The practice of geography

Frontispiece: Rikka Ikebana by Sen'ei Ikenobo, 45th Headmaster, Ikenobo School, Box 36, Nakakyo, Kyoto, Japan

The practice of geography

Anne Buttimer
Clark University,
Worcester, Mass. USA

Longman
London and New York

Longman Group Limited
Longman House, Burnt Mill, Harlow
Essex CM20 2JE, England
Associated companies throughout the world

Published in the United States of America
by Longman Inc., New York

First published 1983

British Library Cataloguing in Publication Data
Buttimer, Anne
 The practice of geography.
 1. Geography
 I. Title
 910   G116

 ISBN 0-582-30087-8

Library of Congress Cataloging in Publication Data
Buttimer, Anne.
 The practice of geography.

 Bibliography: p.
 Includes index.
 1. Geographers—Biography. I. Title
G67.B87 1983   910'.092'2 [B]   82-13091
ISBN 0-582-30087-8

Printed in Hong Kong
by Astros Printing Ltd.

To my father
Jeremiah Buttimer
the best
story-teller
of them all

Contents

List of illustrations   xi

Acknowledgements   xii

1 Introduction   1
Geography, experience, and expertise   1
Why autobiography?   3
Story-telling: knowledge and understanding   5
Adventures in dialogue   8
   Phase one: 'insiders' and 'outsiders'   8
   Phase two: geographic thought and milieu   9
   Phase three: a pedagogy of the impressed   11
Making geographical sense of the world   12
Notes   16

Clarence J. Glacken
2 A late arrival in academia   20

Selected readings   33

Aadel Brun-Tschudi
3 Worlds apart   35

Communication – dialogue – creativity by Jan Hesselberg   42
Selected readings   43

William R. Mead
4 Autobiographical reflections in a geographical context   44

The Vale of Aylesbury   44
A feeling for words   47
Ends and scarce means   48
Prelude to Finland   49
Canadian interlude   50
The Liverpool experience   51
Finland – *ad infinitum*   53
The Scandinavian connection   56
Rêveries du promeneur solitaire   57
Notes   59
Selected readings   61

Interlude One   62
Note   65

5 The environment of graduate school   66

A discussion chaired by Marvin W. Mikesell with Leslie Hewes,
Preston E. James, Clyde Kohn, and E. Cotton Mather
Note   79

John B. Leighly
6 Memory as mirror   80

Notes   88
Selected readings   89

T. Walter Freeman
7 A geographer's way   90

Notes   102
Selected readings   102

Ilmari Hustich
8 An autobiographical sketch of the 'life-path' of a geographer   103

Notes   111
Selected readings   112

Interlude Two   114

9 French geography in the 1940s   119

A simulated discussion designed by Claude Bataillon with Pierre Birot,
Jean Dresch, Henri Enjalbert, Pierre George, André Meynier, Pierre
Monbeig, and Louis Papy
Notes   140

Jacqueline Beaujeu-Garnier
10 Autobiographical essay   141

Career success   142
  An accidental vocation   142
  University career   143
  The Parisian years   144
A succession of interests   145
  Confidentially   145
  Different and successive interests   145
  Parallel pursuits   147
Development of geographical thought   149
  Geography as 'information'   149
  Modernism in geography   149

Systematic reasoning 150
Selected readings 151

William William-Olsson
11 My responsibility and my joy 153

Family 153
School years 1913–21 156
Undergraduate studies and teaching 156
Research 157
Social context and work environment 161
Creativity 162
Geography and research 164
Notes 166
Selected readings 166

Hans Bobek
12 Some comments towards a better understanding of my scholarly life-path 167

Introduction 167
Early stimuli at home and at school 168
University and first scholarly activity in Innsbruck 169
Berlin (1931–39/40) 171
The war years (1940–45) 173
The immediate post-war period (Freiburg i. B., 1946–48) 174
Professor (1949–71) and Emeritus in Vienna 175
Concluding remarks 180
Notes 183
Selected readings 185

Interlude Three 186

13 American geography in the 1950s 196

A discussion chaired by George Kish with Duane Knos, Fred
Lukermann, Richard L. Morrill, and William Pattison
Notes 208

Gerrit Jan van den Berg
14 Some personal reflections on geography 209

Introduction 209
Search for a workable and pragmatically fertile doctrine in (human)
geography 210
My education and training in geography and my experiences with
geography in practice against the background of my life-line 213
Crises and doubts as preconditions for the conception of new ideas
in answer to new challenges 219

The ecology of my thinking on geography and planning 222
Conclusion 223
Notes 223
Selected readings 223

Wolfgang Hartke
15 An interview with Wolfgang Hartke 225

Notes 236
Selected readings 236

Torsten Hägerstrand
16 In search for the sources of concepts 238

Preamble 238
Childhood years 240
Grundzüge der Länderkunde 243
Ten thousand lives 245
Teachers, friends, and books 246
The first steps abroad 250
Ordinarius and planner 251
Graduate student once again 253
Notes 256
Selected readings 256

Appendix
A.1 Participants in the international dialogue project 257
A.2 Alphabetical list of contributors to this volume 258

Appendix B
Highlights of the decades, 1900–80 in nine countries 261
Selective bibliography 275
Index 289

List of illustrations

Frontispiece   Rikka Ikebana by Sen'ei Ikenobo

Fig. 2.1 Clarence J. Glacken, USA   20
Fig. 3.1 Aadel Brun Tschudi, Norway   35
Fig. 4.1 William R. Mead, England   44
Fig. 4.2 The Vale of Aylesbury   46
Fig. 5.1 (Left to right) Professors Clyde F. Kohn (Iowa), Leslie Hewes
(Nebraska), Marvin W. Mikesell (Chicago, Chair), E. Cotton Mather
(Minnesota), and Preston E. James (Florida)   66
Fig. 6.1 John B. Leighly, USA   80
Fig. 7.1 Walter Freeman, England   90
Fig. 8.1 Ilmari Hustich, Finland   103
Fig. 9.1 Claude Bataillon (Toulouse, organizer), Jean Dresch
(Paris VII), André Meynier (Rennes), Louis Papy (Bordeaux)   119
Fig. 10.1 Jacqueline Beaujeu-Garnier, France   141
Fig. 11.1 William William-Olsson, Sweden   153
Fig. 11.2 Perspectives on the world   160
Fig. 12.1 Hans Bobek, Austria   167
Fig. 13.1 (Left to right) Professors Bill Pattison (Chicago),
Duane Knos (Clark), George Kish (Michigan, Chair), Fred Lukermann
(Minnesota), Richard L. Morrill (Seattle)   196
Fig. 14.1 Gerrit Jan van den Berg, Holland   209
Fig. 14.2 Career profile: Gerrit Jan van den Berg   218
Fig. 14.3 The ecological impacts on my thinking about geography and
planning   220
Fig. 15.1 Wolfgang Hartke, Germany   225
Fig. 16.1 Torsten Hägerstrand, Sweden   238
Fig. 16.2 Migration flows between Asby and other Swedish
communities, 1840–1944   249
Fig. 16.3 Career profile: Torsten Hägerstrand   255

Appendix A.1 Participants in the international dialogue project
(diagram by Torsten Hägerstrand)   257
Appendix A.2 Essayists and discussants   258
Appendix B. Highlights of the decades (1900–80) in nine countries   261

Acknowledgements

It would take virtually another volume to recount the story of all those persons, places, events, and institutions which have contributed in various ways to this five-year venture. Mention can only be made of a few. First, of course, is Torsten Hägerstrand, without whose energy and support the whole idea might have remained a pious hope; at least it might have taken an altogether different form. Lund University provided accommodation and material assistance during the first and second phases of the project, 1977–79. The Swedish Committee on Future Studies (SALFO) and the Riksbankens Jubileumfond provided generous financial support for office equipment, video interviews, the Sigtuna seminar, and enabled us to avail of the services of Ingela Gerdin, Christina Nordin, Jim Sellers, and David Seamon during 1978 and 1979. Susanne Krüger, Gunborg Bengtsson, and Eva Särbring continued to offer invaluable support and help right up to the last minute delivery of the manuscript for this book. Clark University hosted the third phase (1979–81) and indeed a debt of gratitude is due to Alice and Milton Higgins who have continued to offer moral and material support for the dialogue effort; to President Mortimer Appley, to the Leir Fund and the Tashahiki Fund at Clark University which supported travel and research assistantships; to colleagues in various departments who continue to support the effort in international dialogue; and most especially to Len Berry and the Graduate School of Geography where experiments in graduate and undergraduate courses were possible and where this book eventually took shape. It is to my colleagues and students in the School of Geography that I feel especially grateful: to Bob Kates who first offered critique and advice, to Bill Koelsch who carefully read successive drafts of my 'interludes'; to Martyn Bowden, Susan Hanson, Harry Schwartz, Duane Knos, Gerry Karaska, Harry Stewart, and the late Dan Amaral who have afforded helpful suggestions; to my students, particularly Izhak Schnell, Rudi Hartmann, Anne Godlewska, Marianne Jas, and so many others who devoted so much time and effort to our dialogue experiment; to Jane Kjems who translated texts and hosted video presentations, to Lynn Frederiksen who carefully edited this entire manuscript, to Terry Reynolds and her staff at the typing pool who have been so generous, to Madeleine Grinkis, secretary, who has maintained such a range of correspondents throughout the world: to all I acknowledge deep gratitude.

Beyond these two 'bases' – Lund and Clark – a wide range of individual persons have offered critique, help, and advice. Among these I would like to acknowledge particularly Olavi Granö (Turku), Christiaan van Paassen (Amsterdam), Dietrich Bartels (Kiel), Shalom Reichmann (Jerusalem), Paul Claval and Philippe Pinchemal (Paris), Egon Matzner and Elisabeth Lichtenberger (Vienna), David Stoddart (Cambridge) and Ron Johnston (Sheffield),

Richard Harrison and David Livingstone (Belfast), Gordon Davies (Trinity, Dublin), Hallstein Myklebost and Jan Hesselberg (Oslo), Gunnar Olsson and Staffan Helmfrid (Stockholm), Hans Aldskogius (Uppsala), Franco Ferrario (Capetown), Jorge Gaspar (Lisboa), Keiichi Takeuchi (Tokyo) and Minoru Senda (Nara), Peter Nash and Leonard Guelke (Waterloo), Allan Pred (Berkeley), Nick Helburn (Denver), Dick Morrill (Seattle), Marvin Mikesell (Chicago), Cecile Struegnell (University of Massachusetts), and Ligia Parra (Bogota).

I also wish to acknowledge formally those studios in many places which offered patient and competent services, sometimes free, for recording interviews. Those involved in this book include Lund University's AV-centralen where the interviews with Mead and Hägerstrand were recorded, the Norsk Rikskringkasting in Oslo for the interview with Aadel Brun Tschudi, the Österreichischer Rundfunk und Fernsehen (DRF) in Vienna for the Bobek interview, the Audio Visual Services at Manchester University for the interview with Freeman, the Service Audio-Visuels of the Centre National de la Recherche Scientifique in Paris for the interview with Beaujeu-Garnier, the Läromedelscentralen at Stockholm University for the interview with William-Olsson, the University of California, Berkeley Audio Visual Service for the interviews with Glacken and Leighly, the Audio Visual Studio at University College, Cork, for the interview with van den Berg, and the NETCHE Studios at University of Nebraska, Lincoln, Nebraska for the discussions 'Environment of Graduate School' and 'American Geography in the 'fifties'. Thanks also to *Hérodote* for permission to translate and abridge the *Table Ronde Imaginaire* of Claude Bataillon which is printed here as Chapter 9.

Finally I offer sincere thanks to Sen'ei Ikenobo, 45th Headmaster of the Ikenobo School who has personally designed the frontispiece: Rikka Ikebana.

Anne Buttimer
Clark University
Worcester, Mass. USA

January 1982

Chapter 1

Introduction

Geography, experience, and expertise

'I have understood that Sweden's government will help us as much as it can, but they do not get it quite clear how it actually is, our life and our situation, because the Lapp cannot explain it precisely as it is. And the reason for that is this: when the Lapp enters into a room, then it is not so good with his reason, when weather cannot blow against his nose. His thoughts do not fly when there are walls around him and roof over his head. And neither is it good for him to be in the thick forests when weather is warm. But when the Lapp is on the high mountains, then he has a quite clear reason. And if there was a meeting place there, on some high mountain, then the Lapp would be able to explain his situation quite well.'

Johan Turi, *Muittalus Samid Birra. En bok om lapparnas liv.* Stockholm. Wahlström & Widstrand, 1917.

It was early in the 1900s, in the Lapp village of Jukkasjärvi, that one Johan Turi used to wonder why the world did not really understand his people and their way of life. Should he try to write a book 'in which everything could be explained about the life and circumstances of the Lapps?' (ibid.: i). He had no formal schooling and to his fellow Lapps the idea of a book was ludicrous, impertinent, and anyhow would not earn bread. He tried Finnish, a language deemed socially superior to Lapp and which he knew better than Swedish, but then Finnish did not really allow him to describe some essential features of the Lapp world; his thoughts could fly more swiftly in his native tongue. Then one day in 1904, Emilie Demant Hatt, a Danish school teacher, came to Lapland. She learned and loved its language and folkways, and met Johan Turi. She returned in 1907 and spent an entire year listening and encouraging him to write down, sketch, and tell her all he could. 'His language was rich in images, beauty and grace in the telling,' she later wrote, 'he jumped from subject to subject and opened for me the full mystery of the life and thought of his people... it was a record of human life journeys, witchcraft, and strange happenings... the saga from barren wilderness sounded in my ears, strange pictures were gliding before my eyes. When Turi spoke I asked him to write it down and this he did – unfortunately (the written word) is only a weak reflection of the style and brightness of his oral telling.' (ibid.: xii) A Danish and Lapp version of *The Lapps and Their Land* appeared in 1910, courtesy of Hjalmar Lundbom, President of the Luossavaara-Kirunavaara Mining Company, and a Swedish version was printed in 1917.

The parable of Johan Turi sets the tone for this volume. In these later years of the 1900s how many of us still – students, researchers, or natives of 'developing' societies – seek to explain our situations to those who would theorize and plan for us? In *Polis* there is still much ado about problems of consultant expertise and its relationship to lived experience, in *Academy* much astir about specialization and fragmentation of knowledge, communication and scholarly creativity. Like Emilie Demant Hatt, I am a teacher and find that few things can beat a good story for evoking curiosity about such questions, opening windows toward understanding other worlds and a more critical perspective on one's own. For some years now I have listened to older colleagues in various lands, shared some of my own worries with them, and have encouraged some to write down the stories of their own life experiences. Unlike Turi, of course, most of my *interlocuteurs* felt quite at ease with the written word, even if for many, the native tongue might have allowed their thoughts to fly more easily. What I present here, in English, are stories about the practice of geography[1] – individuals telling about their own experiences and the contexts in which their life journeys unfolded, as well as group discussions on particular periods and settings. Those included here have been selected from a much wider range of essays, interviews, and discussions held during the past five years in the course of an international 'dialogue' effort which involved participants from diverse countries and disciplines.[1a] The process continues and only a summary sketch of its rationale, procedures, and some of the lessons it is affording, can be outlined in these introductory pages.

The sequence in which authors are presented defies the stereotypical rubrics of national school, chronological age, or even thematic research interest. It resembles a flower arrangement (*Ikebana*) rather than a logically defensible classification, and is intended to evoke appreciation for the uniqueness of each person as well as the diverse strands of similarity and contrast which make geography such a complex and exciting venture. Each author is introduced individually and is positioned in a manner which seems to me at least to highlight his or her particular style of practice. Attached to each essay is a short list of published works, selected by the authors themselves, so the reader may explore connections between that author's life experience and his or her published writings.

Interspersed between individual accounts are group discussions on particular settings: 'The environment of graduate school' (Ch. 5), 'French geography in the forties' (Ch. 9), and 'American geography in the fifties' (Ch. 13).[2] At these interludes, I offer a personal interpretation of the contexts in which discussions are set, illustrating three themes of *meaning, metaphor*, and *milieu* which seem helpful in elucidating the drama of geographic thought and practice, like the three main elements of the Ikebana provide keys to the whole composition. These comments vary in length and detail reflecting my own level of familiarity with the settings. For those who wish an independent rendering of context, the 'Highlights of the decades' (1900–80) in each of the nine countries from which our essayists come, are supplied in Appendix B, prepared by colleagues other than the essayists.[3] All essays and discussions have been prepared especially for this volume with one exception, viz., Chapter 9, which is an abridged translation from the previously published 'simulated round-table discussion' by Claude Bataillon.

Ideally each essay should be read and contemplated as a unique story and each discussion considered against the background of its historical context. The reader will no doubt construe what is written there through the filters of personal experience and a priori intellectual predilections. He or she may be prodded to re-examine such predilections and evaluate them in the light of these accounts. For herein lie the primary goals of this book: to facilitate understanding of the work of those authors whose life journeys are described here, and some critical perspectives on one's own life and work. There is a more ambitious hope also. It invites curiosity about the texts of geography and the worlds they attempt to describe, about the diversity of keys in which disciplinary thought and practice have been played during the twentieth century. It invites the reader to become so involved in the puzzle of interpretations that anxieties over subjectivity and objectivity recede and one engages in a shared intersubjective task to clarify relationships between *Geo*graphic thought and practice and the lived *geo*graphical experiences of humankind.

Why autobiography?

Wherefore, one might ask, such a Socratic adventure? What ends could such experientially grounded reflection, critical self-awareness, and mutual understanding serve? The proverbial cup can be seen as half full or half empty. One reads that our twentieth century has witnessed the political manipulation, trivialization, and amnesia of its intellectuals, the triumph of materialism and tyrannical power; one also reads about the indomitable courage, altruism, and creativity of its many individuals. In each person's life echoes the drama of his or her times and milieu; in all, to varying degrees, the propensity to submit or rebel. Through our own biographies we reach toward understanding, being and becoming.

Some, schooled in particular disciplines, accumulate general knowledge about the world. One of the central motifs in any scholar's life story is the interplay between disciplinary *knowledge about* reality and growth in *understanding of* lived experience with reality. What he or she may write in books and journals yields an image of discrete knowledge products, but may yield little understanding of the intellectual process unfolding within that person's life. Many scholars, too, have extended their work beyond library, classroom or laboratory, applying the results of their research to the resolution or alleviation of public problems. Worries often arise when applied scientific expertise, backed by political fiat, fails to solve problems in the lived world. Turi's frustrations over the allegedly benign efforts of the Swedish government were certainly not unique. There is scarcely any branch of applied scholarship today where tensions have not become apparent between voices of consultant expertise and the taken-for-granted worlds of lived experience[4].

These were some of the general concerns about geography which entered, explicitly and implicity, in the original invitation to these essayists: relationships between knowledge and understanding, between experience and expertise, the dream and reality of applied projects and movements, the significance of context in promoting or inhibiting scholarly creativity. These same concerns have no doubt biased perceptions, excited curiosities, and triggered whatever learning experiences my students and I have enjoyed at Clark University where we studied the essays and the published works of these authors.[5]

'The limits of my language are the limits of my world', and each 'form of life' has its own 'language game', some linguists would argue. (Wittgenstein 1922: 5.62; see also Chase 1954; Whorf 1956). How then may one articulate and defend any generalizations at all about worlds of human experience? How may the world's diverse civilizations 'language' their own experiences and still reach toward an understanding of others? A cardinal issue indeed for geography whose descriptions of the world offer a dramatic story in *interpretation*. Some no doubt reflect the scholarly traditions and lived experiences of those nation states most prolific in the production of textbooks, but do they 'get it quite clear how it actually is' with other worlds of experience? To generate some dialogue over diverse interpretations, as Plato did with parables, might indeed offer insight into problems which stir in Academy as well as Marketplace: how may practitioners of various fields communicate with one another and reach toward a more holistic understanding of reality (Habermas 1979; Boulding 1980; Higham 1980)?

The central issues of self- and mutual understanding which are raised in this book touch on a vast arena of concern for both scientist and humanist: how is the world construed, interpreted, and 'explained' by practitioners of particular academic disciplines (Foucault 1971, 1980; see also Gadamer 1975; Bourdieu 1978; Rorty 1979; Szanton 1980)? Is there a 'geographical' sense of reality which could be distinguished from that of the sociologist, psychologist, poet, or historian (Redfield 1948; Sauer 1956; Rose 1969; Odén 1975)? If so, how does such a geographical sense emerge, how is it articulated, disciplined, and practiced? How do its prose, its maps, and its models match or mirror the geographical experiences of humankind? In the hermeneutical task implicit here, it is hoped that a communication space may be opened which would allow for sustained dialogue among diverse interpretations of the world.[6]

An autobiographically based reflective approach seems most appropriate as a beginning step (Watson 1976; Weintraub 1978). And geography lends itself admirably to the threefold interpretive task of hermeneutics, viz., expression, explanation, and translation (Palmer 1969). For most of the authors in this volume, for instance, the appeal of sensory (visual, kinetic, auditory) perception, affective bonds to place and region, as well as imaginary experiences gleaned from story books, stamp collecting, and musing on maps, all contributed to their choice of geography. In their 'schooling' years, traditions of thought were transmitted orally, whether in field excursions, laboratories, seminars, or their sequels. The cultivation of this geographical 'sense', which I call 'choreographic awareness' may have roots in childhood experiences of places, events, and people. To understand such roots may be a first step in understanding the more formal languages of description and explanation which are part and parcel of the discipline's prose.

Autobiographies can also, of course, shed light on essential questions about the history of the field, and the social construction of its thought and practice, thus offering a critical complement to conventional approaches which rest on archival records. Many colleagues, particularly before mid twentieth-century, may have devoted so much time and energy to teaching, field work, administration or non-academic employment that they had neither the time nor inclination to write their ideas down. From an international perspective, too, the hegemony of a few vernacular languages, vicissitudes of editorial policies, power relations within and among invisible networks of scholars, all probably have imposed a selective screening on what has found its place in documentary archives. And within any school the published record may reflect oscillating criteria of status, orthodoxy, and 'relevance' quite as much as the logical or epistemological preferences of any particular individual. But finally, there may be facts and feelings about the learning process itself and the challenge of articulating one's discoveries which cannot ever be incorporated in conventional research reporting or scholarly books. Most of those included in this volume, for example, have published a significant amount and have been recognized leaders within particular fields yet nearly all welcomed the opportunity to reflect on threads of continuity in their own intellectual journeys. Many felt that their published work consisted mostly of responses to short-term demands such as reports on particular pieces of research, and that textbook and journal editors had usually welcomed the fruits of analytical endeavor rather than reflective insight or personal views. Some welcomed the opportunity to question what they considered to be distorted stereotypes and unfair commentary on their writings.

Why not, then, invite senior scholars to share insight from their own life experiences? Freed from those psychological and political constraints which normally surround institutionally defined roles and agenda, would they not be in a better position now to offer information and reflection on the oral history of the discipline? At least they can offer their own interpretations, tell their own stories, and let us see what may be learned from the listening.

Story-telling: knowledge and understanding

I was born in a land of story-tellers. The knack of telling a good story is still a cherished art, calling for memory and imagination, dramatic skill and body language in the telling. It reaches its highest appeal when someone speaks of his or her own experiences. Like its ancestral prototype in the epic or saga, the story addresses practical concerns of everyday life and celebrates the splendor and magic of the faraway. It thrives in places and periods where many life experiences meet, as in the medieval *atelier*, forge, or trading post. 'If peasants and seamen were past masters of story-telling,' writes Walter Benjamin, 'the artisan class was its University. In it was combined the lore of faraway places, such as a much-travelled man brings home, with the lore of the past, as it best reveals itself to natives of a place' (Benjamin 1969: 85).

If imaginations could fly to the pre-literate, pre-disciplinary eras of Western intellectual history, one would undoubtedly see the origins of historical and geographical curiosities in story-telling. The names of Herodotus, Homer, Virgil, Plutarchus, and others may be convenient symbols for a tradition of life quite as much as specifications for particular authors (Vernant 1962, Detienne 1973, Foucault 1977). Turi's account of reindeer hunt and witchcraft, kidnap and sabotage, revolves around *dramatis personae* which are not actors *qua* subjects; rather they are beliefs and practices, events, animals, plants and humans, all part of a life drama attuned to this very special milieu of the North. There is a moral ('ideological') implication, too, in the story: it is that the Lapp, over centuries of experience, had developed a rational way of ordering society and milieu, a *genre de vie* bolstered by myth and practice fostering a sense of cultural and geographical identity. The definition of a political boundary between Norway and Sweden posed an intrusion on Lapp geography and heralded the arrival of unfamiliar *dramatis personae* like textbooks on European history, monarchies and Christianity, as well as schools and clinics to serve 'educational' and 'medical' interests which had previously been accommodated as an integral part of life and death.

The tradition of story-telling, for perhaps most of human history, has served not only to sustain human efforts to live poetically – to create and cultivate meaning, rationality, and wisdom in modes of living on the earth – but has also been the catalyst and wellspring for community and cultural identity (Arendt 1968; Parry 1971; Lord 1971; Cavell 1980; MacIntyre 1981). Once stories are committed to print a profound change occurs. The novel and fictional short story, like the treatises of history and geography, may now become the work of a particular author, to be enjoyed privately by an individual reader; the story yields up its social and catalytic role to the vicissitudes of authorship, readership, medium and literary criticism (Benjamin 1969; Foucault 1977). Wisdom of life is no longer guided by proverb or epigram; it becomes the *raison d'être* for specialist castes of philosophers, psychologists, economists, and engineers. Maps and charts, garnished in appropriately disciplined prose, replace the journey and field observation of geographical experience; proper history supplants the story-teller.

Throughout the nineteenth and twentieth centuries, however, scholars have frequently sought insight from oral history and autobiography (Garraty 1970; Toynbee 1975; see also Gadamer 1975; Rabinow and Sullivan 1979). In German historicist writing one finds an emphasis on *verstehen* (empathetic understanding) a mode deemed more appropriate for the human sciences than *wissen* (scientific knowledge). Wilhelm Dilthey claimed that autobiography epitomized the fundamental challenge of the human sciences in that it dramatizes, for hearer as well as for teller, those dialectical tensions between internal and external forces, between the synchronic and the diachronic, between the unique and the general – those same tensions which have set the humanities and sciences apart in Western academic traditions (Dilthey 1913; see also Habermas 1968: 150–152). Biographies and autobiographies abound.

Scholars from diverse disciplines have bequeathed a rich legacy of personal stories (Horowitz 1969; Heisenberg 1969; Schweitzer 1949; Dalenius *et al.* 1970; UNESCO 1973; Baum 1975; Chargaff 1978; Kazin 1979). Geographers, too, have now and then revealed insight into their own experiences (Løffler 1911; Schouw 1925; Siegfried 1952; Bingham 1959; Hettner 1960; Taylor 1958; Christaller, 1968; Schmieder 1972; Mill 1951). Yet somehow the harvest has not been reaped. Autobiography is still generally regarded as a literary form (Bruss 1972; Zweig 1976) or at least the proper domain of the humanist (Gunn 1977) and 'foreign' to most fabricators of theory and paradigmatically minded scientists.

Diverse attempts to forge links between the unique and the general within a basically *verstehen* approach to truth have tended to focus on method (Gadamer 1975), on ideology (Habermas 1968) or on epistemological foundations (Rorty 1979; Schrag 1980). All reveal how deeply ingrained are those habits of thought inherited from Descartes and the Enlightenment where subject and object, unique and general, 'insider' and 'outsider' confront each other across impenetrable walls of language and cultural tradition (Wittgenstein 1960, 1969; Habermas 1979; Dallmayr and MacCarthy 1974). Today such bifurcations are suspect; the human subject as author of consciousness is questionable and the very idea of epistemological foundations for knowledge regarded by many as futile and illusory (Foucault 1969, 1971; Rorty 1979). In geography itself, a cacophony of '-isms' battle on issues of knowledge, language, and power (Olsson 1979; Gale and Olsson 1979; Smith 1979); opinions clash on what enquiry processes might be appropriate for delivering credible statements about reality (Sack 1980; see also Sellars 1964).

Conflict and controversy also surround interpretations of intellectual history. Idealist typologies, e.g., those of Hegel and Collingwood, assume that thought becomes progressively more refined and truthful as scholars build on knowledge accumulated by past generations and celebrate the onward march of paradigms (Popper 1961; Lakatos 1974; Johnston 1979). Others, of materialist bent, would imply that external conditions such as politics, economics, and technology are the fundamental shapers of thought (Bernal 1939; Needham 1964; Nakayama 1972). To the much-celebrated debate between 'internalist' versus 'externalist' theories of intellectual history (Mendelssohn *et al.* 1977; Lilley 1953) structuralists and others would cry 'a plague on all your houses' and argue that language is the key to the structures of thought (Levi-Strauss 1966; Jakobson 1960; Bourdieu 1978). Western knowledge and practice might more appropriately be described in terms of an archipelago of epistemic structures where discontinuities, ruptures, and major shifts are more characteristic than broad lines of continuity (Foucault 1969, 1971; see also White 1973a; Claval 1981). And in this cacophony of competing interpretations of history there are those who find in dialectical reasoning the Ariadne thread for deliverance from polarized and bifurcating legacies (Kuhn 1962; Elzinga 1980; Blume 1977; Young, 1973). A neophyte might well observe that each interpretation yields its own array of partial insights and that none merits credence until tested in the light of experience and practice.

It can scarcely be denied, for instance, that the nation state has been one of the most powerful sponsors, mediators, and audiences for geographic effort. Does this mean that all geographic thought has borne the stamp of 'mistress' to national interests, that the maps and matrices of spatial analysis are but the product and instrument of externally defined objectives? One wonders if Johan Turi worried about such questions before accepting Emilie's editorial help and the financial support of Lundbom. Strange how 'mapping' information which was intended by one generation as gratuitous 'service to the fatherland' can get construed by a later generation as prostitution to state interests (Eriksson 1978).

The legacy of 'either-or' categorical thinking has fostered tendencies to separate and often polarize these distinct dimensions of reality, beclouding the immense variety, ambiguity, and often paradoxical nature of lived experience. Autobiographically based career stories in geography demonstrate, explicitly and implicitly, how necessary it is to maintain an open attitude toward both the unique and the general, and suggest what a loss it would be if a pluralism of styles could not be encouraged. If regions and places could tell their own stories, would the same tensions not be apparent (Blythe 1969, 1979; Parker 1978; Helias 1975)? The lived account, in either case, would dramatize geography's perennial ambivalence *vis-à-vis* sciences and humanities. A discipline devoted to the study of the earth as home for mankind should surely be a house of many mansions, as ecumenical in spirit as the *oecumene* itself could inspire.

Adventures in dialogue

Phase one: 'insiders' and 'outsiders'

It was in an existentially critical mood that this project began. The romantic and idealist tone of historicist prose, too, seemed appealing. With the spirit of writers like Dilthey, Schütz, and of course Saint-Exupéry, I suggested: 'Let the authors speak for themselves . . . let us listen to their life experiences and look critically at our own taken-for-granted worlds of thought and practice.'[7] For some years I had found that dramatizing the contrast between the worlds of insiders and outsiders was both pedagogically effective as well as analytically exciting.[8] When it came to the history of geographic thought of course it was easy to play a similar tune: there was the world of the teacher and field researcher, who actually *did* geography, and the world of the annalist and commentor, who recorded its history. Should the authors of geography themselves not be the best sources of insight into relationships between thought and context, into what might constitute a 'geographic sense' of reality, and how, in fact, the discipline developed? I claimed that conventional methods of research, relying as they do on documentary evidence (i.e., already published works) could only yield an opaque, outsider view of history.

The aim, however, was not to construct alternative histories, or for that matter to question the validity of archival research. It was rather to open up a process which could facilitate better communication and mutual understanding not only between generations, but also between the increasingly specialized branches of geographic effort today, as well as between geographers and colleagues in other fields. I had reason to believe that such a process could best be encouraged if there were some concretely grounded stories to share orally, in so far as possible, as springboards for dialogue. For this approach had worked so nicely during the spring and summer of 1976 when I was invited to lead a seminar at Lund (Sweden) on issues of knowledge and experience. Participants from over fifteen fields would meet weekly to discuss common readings as well as to share insights from their own experiences. I would play 'outsider' and invite them to explain Swedish life and thought to me; I would tell stories from my own worlds of experience and listen to their responses. It was at once a thrill to learn so much that was new, and more often a shock to discover how naive some of my own presuppositions were. One general result, shared by many of the sixty to eighty individuals who participated, was that the experientially grounded approach could indeed facilitate the discovery of common denominators across disciplinary boundaries.

Why not then try a similar approach within geography? To look at the recent history of the field through the accounts of senior colleagues, and at the same time look critically on our own practice of geography?[9] The fundamental aim of this first phase, then, was to open up dialog over diverse interpretations, to generate questions rather than to deliver answers. In Socratic and Idealist vein, I wish to demonstrate the need for pluralism of research style and personal vocation in the field, and to raise questions about contextual as well as cognitive dimensions of its thought and practice.

It was from Sweden, and particularly from Torsten Hägerstrand, that the most positive response came. From his many anecdotes, sketches, and interpretations of events, he gradually clarified for me his own thought style and world view which in so many ways contrasted with mine.[10] In 1977 I was invited again to Sweden to study problems related to integration in geography and to work at promoting better communication between humanists and social scientists. Now the opportunity came to render those very general aims into the form and language of a research project. One would proceed inductively, without promise of clear-cut generalizations or products, but there still was need to outline a conceptual framework, some general foci of enquiry, and appropriate research strategies.

Phase two: geographic thought and milieu

Through discussions, correspondence, and meetings both formal and informal with colleagues from several countries Torsten Hägerstrand and I worked toward discerning key themes which could be explored via autobiographical interviews with individuals.[11] Three identifiable clusters of concern emerged as viable bases for designing such interviews: (1) Creativity and Milieu; (2) The Social Construction of Geographic Thought and Practice; and (3) The 'Dream and Reality' of applied geography. We also planned to make video recordings of at least some of these interviews and to use these as catalysts for small group discussions so that a further question could be examined: (4) Media and Message in the articulation of ideas.

One of the first exercises was a seminar on creativity and context which was held at Sigtuna in Sweden in 1978 where forty-four senior scholars from a variety of disciplines were invited to share insight from their own career experiences.[12] Geographers numbered about one-third of this group and, by and large, those who had participated in the 1976 seminar probably felt more comfortable with the agenda than those who had not. Unlike more conventional approaches to the study of creativity, where emphasis tends to rest on factors of personality and motivation, or 'critical moments' (Maini and Nordbeck 1975) our approach directed attention to context: participants were invited to reflect on the significance of places, people, networks, events, and general intellectual milieux for their own creative work. Some fascinating discoveries were made by individuals about their own lives, and indeed some of the essays submitted during the subsequent year have helped us enormously in framing questions for our interviews.

It is only fair to admit, however, that the question of creativity did not evoke an unequivocally positive response. At the Sigtuna meeting, and indeed on several other occasions, opinions on this approach ranged from wholehearted enthusiasm to outright indignation. To some of the participants at Sigtuna the agenda appeared not only unconventional but actually offensive, a preposterous intrusion on their private lives. Given the fact that there was not adequate time to explain the entire philosophy underlying the project, it is not surprising that some would construe it as narcissistic, reflecting the virtually 'cultic' character of the 1970s rhetoric on the Self (Hougan 1975; Schur 1976; Lasch 1978). Other cultural and historical circumstances no doubt made 'existential jargon' particularly repugnant to Scandinavian scholars (Fromm 1968; Jung 1957; Adorno 1973; Jacoby 1975). The commonest objection, however, stemmed not from philosophical critique but from convention: this proposal was impertinent and probably would not earn bread.

During the years 1978–81 a number of interviews were recorded on videotape: conversations which invited individuals to share freely reflections on their own career experiences.[13] Nobody was asked to *represent* any particular school of thought, ideology, or national tradition; each was invited to *present*, in the language and form which seemed most appropriate, the account of his or her own career. The video format, for all its limitations, seemed to offer an opportunity to dramatize the effects of media on the language and speech of scholarly communication. Viewers could compare the oral versus written modes of self-presentation and discern which medium seemed more appropriate for the communication of particular types of insight. This has certainly given rise to many questions about the construction and interpretation of disciplinary history and its actual practice, as well as a keen awareness of language and symbolic communication.

Phase Three began when these interviews were shared with audiences of students and colleagues in various places, and especially at Clark University, as opportunities were available. Viewers were first asked to write down their responses to what was seen and heard, before engaging in discussions.

Summaries of questions were then communicated back to the individual and he or she was invited to write an autobiographical essay. In a sense, then, the essays contained in this book are already the product of some minimal dialogue varying from case to case depending on opportunities – and are still potential catalysts for further dialogue.

Phase three: a pedagogy of the impressed

One of the underlying assumptions of the entire effort was that an essential prerequisite for dialogue was the possibility of some community context within which the same participants could share views over an extended period of time. Graduate seminars at Clark University, during academic years 1979 to 1981, provided such a context. Another assumption was that the dispositions necessary for dialogue oriented toward mutual understanding would require a free interplay of emotion and reason – to *care about* a subject was as important as wanting to *know about* it. One guessed that when senior scholars would speak and write about their own practice of geography, they would bring into focus particularly those subjects about which they cared deeply, for whatever reason. Students also presumably would care about discerning a direction for their own prospective practice in the field and this would lend energy to the intellectual curiosity not only about geography as a field of knowledge but as a vocation to life. Fortuitously indeed many of my students were polylingual and had come from very different cultural backgrounds – Hebrew, German, French, Swedish, and Russian. From the beginning there was an experientially grounded sense of the nuances involved in language, and parallels between cross-disciplinary and cross-cultural communication.[14]

What we experimented with was a juxtaposition of these two agenda which have conventionally been regarded as separate and distinct: in-depth understanding of the particular case, and knowledge of general patterns, without denying the integrity of either quest. Through correspondence with the authors, limited as some of it was, we came to regard this institutionalized antinomy of so-called humanistic and scientific perspectives as part of a dramatic story in the socialization of Western thought rather than a logically defensible set of rubrics for the acquisition or transfer of knowledge. We began to appreciate how these perspectives could complement and enrich one another, if only to elucidate the myriad ways in which they both have become imbued with Western and national values, and marked by the vicissitudes of Euro-American history. One of the sometimes painful lessons learned from various attempts to establish common denominators among our authors, for example, was that we were imposing our own preconceptions, striving to justify and elaborate theories with which we temporarily felt comfortable, establishing rubrics and labels with which to classify these 'objects of research'. Microcosm of Western intellectual (and political) history?

It was during this third phase that tensions between the preoccupations of the first phase (personal creativity, insiders, and outsiders) and those of the second (social structuration of thought) became dramatized. A great deal of time and energy was devoted to the construction and analysis of 'career profiles' which would show, as comprehensively as possible, all potentially relevant aspects of an author's thought and milieu. The enormity of this challenge became clear when students sketched their own experiences.[15] Eventually we concluded that the primary value of such schematization was for an individual who wished to reflect on his or her own experience. At least for this sample, chosen from such widely diverse contexts, there was not much to be gleaned from a comparison of profiles. They did, however, serve the heuristic function of dramatizing the dialectic of particular story and general hypothesis.

What emerged most convincingly during this third phase was the need for a more explicitly hermeneutic approach, a heightened curiosity about language and symbolism, and the open challenge of discerning meaning and metaphor in the work of our authors. It would be inappropriate indeed to elaborate on the entire process through which we sought hermeneutic understanding of the texts which follow. It is better to allow the reader to enjoy the uniqueness of each essay first, to leave imaginations free to discover other themes. In the interests of further dialogue, however, let me summarize my own perceptions at this stage, of themes around which one could develop appreciation of an author's style of geography, and a contextual understanding of disciplinary thought and practice during the twentieth century.

Making geographical sense of the world

'In any field not yet reduced (or elevated) to the status of a genuine science thought remains captive of the linguistic mode in which it seeks to grasp the outline of objects inhabiting its field of perception.' (White 1973b: xi)

When a geographer wishes to make sense of that vast panorama of fact and fiction, pattern and event, on the surface of the earth, what goes on in his consciousness? How does a geographic sense of reality emerge, and how does it distinguish itself from that of the geologist, poet, painter, or historian? Even if one could place in parentheses, for the sake of speculation, the accidents of institutional history and the rituals of the discipline era, it seems worth while to look closely on this question.

What is suggested in the following essays, and indeed in several of the interviews not included here is that prior to any explicit 'figuring out' of any explanation of pattern or process, there is a prefiguration, an intuitive grasp of how the world is and how the story of its landscapes should be told. Such prefiguration, which I call 'choreographic awareness' can best be understood as a poetic rather than a calculative exercise; it seems to involve moral, esthetic, and emotional commitments which are related to lived experience and which underpin a scholar's eventual choice of epistemological orientation and *style of practice*. Through childhood experiences of particular places and journeys, family and school, a person moves through a process of discerning meaning and order in his or her surroundings. Elements of such early impressions seem to have an enduring influence (positive and negative) on subsequent ways of construing intellectual and practical challenges in geography. What seems to distinguish the geographical sense of reality for most of these authors is the attempt to grasp both synchronic (spatial–structural) and diachronic (temporal) aspects of world reality at once. For many, this very general awareness does not begin to crystallize into formal modes of explanation until university years, and most often around a Ph.D. dissertation.

During these early years, too, individuals manifest significant differences in how impressionable they are to external versus internal stimuli. The moral and pragmatic implications of how an individual is to relate to society often reflect themselves in how a scholar will negotiate his own intellectual convictions with those prevailing within the discipline. 'Self' versus 'system' emerges as a cardinal issue in the evolution of a scholar's style of practice.[16] Individuals vary so much on this issue: some perceive themselves as 'loners', prevented from much collegial interaction by the burdens of administration, direction of one-man departments, or the duties of non-academic employment. Some, socialized in milieux more open to individuality and personal creativity, rejoice in their uniqueness *vis-à-vis* prevailing trends. Others speak of the thrill of belonging to a new movement in which they participated with passion and zeal, eager to win fresh recruits. And some bother little about social context, focusing attention on ideas and their practical or theoretical appeal.

At the level of prefiguration, thus, two essential considerations seem to be (a) choreographic awareness, and (b) relationship to social reference system. Both seem to have an enduring influence on the style of practice which, like ballet, can accommodate melodies of varied key. It is on the basis of these melodies that some common denominators of style can be observed among our authors. For me, it seems, there are at least three major interlocking voices which could be discerned in each author's story: (1) *Meaning*: values and convictions expressed concerning thought and life and the practice of geography; (2) *Metaphor*: key modes of symbolic expression and modes of argument ('explanation') evident in autobiography as well as in published works; and (3) *Milieu*: physical, historical, social, linguistic, and political contexts deemed significant in the thought and experience of an author.

Meaning and values appear to be unique, their sources vaguely discernible, but their significance for an author's work is undeniable. Professionally, for example, some perceived their primary vocation to be one of becoming inspiring teachers, others to expertise in problem-solving and societal relevance, others to the poetic art of evoking meaning in landscapes and the sense of place, others still to promoting organizations and networks of scholars. Values change, assume different expressions, and new ones appear in various combinations in the course of a lifetime; whether ascribed to family or societal influence or to personal choice, they constitute a kind of 'red thread' of meaning in an author's practice of geography. It is on this level of meaning and vocation that students today can often find the most direct bases for understanding the work of colleagues in another generation.

There are other dimensions of meaning, of course, which may be discerned in the published work of an author as well as from autobiography. Ideological (moral) stances on the nature of society and change, for example, vary from conservative to radical, and such commitments no doubt influence a scholar's preferences for particular types of explanation over others (Mannheim 1946; White 1973b; Young 1973). Categories of nineteenth- and early twentieth-century 'ideology', however, seem to be oddly inappropriate for those scholars whose careers spanned the Second World War. Even more legible for the careful reader are the symbolic meanings of the prose style adopted by an author (Frye 1957; Jakobson 1960; Burke 1969). Some may express a romantic conception of the world in variations of comedy or tragedy; others prefer satire, anxious to unmask the ultimate inadequacy of world visions displayed by the others. Such elements of prose and narrative style enter into that process of *symbolic transformation* of 'raw data' to disciplinary language and it is in this process that meaning and metaphor (values and cognition) meet (Langer 1957; Merleau-Ponty 1973; Ricoeur 1975; Pribram 1980).

Metaphor, as used here, is a shorthand way to identify cognitive style and mode of argument: that of an author as well as those which held dominance within geography at different times.[17] At one period, for example, many practitioners saw the world as a set of maps, a mosaic of spatial or regional patterns, while at another time one looked for mechanisms underlying the spatial or functional organization of phenomena on the earth. Earlier one spoke of the earth as organism – each place, region, niche, and nation seen as an organic unity, a microcosm of some global whole. And at most times and places there were voices insisting on a view of the world as arena of unique events, each one to be interpreted in its own context.

These key metaphors: 'map', 'mechanism', 'organism', and 'arena' are, of course, global symbols which serve to illustrate rather than exhaust the full range of geographic styles.[18] To look at geography's claims to truth via root metaphors serves several valuable functions: it can demonstrate the essential and partially irreconcilable differences in basic world view which have characterized particular definitions of the field. It can also defend pluralism, in that each root metaphor projects its own constellation of claims to truth, and none can be evaluated or understood in the categories of another. Thus attempts to impose one dogmatically on the field as a whole, dismissing deviants on inadequate and inappropriate terms, can be seen as a futile imperial exercise, however worthy the alleged cause such as 'integration', 'identity' or 'analytical efficiency.'

Individual geographers cannot, of course, be identified unequivocally with any one of these global metaphors; most scholars have moved quite freely among different modes of argument. At times when an individual's own preferred style is patently out of tune with what prevails in the discipline, much can happen; in fact, a new creation may emerge (Fleck 1935; Kuhn 1962).

Milieu, as used here, adumbrates the physical, social, and ideological setting within which research is conceived and executed, the material objects to be investigated, and the audience with whom results are to be communicated whether in teaching, in applied work, or in publication. It is thus more amenable to definition and analysis than either meaning or metaphor. The salience of particular aspects of milieu varies, of course, not only from one individual to another, but between phases of any one person's life; the only justifiable generalization is that one should always regard milieu both synchronically and diachronically. In childhood, for instance, the immediate family setting and home area may impress particular conceptions of reality in a lasting way. During school years peers and group activities may foster certain types of curiosity and teachers may stir imaginations about places and events far and near. During university years the institutional milieu and courses may further channel one's image of geography, and external (political and social) events may evoke commitments to practical and theoretical directions. Each event in a scholarly life journey affords occasion for redeploying previous insights and redefining the nature of one's future commitments. One interesting feature of the essays which follow is that there is sometimes little congruence between an author's implicit or explicit 'ideology' and the significance ascribed to milieu in his or her career story. Neither is it always evident that individuals choose their intellectual orientations from lived experience: city children do not always choose urban geography, nor do farm or rural children choose agrarian interest! Quite a few cases of the opposite can be seen.

Under the rubric of milieu, as the term is used here, one encompasses all relevant aspects of the existential context within which meaning and metaphors are selectively negotiated. Chief among these, of course, is language which, in so many ways, seems to have a life of its own (Wittgenstein 1960, 1969; Foucault 1977). It not only affords opportunities and constraints in the definition and analysis of problems, but also the potential 'peer' context for discussion and evaluation as well as the audience of potential readers. Would that all our authors could have written in their own vernaculars!

Meaning–Metaphor–Milieu: each theme fans out into a distinct array of insight and possible common denominators among our authors. The trilogy itself warrants an entire treatise. A careful attunement to these three melodies can help one to appreciate the dramatic uniqueness of each life story: meaning can open doors for interpersonal understanding; metaphor and milieu can shed light on general issues of ideas and context. Bearing all components in mind, one can appreciate how limited are those generalizations about the history of geography which are based on *thought* alone, or on *context* alone. Meaning and metaphor afford antennae for picking up the treble chords of an author's story, milieu the base. But the analogy limps. Milieu can be seen as playing a dominant refrain, not only in suggesting themes for research but also in guiding the choice of language and methods to be used (Ravetz 1971; Teich and Young 1973). In fact, it is only when one frames a focus on particular milieux, e.g., on the 1930s, 1940s, or 1950s that the dynamic interplay of meaning–metaphor-milieu can be unraveled.

In a life story, however, particularly if one supplements it with an in-depth study of an author's published works, there are these other questions which can be explored, e.g., elements of prose and narrative style which provide continuity and consistency in the author's way of rendering an account of the world.

The reader, too, brings a set of meanings and values to the encounter.

Suppose a conscientious social activist 'meets' an author whose highest ambition is logical clarity or poetic edification? How, in this encounter, can there be understanding? The ground on which these two types of perspective meet is the *text* itself (audio, visual, or print) and that symbolic mode offers the challenge to an exercise in translation. In this challenge one's a priori biases and predispositions have to be stretched, such as in the process of learning a. foreign language (Gadamer 1975). One has to discover the rules of this translation game by playing it, and all those efforts of hermeneutical scholars to find the appropriate method show how futile it is to pin the process down. Let me illustrate: it is no longer the personalities of Johan Turi or Emilie Demant Hatt which occupy the forefront of my attention, nor is it my own difficulties to empathize with the Lapp because of an instinctive wish to force-fit the Lapp story into the a priori categories of my own experience, but rather my enormous curiosity about *The Lapps and their Land*. What is given to historiographic scrutiny is, after all, the archival record, written texts, and the puzzle we can all share is that of our collective inheritance of symbolic transformations from the 'realities' of landscapes and events on the earth's surface to the prose of geography. The face of the earth offers the *explicanda* (what is to be explained), texts reveal how these have been interpreted by observers of different glance. We have the opportunity for a double kind of hermeneutic, then, which could trace the circle of experience and expertise on the one hand, and the dynamic interplay of thought and landscape on the other.

'Det gäller att väcka aptiten, inte att mata' (It is a question of whetting the appetite rather than spoon-feeding it), a Swedish high school teacher was wont to say. No doubt each essayist in this book felt there was much more to say. To minds bemused and bombarded by a priori hypotheses concerning intellectual history, and ideologically fixed stances on the social construction of knowledge and practice, this selection of stories offers the voices of lived experience. To the monological style of discourse which has come to characterize this functionally specialized era of scholarship, these stories offer an invitation to dialogue. Among the few unequivocal results of the effort to assemble this collection is that dialogue can indeed lead to greater self-awareness, mutual understanding, and the dawning of a more emancipated attitude toward geography as a vocation to life.

Notes

1. The term 'practice' should be understood throughout as 'thought and practice'. It is assumed that professionals in geography seek not only characteristic styles of description and explanation of phenomena and make certain claims to truth, but also express artistic, moral, and esthetic choices in their modes of writing, teaching, and applied work.

1a. See Appendix A.1

2. The first and third of these discussions are edited transcripts of videotaped conversations recorded at the NETCHE Studios at Lincoln, Nebraska (April 1978) and printed here with permission. The second is an abridged translation of Claude Bataillon's (1980) 'Table Ronde Imaginaire sur la géographie universitaire Française, 1930–1940', *Hérodote*, **20**, 116–15, printed here with permission.

3. I wish to acknowledge gratitude to Marvin Mikesell (Chicago), who offered this suggestion and prepared the profile of American geography, as well as to those colleagues who prepared profiles for their respective countries: Ronald Johnston (England); Olavi Granö (Finland); Paul Claval (France); Elisabeth Lichtenberger and Dietrich Bartels (German-speaking countries); Christian van Paassen (Holland); Hallstein Myklebost (Norway), and Staffan Helmfrid (Sweden).

4. The terms 'expertise' and 'experience' derive from the past and present participles of the same Latin root, *experiri*, to try, to learn from one's trials. 'Experiment' is a noun which, like 'expert', refers to past events, whereas experience is present and continuing. In so far as expertise is built upon successful experiments from which intellectual generalizations are derived, is it not inevitable that expertise plays a conservative role in the reproduction of scientific knowledge? The language and speech of the expert, moreover, be it sectoral or disciplinary, is also socially constructed, hence can only serve as a good medium for the communication of experience to the extent that societies can interpret words and symbols in a common manner.

5. I wish to acknowledge the time, energy, and positive support afforded by colleagues and students at Clark University throughout academic years 1979–81, in particular Bill Koelsch, Harry Stewart, Duane Knos, Jonathan Bordo, Tamar March, and the late Dan Amaral; Izhak Schnell, Rudi Hartmann, Anne Godlewska, Ute Dymon, Göran Nilsson, Murdo Morrison, Michael Enbar, Marianne Jas, Maxine Grad, Uwe Hermann, Jonathan Cheney, Barry Sillik, and several undergraduate students.

6. Hermeneutics (from *hermeneuein* 'to interpret') could be defined in a very general way as 'methodological principles of interpretation and explanation'. The origin of the term is frequently ascribed to Hermes, the wing-footed messenger of the gods, whose task it was to communicate what was beyond human understanding in a form that human intelligence could grasp. In Greek thought Hermes is associated with the discovery of language and writing. Traditionally, the interpretive task of hermeneutics involves three main directions of concern: (1) oral recitation; (2) reasonable explanation; and (3) translation (Palmer 1969) and each cultural tradition brings its own perspectives on this endeavor. Far from espousing a particular positive doctrine, however, hermeneutical scholars display diverse approaches, each of which offers partial yet coherent accounts of culture, language, and practice. Each seeks to restore an integrity to thought which has been lost by the overemphasis on general doctrines. There is a hermeneutics of religious texts (Bultmann, Schleiermacher, Rahner), of literature (Derrida, Barthes), of art (Croce, Gombrich), of science (Heidegger, Habermas, van Fraassen), of language and culture (Blumenberg, Arendt), of human understanding (Dilthey, Gadamer), and others. National traditions are also discernible. Gadamer's hermeneutics of language continues the work of Heidegger and raises issues germane to those of the critical social theory of Habermas and Apel. So-called French structuralists and the semiology of Barthes redeploy the cultural, linguistic, and psychoanalytical components of Marx, deSaussure, and Freud (among others), with a philosophical orientation traceable to Descartes. Rorty attempts to connect threads of English-speaking philosophy to recent continental thought. If there is one unifying theme which connects hermeneutics with other movements in recent Western thought, it is probably its turn toward the social – an effort to show that philosophical concerns over rationality, value, and action, can be best understood in the context of 'life forms' (e.g., Wittgenstein 1960, 1969) and to remove them from the autonomous discursive domain of philosophical doctrine. For a geographer (or geosopher) this turn may appear to be long overdue. Attempts to relate geographic thought and practice to the life experiences and landscapes of humankind could unquestionably be construed as a hermeneutical task.

7. See my 'Values in geography' (1974) *Resource Paper* No. 24, A.A.G., Wash. DC, and 'Grasping the dynamism of life world' (1976) *Annals*, A.A.G. **66**, 277–92.

8. Buttimer, A. and D. Seamon (eds), (1980) *The Human Experience of Space and Place*. Croom Helm, London.

9. The main arguments were briefly summarized in a memo to the Leningrad Congress of the International Geographical Union's Commission on the History of Geographic Thought, 'Proposal concerning philosophical critique and improved communication within geography' (1976) and more formally outlined at Edinburgh (1977) in 'On people, paradigms, and 'progress' in geography' (circulated first by Institutionen for Kulturgeografi och ekonomisk geografi vid Lunds Universitet, *Rapporter och Notiser*, **47** (1978) and reprinted in D. Stoddart (ed.) (1981) *Geography, Ideology and Social Concern*. Basil Blackwell, Oxford, 81–98).

10. Hägerstrand, 'Commentary' in Buttimer (1974), pp. 50–4.

11. In June 1977 Professors Olavi Granö (Turku), Dietrich Bartels (Kiel), Christian van Paassen (Amsterdam) joined Torsten Hägerstrand and myself in preliminary discussions about the project. See our 'Invitation to dialogue' (1980) *DIA Paper* No. 1, Lund and my 'Reason, rationality and human creativity' (1979) *Geografiska Annaler*, **61B**, 43–49.

12. Buttimer, A., Seamon, D. et al., (forthcoming) 'Creativity and context. Report on a seminar at Sigtuna. Professors William William-Olsson, G. J. van den Berg, Wolfgang Hartke, and Aadel Brun Tschudi were also present at the Sigtuna seminar. See also Walter (1980–81).

13. Transcripts of these interviews, and complete lists to date are available from Torsten Hägerstrand, University of Lund, and from Dialogue Project c/o Graduate School of Geography at Clark University.

14. The specific agenda for these seminars varied depending on the size and composition of the group. In general, however, each student was invited to pursue three distinct lines of enquiry and reflection: (1) discernment of his or her own values and the circumstances associated with his or her choice of geography as university subject; (2) in-depth study of one particular author's published works, autobiographical essay and video interview, as well as gaining some familiarity with the academic and social context of that author's career; (3) leading a discussion on one particular theme, e.g., thought and milieu, social structuration of thought, sense of place, and others, in the work of all authors then being considered in the group.

15. Some authors actually sketched their own life-paths, and those of Hägerstrand and van den Berg are included with their essays.

16. Ludwig Fleck's concepts of *Denkstil* (individual thought style) and *Kollektivstil* (collective thought style) illustrated well the social as well as epistemological aspects of scientific thought and practice (Fleck 1935, trans. 1980). The relativist underpinnings and contextual emphasis in this approach have, of course, been more widely popularized in the notion of 'paradigms' and 'scientific revolutions' (Kuhn, 1962).

17. Metaphor is used here in a very general sense, referring to the characteristically human propensity to transpose symbolically the unknown, or hitherto unfamiliar, into more familiar forms (Cassirer 1955; Langer 1953, 1957). Emphasis, however, rests primarily on the cognitive claims of metaphor, its capacity to generate paradigms and models of research (Black 1962; Ricoeur 1975; Pepper 1942). As literary trope, of course (Jakobson 1960; Levi-Strauss 1966; White, 1973b), metaphor could be regarded as serving an essentially representational function, pointing to relationships between parts and the whole. In this it can be distinguished from *metonymy*, which is reductionist in character and points to relationships among parts, as well as from *synecdoche*, which is integrative and suggests microcosm–macrocosm relationships, and from *irony* which is dialectical in character, skeptical of such analogical leaps as the others attempt, and thoroughly relativist in its ethical orientations. For excellent illustrations of the relevance of metaphor in geography, see Livingstone and Harrison (1981) and Berdoulay (1980).

18. The initial inspiration for this approach to geographic thought cane from

Stephen Pepper's *World Hypotheses* (1942). He claimed that there were really only four 'relatively adequate' hypotheses about the nature of the world in Western science, literature, and art: formism; mechanism; organicism; and contextualism. Later he qualified this and added another potential world hypothesis, but his arguments in the initial volume do appear to have a worthwhile message for historians of geographic thought. Our 'mapping' seems to correspond rather well to the tenets of formism, our 'functional systems' to classical mechanism, 'organism' to organicism, and 'arena' to contextualism. The so-called root metaphors underlying each of these world views are not treatable in the categories of the others. See my "Musing on Helicon: Root Metaphors in Geography" GEOGRAFISKA ANNALER, Vol. 64^2, (Nov. 1982), 89–96.

Chapter 2

Fig. 2.1 Clarence J. Glacken, USA

From the American Far West's Bay Region, with its 'easy-going and tolerant humanity', comes Clarence Glacken, whose *Traces on the Rhodian Shores* is a classic work. How many readers know that the author of this highly intellectual treatise was an eager voyager himself, tracing the shores of Pacific and Mediterranean, marveling at Norwegian *seter* and *fjord*, reliving history in the *Jardin des Plantes*, the streets of Istanbul and Port Said . . . and on to Asia via slow train from Beijing to Nanjing, eventually to discover the ingeniously woven textures of Japanese village life and landscape. It is this extraordinary combination of intellectual vision and sensitivity to the concrete that makes Clarence's contribution so unique.

To academic geography he is a 'late arrival', but his awareness of geographical interests comes early, not only from travel and reading, but also through first-hand experiences of human trauma in the Dust Bowl era, when he served with the Farm Security Administration in relieving the plight of migrants to the valleys of California, and became acutely aware of ecological problems associated with soil erosion.

What better route to follow in weaving all these disparate interests than a lifetime of research and reflection on the history of environmental ideas – those latent but powerful streams of consciousness and conviction which articulate themselves in landscapes and livelihood? Evocative indeed is Clarence Glacken's invitation to critical self-awareness in the Western world and an open, humble attitude toward other traditions of thought and life.

A late arrival in academia

Looking back in later life on the earliest years, one must be careful to avoid a teleological view. Hindsight can be an illusion, and we are apt to see how earlier events, experiences, and feelings (whose later significance was not apparent at the time) contributed in an orderly development to its culmination in a life's work.

My earliest recollection of what I wanted to be was a streetcar conductor. I regarded him as an important personage, not on the exalted level of a train conductor, but within the range of my aspirations. There were, however, several interests in early life which I later perceived to be geographical and historical. Sacramento, the capital of California, where I was born on 30 March 1909, is an historic city, perhaps not by Old World standards, but certainly by American ones. Its early days were closely associated with the discovery of gold, and it was a supply center for the mines. In my boyhood, I often roamed through Sutter's Fort and as an adult marveled at its furnishings; with only a slight suspension of belief, they seemed to come out of a Neolithic site. I visited the mill near Coloma where Marshall discovered gold, and took a picture of the monument erected to him with by Brownie box camera. Sacramento was also the western terminus of the Central Pacific, the first transcontinental railroad in the United States.

I came from old Sacramento families on both sides. My maternal grandmother was born in a hamlet long since absorbed into the city; my paternal grandmother, born in St Louis, had come across the plains, when she was nine months' old, in a covered wagon, and both grandfathers had settled there as young men. Both of my parents were born there.

Like most American children of the time, my introduction to geography took place in primary school (with many it has remained there). Ours was what the English call bays-and-coves, and what I call county-seat geography. I do not say this in disdain because it is a good thing to learn early in life the names of capital cities, principal rivers, highest mountains, largest lakes and to know where they are. Our text had interesting if conventional illustrations like the railroad station at Oakland, the State Capitol building, and an orange grove in southern California, no doubt long since disappeared, sacrificed to developers.

My elementary school teacher liked geography, and periodically we had a happy variant of the spelling bee. We would line up, she would snap out a place name, and if the student targeted could not point it out correctly on the wall map, he or she had to sit down. One of these places was the Gulf of Carpentaria. I do not remember whether I kept standing, sat down, or was even called upon, but I have never forgotten where it is.

I remember early fascinations that could be considered geographical. Sacramento is located near the confluence of the Sacramento and the American rivers (on whose headwaters gold was first discovered in California), and I often experienced a thrill in seeing the waters meet. It was also exciting to cross boundaries. It was customary, as it still is, for signs to be posted at county lines, and there was something dramatic in crossing from one to another.

The strongest stimulus in early years to an interest in geography was stamp collecting, a period which did not carry over into adult life. I had the usual beginner's Scott's blank stamp book, and on summer evenings especially, I rode a circuit on my bike to my friends' houses engaging in the delicious occupation of trading. Gradually I became friendly with hundreds of places throughout the world, including the Azores, Bavaria, Lichtenstein, San Marino, Bosnia and Herzogovina, German East Africa, the Sudan, Luxembourg, French Indo-China, Heligoland, Martinique. Some stamps enchanted me with their beauty, like those from the French colonies in Africa and faces on the stamps of Martinique. I had a beautiful vermillion $5 uncancelled stamp from El Salvador. When I was fifteen, my grandfather, a lifelong, avid stamp collector, presented me with a huge album which I still have. It has thousands of stamps, and along with my own later additions, considerably increased my knowledge of places.

I was most fortunate with my high school and junior college teachers, several of whom I remember with great affection. They were not scholars, but encouragers, gentle but firm correctors of incorrect English. The book that stimulated me most in my junior college years was Isaiah Bowman's *The New World. Problems in Political Geography*, first published in 1921. When I studied it in the late 1920s, the First World War and the Treaty of Versailles were still fresh in many minds. That was what Bowman's book was about, the new world created by the defeat of the Central powers and by the Treaty of Versailles. I was and still am impressed by the wealth of maps, some in color, some general, some detailed, some with distributions like nationalities, languages, and religions.

When I was about to graduate from the Sacramento Junior College, a friend, a young instructor, himself a graduate of the University of California at Berkeley in English, said I should take a course from Professor Teggart. In high school and junior college I had the usual courses in American history and modern Europe and had done some reading on my own, but I had no realization of what I was soon to discover, that there was such a field as the history of history and the history of ideas.

In 1928, when I entered as a junior, the University of California at Berkeley already had an international reputation with scholars and scientists of world renown. It was then, as now, a cosmopolitan campus with many students from various parts of the world. In my junior year, I cultivated my early interest in Asia and the Pacific by taking a course from a Chinese professor on Far Eastern politics and history.

I enrolled in Frederick J. Teggart's course, *The Idea of Progress*. It was a two-semester sequence, a revelation, because I then realized the importance of the history of ideas. I knew there were histories of science (mostly organized around the history of discoveries and inventions or the lives of scientists, not around scientific ideas) and of literature. I had myself read with great pleasure Taine's *History of English Literature* in translation.

In Teggart's course we learned about cyclical theories, ideas of a golden age, and eternal recurrence in the ancient world; the providential interpretation of history in the Middle Ages; and the full-fledged emergence of the idea of slow, gradual, continuous, and inevitable progress in modern times. We read Bury's *The Idea of Progress*, and a stunning collection of readings from the sources which Teggart had assembled.

His stature as a scholar intensified my interest. He was an eloquent lecturer and wrote beautiful, simple English innocent of jargon. (The jargonification of the English language was then in its infancy.) *The Prolegomena to History* and *Theory of History* are distinguished scholarly works. *Theory of History* (1925) I still regard as an outstanding and fundamental analysis of the nature of history and of historiography.

In my senior year, I became a major in the Department of Social Institutions of which Teggart was then the chairman. On the staff was a young assistant professor, Margaret Hodgen, who became a lifelong friend. She died, still working, in 1977 in her eighty-seventh year. She had read widely in the history of history and of the social sciences, was deeply interested in anthropology and its history, in cultural diffusion, the nature of social change, and cultural inertia. I took several courses from her, and she introduced me to many classics in these fields and to environmental ideas beginning with the Hippocratic corpus and including the French possibilists and Huntington.

I have often been asked whether in these undergraduate years I took any geography courses. I did not. I had heard of Carl Sauer, but I had no time for anybody but Teggart. I did not even bother to take courses from two world famous anthropologists, Kroeber and Lowie.

In these years I read two books which, in addition to the *Theory of History*, had a profound effect on me. The first was Edwin Arthur Burtt's *The Metaphysical Foundations of Modern Physical Science* (1925). It was not about the science of men like Galileo, Kepler, and Newton, but about their metaphysical ideas, the assumptions behind their discoveries, their religious beliefs. Burtt was particularly interested in the thesis that the Copernican triumph over the Ptolemaic conception had the effect of making the human feel lost and orphaned in the vast universe. It was a history and philosophy of science based on concepts and ideas.

The second book made an equally profound, but entirely different impression. It was John Livingston Lowes's *The Road to Xanadu*. What is it about? It is about Coleridge, not all of Coleridge, only the Coleridge of *The Rime of the Ancient Mariner* and *Kubla Khan*. (It was only recently when I reread these poems and *The Road to Xanadu* in connection with the work I am now preparing, that I realized that of all the great English poets, Coleridge was the most ardent student of geography. He was a tireless and insatiable reader of travel books, voyages of discovery with particular interests in the polar regions and the tropics.)

Lowes's *Road to Xanadu* is 'a study in the ways of the imagination'. He took both poems, almost line by line, and showed how Coleridge's glorious verses had come out of his unconscious mind. He had received and transformed mundane phrases or commonplace observations in the books he had been reading. I had long been interested in the nature of human creativity, not in the work of psychologists, but in reading personal accounts of creative people of how their ideas and inspirations had come to them. It was consoling to have been shown, by such painstaking documentation, that even a young poetic genius like Coleridge needed something to work on; he could not and did not create out of thin air.

The early 1930s were bleak years for young people, and the details of my life then are of little relevance here. One period is pertinent. In Franklin Roosevelt's first term, The Farm Security Administration, along with many new federal agencies, was established. I secured an appointment at the Regional office in San Francisco and for almost a year alternated between living there and traveling throughout the Central Valley of California.

It was my second period of residence in the 'The City', as almost everyone living in northern California called it. In my early teens, my mother's cousin had given me a lesson in the proper respect for place names. She sternly warned me against *ever* calling it Frisco. San Francisco then was far different from what it is today. The quarters now given over to a wide variety of 'adult' entertainment were then known, and affectionately, as Little Italy. There were many genuine reminders, including smells, of China as I walked down Grant Avenue through Chinatown, on my way to work. Fisherman's Wharf was a fisherman's wharf.

I do not know when my interest in cities was aroused, but it was surely intensified by living in San Francisco. I had often visited it in younger years and had taken the overnight riverboat ride, enchanting in moonlight, from the ferry building to Sacramento. At that time the Sacramento River was the Mississippi of the Hollywood film masterpieces.

Part of the program of the Farm Security Administration was to establish a series of camps in selected agricultural areas in California where migratory labor was employed most intensively, like the Imperial and Salinas Valleys and the Central Valley (Sacramento and San Joaquin) from Redding in the north to Bakersfield in the south. My job was to visit proposed sites, study the facilities like schools, churches, markets, transportation, crops, and similar matters that might have a bearing on decisions in establishing the camps. In these travels I became acquainted with many people (many, many, more in welfare administration in the late 1930s after returning from my travels) from Texas, Oklahoma, Arkansas, and other midwestern and southern states. These were the kinds of people that furnished Steinbeck materials for *The Grapes of Wrath*. One result of this and the subsequent experiences was that I became intensely interested in soil erosion on the plains, migration, and re-settlement. I prepared several reports, modest performances, factual, with statistics ranging from average daily attendance in schools to the going wage for agricultural labor.

Since my early teens, I felt provincial, limited in experience and outlook, notwithstanding the cosmopolitan atmosphere of the university and San Francisco. I had an intense yearning for travel and determined to be denied no longer, decided, once and for all, that I would do so. I had saved money, foreign exchange was favorable, and fares on ocean liners were remarkably cheap in tourist and third class.

I grew up and lived during the early part of my life in the age of the railroad and the ocean liner, not the world of airports which even now I find strange and disembodied. There is nothing disembodied about a train trip from Oakland or Los Angeles to Chicago, from Rome to Paris, from Port

Said to Cairo, nor is there anything disembodied about going through the Suez Canal on a great ocean liner or sailing out of an international port like San Francisco with hundreds of people on the dock waving good-bye while the bands play.

This is not the place for an account of my travels, which lasted from February to December 1937, and included Japan, China, French Indo-China and Angkor Wat in Cambodia, Egypt, the coast of Asia Minor, Cyprus, Greece, Turkey, the USSR, Finland, Sweden, Germany, Austria, Italy, France, and England. I would like, however, to make a few general remarks about these experiences.

I did not go as an accomplished trained observer, but as an ordinary young traveler with an A.B. My experiences intensified what my studies had suggested: the profound importance of cultural differences. In cycling in Cyprus, long before the terrible days to come, I knew by the minarets and churches whether I was coming to a Turkish or a Greek village. Shepherds I saw while crossing the Caucasus seemed modern representatives of a primordial way of life. The miles upon miles of tomb mounds on the train trip from Beijing to Nanking were silent, ancient testimonials closely linked to the Chinese family system.

In retrospect, I look upon my travels as a species of field work. There are many different conceptions of it, the most popular being the contrast between the library and outdoor tramping. In the light of subsequent interests, this was field work in an entirely different sense, a preparatory experience.

I do not think I would have ever developed my intense enthusiasm for the history of ideas without it. It would have been a world of abstractions. When I later studied the history of environmental ideas, especially the influences of climate, I remembered vividly many Mediterranean scenes or similar ones that had inspired such theories since antiquity. When I read Marsh's *Man and Nature* much later, I had a deep personal feeling for areas he had written about – parts of Italy, treeless slopes of Greece, goat grazing in Cyprus, the shores of Asia Minor.

Whether travel broadens one, I am not prepared to say; some think it only freezes one's prejudices and is a great stimulator of inaccurate observation. Perhaps it all depends on the person, but I have a mind that needs the concrete, the mountains, rivers, mirror lakes, cities, and markets. When I read about ideas and their history and they are related to places I know both become more vivid, more full of meaning.

Travels like mine are powerful reminders of the limitations of theory, of the pitfalls of models, of what in the 1960s became by some a virtual apotheosis of nomothetic thinking. The Romantics have been much maligned but places I visited in my life were also visited by Goethe (a Romantic of sorts), Byron, Shelley, Cooper, and Chateaubriand. For three golden days I did little more than walk in Istanbul. On my third-class railroad trip from Beijing the train stopped every few miles. Peasants with their luggage and their chickens got on or off, there was opportunity to buy boiled eggs or flowers for tea. A young mother would gently seat her child on the spittoon in the crowded railroad car when the calls of nature demanded it. All this has intensified earlier as well as subsequent feelings in me, a respect for individual experience, the discrete, the concrete, and a critical attitude toward theory, especially Western theories arrogantly striving to be ecumenical.

I have never forgotten these lines in Goethe's *Faust* which I read in my early twenties, when Mephistopheles tells the student,

> *Grau, teuer Freund is alle Theorie,*
> *und grün des Lebens goldner Baum.*
> [*Grey, dear friend is all theory,*
> *And green, life's golden tree.*]

About three months after the death of my first wife in 1941, I was drafted into the US Army, and save for a short period, served until April 1946. It is not necessary to go into these years except for the later period. I was selected as an officer-student in the Civil Affairs Training School, specializing in Japan, at the University of Chicago. The director, Fred Eggan, was already a well-known and highly respected anthropologist, and we had distinguished professors of the university as instructors. The burden of imparting Japanese culture was left to John Embree whose book, *Suye Mura*, became a Bible to us It is set in Kumamoto prefecture, in Kyushu, and Embree systematically guided us through all aspects of village life. We became good friends, and later I felt greatly indebted to his book when I lived in three Okinawan villages and subsequently wrote *The Great Loochoo*.

In addition to the formal lectures, we spent three hours a day six days a week in classes in spoken Japanese which I later put to use in Korea and in the Okinawan villages. The book which now stands out most in my mind from this period is Sir George Sansom's *Japan. A Short Cultural History*, in my opinion a masterpiece of modern historiography. I was particularly fascinated by his masterly account of the Tokugawa Shogunate. Many years later when I had joined the geography department at Berkeley, I met Sir George at a conference at Stanford. He was then quite old, a man of almost unbelievable grace and charm.

Most of my final year in the army was spent not in Japan but in Korea where I was deputy director of the Bureau of Health and Welfare in the military government. I especially enjoyed Kaesong, just south of the 38th parallel, the older parts of Seoul, and the cities of Taejon, Taegu, and Pusan, but what impressed me most was the extent of the deforestation in South Korea.

I had remarried and now had two children. After my release, I concluded the only occupation in which I could find real and enduring satisfaction was as a university professor. My wife generously supported me in this decision. Accordingly I worked for my doctorate at the Isaiah Bowman School of Geography at Johns Hopkins. I knew what I wanted to do. I wanted to write a history of ideas on a subject involving culture and environment. I had specific ideas in mind, some left over from pre-war years in the Central Valley, some from what was occurring after the war, especially the United Nations

Scientific Conference on the Conservation and Utilization of Resources (1949).

I must go back a moment to explain that my wife, Mildred, during the war accepted a job in Washington in the Office of the Director of Inter-American Affairs, and I was at the Adjutant General's Officer Candidate School at Fort Washington on the Potomac. My wife became an assistant to William Vogt, and I met him through her. The three of us were close friends thereafter. Although we did not see one another often in later years, I was much impressed by the studies of environmental destruction in Latin America which he had done for the Pan American Union and his *Road to Survival* (1948). His aim in that book, he once told me, was to make people in power, the politicians, and leaders in government aware of the dangers of continuing increasing world population growth and the worldwide destruction of the environment that was taking place, exacerbated by the increasing population. These developments encouraged me also to delve into the history of ideas about such subjects. William Vogt died on 11 July 1968, and I was honored to write an appreciation of and a tribute to him in the *Geographical Review*, **59**, No. 2, 1969.

My doctoral dissertation, *The Idea of the Habitable World*, was a history of four interlocking ideas, roughly from the middle of the eighteenth century to the post-war years: population theory; ideas concerning the influence of the environment on culture and the reverse; concepts of the soils; and modern ecological theory. It was not published, and I am happy that it was not. I am not ashamed of it, but after all these years, I have learned much more about the complexities of these periods.

Shortly after receiving my Ph.D., I was invited to take part in a project sponsored by the Pacific Science Board of the National Research Council, the *Scientific Investigation of the Ryukyu Islands*, part of the overall plan being to make ethnographic studies of the various islands. I accepted. There was no academic appointment in sight.

It was on my way to Okinawa that I first met Carl Sauer. I had been stimulated and encouraged by his writings, especially one of his finest up to that time, *Forward to Historical Geography*, in which he expressed his enthusiasm for studies in the history of geographical thought. I had a pleasant visit with him, but do not remember any details of our conversation.

In 1952 I began the study of three Okinawan villages: two, agricultural; one, a fishing village. My Japanese was rusty, but I was able to remedy that fairly soon. At first I stayed in the villages only during the day because of an epidemic of sleeping sickness, but later spent most of the time in them. Work of this kind from time to time could be depressing and discouraging because I often had the feeling (which I gradually overcame) that I was accomplishing little, that I was talking to a grandmother here, and a schoolboy there, and wondered often how I would be able to put together an organized ethnography.

From a personal point of view, the experience was most rewarding. I tried to organize my studies around three themes: the family system (a blending of the Chinese and the Japanese systems); the land tenure system, particularly the division of land and the effects of the inheritance system; and the actual results of these in the use and appearance of the land.

I had a deep respect for the people I lived with in the villages and found them as they were described so touchingly by Basil Hall in *Account of a Voyage of Discovery to the West Coast of Korea and the Great Loo-Choo Island* (1818), not as they appear in Commodore Perry's account (on his way to opening up Japan) poisoned by the prejudices of the missionary Bettleheim. A boy of fifteen often came from Naha to the village to visit his grandmother who could speak only Okinawan, a woman in her late seventies of great beauty and dignity. Occasionally I walked over to the primary school and led the class in teaching the boys and girls how to pronounce English, giving them a few simple phrases to learn. Once I addressed a large assembly of primary school students, and I never knew whether they were laughing because of what I said or because I had mercilessly massacred the Japanese language.

My wife and children lived in Berkeley during my stay in Okinawa. On my return I again called on Carl Sauer. On one visit, he asked me so casually that I did not know whether it was an offer or not, if I wished to join the department. I knew definitely I would when he told me what office I could have. So I began my career at the University of California in the fall semester of 1952.

Much has been written in recent years by geographers and others about places, the sense of place, roots, the importance of the past, and kindred preoccupations. No doubt there are many different reasons for these, arising perhaps out of a feeling of being disembodied (like going on a long plane trip without seeing anything); it is also associated, in this country at least, with nostalgia and love of the artefacts of the past whether they are old chairs or old beer cans, but I think more than nostalgia is involved. Vague as it might be, such feelings seem a recognition that we live in the present because there is a past; that the past creates for good and evil; that civilization is a creation in large part of the past.

My feelings were similar to these when I returned to the university from which I had graduated over twenty-two years ago, but I was on familiar ground. The great buildings, the masterpieces like the Doe Library, Wheeler Hall, California Hall, and the Hearst Mining building were still there. So was the International House where I had lived for a year. I was grateful for the opportunity Carl Sauer gave me to return. I had always felt like a Far Westerner, not in a chauvinistic sense, but the place I knew and felt most at home in was the Bay Region. To me it had an easy-going and tolerant humanity; I felt comfortable because of associations dating from boyhood, because I felt I belonged here, needing no guides to the terrain.

Carl Sauer, a kind, thoughtful and considerate man, believed in letting people follow their own research interests and in allowing them to give courses they wanted to and were prepared to teach. I was delighted when he invited me to give a course on the history of ideas, and he christened it with its university catalogue title, 'The Relations Between Nature and Culture'. It set out to survey the leading ideas on the subject in Western civilization from antiquity to the present. With the exception of leaves, I gave this course every year until my retirement in 1976; and once again in 1978.

The highlight of these early years in academia was my participation in the international symposium, *Man's Role in Changing the Face of the Earth*, held at Princeton from 16–22 June 1955. The idea for the symposium came from William Thomas, Jr, who combined with this inspiration a rare talent for organization and editing, proved again when the proceedings were published by the University of Chicago Press in 1956.

I was most grateful for this opportunity for two reasons: it gave me the chance to formulate the main developments in the history of the idea of man as a geographic agent, as a transformer of the natural world. I was not interested in the actual history of those changes but in the interpretations made of them. Equally important was the opportunity to meet people from various parts of the world who were concerned with the nature of these transformations. In the dedication to the published volume, in addition to the countless, nameless persons who had contributed to the symposium, the name of George Perkins Marsh, who wrote *Man and Nature* in 1864, and on the same theme, was singled out. Sauer called the symposium 'a Marsh festival'.

It had three chairmen, Sauer for 'Retrospect', Marston Bates for 'Process', and Lewis Mumford, for 'Prospects'. I had not met Bates or Mumford before. Bates's touch was so light it could be deceptive because he had a profound sensitivity to and knowledge of cultural and environmental interrelationships. Mumford I had admired (and still do) since my early twenties; I remember vividly reading *Technics and Civilization* (1934), one of the first works, to my knowledge, which dealt in a satisfactory way and in detail with the role of technology in the history of civilization.

I will never forget the symposium. The impact which the published volume, *Man's Role in Changing the face of the Earth*, made surprised many, including me. I used it for years in our course in cultural geography.

Two other conferences in which I participated made a lasting imprint on my research interests and teaching. The first, *Man's Place in the Island Ecosystem*, convened and organized by Ray Fosberg, the outstanding authority on the flora of the Pacific Islands, was at the Tenth Pacific Science Congress in Honolulu in 1961. Taking my cue from Cicero, I wrote a paper on *This Growing Second World Within the World of Nature*, a short history of ideas outlining the ancient and early modern background of modern ecological thought.

In 1965 F. Fraser Darling (who had participated in the Princeton symposium) and John Milton organized the conference on *Future Environment of North America*.

There I met William Vogt again (with the same worries) and Lewis Mumford who gave a stirring closing address. My paper, *Reflections on the Man – Nature Theme as a Subject for Study*, was an attempt to show common underpinnings of the theme and the opportunities there for histories of ideas.

Since my writings on this subject have been concerned with their development in Western civilization, I have frequently been asked whether I have studied ideas of non-Western cultures as well. I have and have found many parallels with Western thought, but have never seriously considered doing scholarly work in them. Particularly in dealing with non-Western thought one must know how the ideas are expressed in the original language. Over the years I have encouraged graduate students from non-Western countries to undertake such studies. Once made sensitive to the possibilities in the history of ideas, they can find rich ores.

For years, in addition to what are accurately and inelegantly called bread-and-butter courses, I offered three regional courses, one on the geography of China, Japan, Korea, and the Ryukyus; another on Western Europe; and a third on Mediterranean Europe. I have repeatedly been grateful for an early interest in the Far East (and in non-Western cultures as a whole) because it has been an insistent and imperious warning against parochialism based on the experience and study of Western civilization alone; this often takes on virulent forms in the study of society. Western models are regarded as inspired illuminations for the rest of the world. As part of this same feeling, I have been distressed by the simplification involved in contrasting the East and West (successor to ideas of the 'unchanging' and the 'mysterious' East) committed by people on both sides. It never seems to occur to them that the differences between China and India might be as great as the differences between either one of them and Western civilization.

My research and writings in the history of ideas have been within the traditions of Western civilization because I know it best and am a product of it, but I have never forgotten that it is a partial, not an ecumenical view.

In 1956, my book on Okinawa, *The Great Loochoo*, was published. During the eighteenth and nineteenth centuries the Ryukyu Islands (the Japanese reading of the Chinese ideographs) were known, in a variety of transliterations, as the Liu Chiu, the Lew Chew or the Loochoo Islands (the Chinese reading of their own ideographs) and since Okinawa was the largest island of the chain, it was known as the Great Loochoo Island.

This was my last venture in the study of an Asian culture. East Asia still interests me intensely. So does Japan. I was, however, too old, had too many academic and family responsibilities to undertake serious further study of the incredibly difficult Japanese written language. The main reason it is so difficult is that the ideographs (the Japanese call them 'kanji', a Chinese character of ideograph) were devised and developed by the Chinese for themselves not the Japanese.

In 1957–58, I was a Fulbright scholar to Norway and had the good fortune to be guided by the late Fridtjov Isachsen, an exceptionally cultivated man, then head of the geographical institute at the University of Oslo, who seemed to know every square foot of Norway, physical and cultural. I was impressed with the deep interest both in the university and the government in the history and preservation of rural life. The eastern coastal cities also attracted me, particularly Stavanger, Haugesund, Bergen, Trondheim and Tromsö in the far north.

My geographical studies centered around forest history, deepening my conviction of the close interrelationship between it and the history of civilization. I was introduced to the *seters*, the mountain chalets (part of the ancient transhumance from the Mediterranean to the Alps and on to Scandinavia) where people went for the summer to watch over the cattle, goats, and sheep, formerly making their cheeses there. Some *seters* are still in use; a particularly fine one was set up in the open-air museum at Lillehammer.

From an esthetic point of view I found deep satisfaction in the beauties of the Norwegian landscapes. One has the feeling that wherever one goes in Norway, one is never far from a fjord or an arm of a fjord, a lake, a waterfall, a mountain, or a glacial valley. It is supreme for seeing mirror lakes, especially on a still morning. The most glorious of these to me was Loenvatn, but Norway is full of beautiful lakes.

What enchanted me in Norway's geography was the variety of glacial landscapes, the glaciers themselves, fjords, cirques, glacial valleys, moraines, striated rocks, and erratic blocks. Each landscape seemed stolen from an illustration in a textbook on geomorphology.

On my return from Norway, I resumed my academic duties. In 1961 I completed a study of Buffon's *Histoire Naturelle*, with special emphasis on his descriptions of environments changed by culture. It was published in 1961 and later parts of it were incorporated in *Traces on the Rhodian Shore*.

Carl Sauer introduced the first offering in cultural geography in the department, a lower-division course planned for freshmen and sophomores. It was continued by Erhard Rostlund, and on his death by me. In the early 1960s, Paul Wheatley, who had joined the department permanently, and I offered a year sequence in the upper division in cultural geography in which we emphasized the non-material as well as the material culture. Among many other topics, Wheatley added materials on gardens, urban origins, non-Western cities, 'feng-shui' (commonly referred to as Chinese geomancy), and we shared lectures in religious geography. I added materials on the history of population theories, the history of ideas, and various attitudes toward the natural world. My association with Paul Wheatley was rewarding and enlightening. We listened patiently to one another's jokes, asides, and *ex cathedra* pronouncements for years, and there was never a harsh word between us.

Traces on the Rhodian Shore was published in 1967. The relation between human cultures and the environments in which they live is a brute fact of human life. Much of human thought has been devoted to this subject, concealed because it appears under different names. As my studies continued through the years, I became convinced that, although many ideas owed their origin to this brute force, three stood out in the history of Western thought: (a) the idea of the earth as a divinely designed planet; hence, the unity and harmony of nature – I argued that it ultimately led to ecological thought; (b) environmental determinants or influences on culture; and (c) human beings as geographic agents, transformers of nature.

Since Teggart's lectures, I knew the powerful influence of teleology and teleological explanation in Western thought, with all the variants like the design argument and the doctrine of final causes. I did not realize until I had studied the matter in depth how all-pervading teleology has been in the history of Western interpretations of nature, either in an extreme anthropocentric form in which the Creation was made for and exists for man or associated with the idea of a chain of being with man at the top, but not necessarily the lord and master of Creation. It is found in Cicero and in John Muir's *Mountains of California.*

I had received a Guggenheim fellowship, and after the book's publication I left for Europe. *Traces on the Rhodian Shore* carried the story to the end of the eighteenth century, and I now planned a sequel, on which I am now engaged, to continue through the nineteenth century to the beginning of the First World War. I decided to broaden the scope to discuss esthetic, emotional, and subjective attitudes to nature. I visited, and with great profit, The Nature Conservancy in London; The International Union for the Protection of Nature and Natural Resources at Morges, Switzerland; and the Swiss Union for the Protection of Nature (Schweizerischer Bund für Naturschutz) at Basel.

At Arles, still with many reminders of the Provençal poet Mistral, walking daily around the Roman arena, I wrote down what I thought were the points to consider in writing such a work.

I am quite sure why this topic attracted me to such a degree that I was willing to work on it for so long a period. Part of the explanation comes from the adventures I have already described. Another reason was that the deeper I studied these themes, the more inter-twining relationships came to the surface. I find this to be even more true in the work I am now engaged on. I have always been interested in works of synthesis; there is a certain security in them; they reverse trends to atomization, give one a feeling of interconnections, hence, of reality.

I was fascinated in seeing how ideas spread through disciplines seemingly remote, ideas like evolution, Platonism, the great chain of being (this was Arthur Lovejoy's great contribution), teleology, and environmental determinism. I wanted to show how these three ideas had diffused throughout wide ranges of thought, and this could not have been achieved by writing short articles or monographs on a single person or topic, a national school, or contemporary trends.

One can see the diffusion of ideas more clearly by studying the ideas, not individuals, periods, or individual disciplines. It was also a challenge to try to find some order in masses of historical accumulations. They must be related. People do not create out of the blue. People must build on what has gone before them. They are not that original.

The most moving experience was in Paris. I visited the Jardin des Plantes, whose history I had studied earlier, as a homage to Buffon. On the way, we stopped at Montbard on the Paris–Lyon Méditerrané line to visit his birthplace. The Jardin des Plantes is bounded on three sides by streets named after great French naturalists, the Rue Cuvier, the Rue Buffon, and the Rue Geoffroy St Hilaire. I walked along the Allé Centrale which leads to the Place

Valhubert, the Gare d'Austerlitz, and the Pont d'Austerlitz over the Seine nearby. At one end is a bust of Buffon, at the other a statue of Lamarck.

I should say a few words about relationships with students and teaching, difficult subjects to speak about generally and briefly. The relationship, even in large classes, can be personal and mutual. Students, perhaps unwittingly, often exaggerate the impersonality of a large university. The relationship between professor and student need not be a gravity flow down from on high. Undergraduate students are fresh, capable of asking most difficult questions, often jaunty, lazy at times, which can be an advantage because a professor's words and writings often are not *that* memorable. An American university professor will assume at his peril that students lack critical abilities. Over the years I have had many friendships with graduate and undergraduate students which have enriched my life immeasurably. Perhaps these are as close as anyone can get to the fountain of youth. Fewer opportunities for these encounters are what I miss most after retiring from active academic life.

In teaching, when appropriate, I used past experiences, especially travels, attendance at conferences, and results of sabbatical leaves, often indirectly, and without reference to the materials, because of the new ways of looking at things they suggested. If I went to a conference and thought it worth while, I told my classes about it. I always made it a practice to put on the blackboard the names of books I had read and found rewarding, and the main sources (other than myself) of the lectures.

Upon reaching the age mandated by the Regents, I retired in 1976, but one of the many blessings of the academic life for those who cherish it is that one does not really retire unless one chooses to. I no longer ask myself whether I have deep regrets about entering academia at an age when most others, starting as young assistant professors, were already full professors because the question no longer has much meaning for me. It is moot, as the lawyers say. I used to try to balance the advantages of a wide variety of experiences I brought with me to the university against the advantages of starting early and having a much longer career. As I look back on it, I think in my own personal case (I am neither generalizing nor advising anyone) that I would have had a more limited and constricted outlook had the second course prevailed.

If I were to summarize my academic experience, I would say that one thing leads to another; that embarking on an extended scholarly undertaking is like going on a long journey – one is bound to pick up a lot on the way; that one can never be too careful. A professor once wrote on a student's term paper that there were too many misspellings. He misspelled 'misspellings'.

Selected readings

1956   *The Great Loochoo: A Study of Okinawan Village Life*. University of California Press, Berkeley, CA, 324 pp.

1956   Changing ideas of the habitable world, in *Man's Role in Changing the Face of the Earth*, William L. Thomas, Jr (ed.). University of Chicago Press, Chicago, pp. 70–92.

1963   This growing second world within the world of nature, in F. R. Fosberg (ed.). *Man's Place in the Island Ecosystem*. 10th Pacific Science Congress, Honolulu, 1961. Bishop Museum Press, Honolulu, pp. 75–95.

1966   Editor's note to Norman Thrower, *Original Survey and Land Subdivision*. Association of American Geographers Monograph Series, Rand McNally, Chicago, pp. v–vii.

1967   *Traces on the Rhodian Shore. Nature and Culture in Western Thought from Ancient Times to the End of the Eighteenth Century*. University of California Press, Berkeley, CA. Paperback edn, 1976.

1970(a)   Man against nature: an outmoded concept, in *The Environmental Crisis*, H. Helfrich (ed.). Yale University Press, New Haven, pp. 127–42.

   (b)   Man and nature in recent Western thought, in *This Little Planet*, Michael Hamilton (ed.). Scribners, New York, pp. 163–201.

1973   Environment and culture, in *Dictionary of the History of Ideas*, Philip P. Wiener (ed.). Charles Scribner's Sons, New York, 2, pp. 127–34.

Chapter 3

Fig. 3.1 Aadel Brun Tschudi, Norway (Jan Hesselberg, Oslo)

'A beautiful person and a wonderful friend.' Aadel Brun-Tschudi has gifted many with inspiration and challenge. Despite failing health and the burdens of other responsibilities, she engaged wholeheartedly in our dialogue effort and contributed much insight indeed from her own lived experience.

Rich and varied has her own life been. Born in China of Norwegian missionary parents, broadcaster and foreign news analyst for Norwegian radio, mother of three, and eventually the first lady professor of geography in Norway, Aadel holds a precious place in the hearts of students and colleagues throughout Scandinavia. Geography was for her a 'calling', albeit a late one. During student years it had offered her a 'sense of purpose' and direction; to it she herself was to offer lessons in cross-cultural understanding and international development, and an edifying example of intellectual integrity and courage, human compassion and warmth.

The following essay was Aadel's final gift before her death in 1980. Time did not permit subsequent dialogue so I have added some reflections by Jan Hesselberg, who speaks for that 'lively group of young assistants' with whom she enjoyed much 'give-and-take' during her later years.

Worlds apart

I was born in a small town as the eldest of seven children, six sisters and one boy, in a small town in Hunan province, west of Lake Tungting, in the subtropical parts of China. My mother went by a small river junk for five days to stay near a doctor in case complications should occur. Upon her return to the place where she and my father were missionaries, they occasionally brought me with them on regular tours of 'out stations' in the mountain district to which they were assigned by the Norwegian mission conference. This, my father held, had a wholesome effect on people in the villages and townships who were then full of misgivings and easily roused by rumors and grievances as well as justified accusations against converts and adherents of the not yet firmly established foreign mission. When I was three or four years old my mother sometimes relied on me for some interpretation in chatting with visiting country women who used their highly localized tongue.

These experiences may well have unconsciously contributed toward a general cross-cultural awareness.

My earliest conscious memories were those of my Norwegian heritage. I simply adored my given name, Aadel, and have kept on loving it all my life: it was a unique name – no one had exactly the same – it set me apart from everybody else. It was derived from my grandmother from whom I had also inherited a coverlet of eiderdown, warm in winter and light as air. The down, I was told, had been collected from nests prepared for the almost naked birds just escaped from their eggs in northern Norway where my grandfather was a parish pastor at the time. The distance made my coverlet all the more precious. The courage of poor fishermen and farmer climbers, the busy stitching of my grandmother's needle in the beautiful soft-colored pattern of the satin cover made this possession something very special too – another proof that I was somehow unique – favored and set apart. My grandmother was highly praised for her ability as a housewife, having to move about between clerical incumbencies in widely different parts of Norway, and for her fortitude during adversities in late years. Looking at her handsome face in the simple, somewhat faded photograph in our sitting room I repeatedly renewed a decision to become like her. Whether this resolution was a reflex of the Chinese attitude to harken back in behavior patterns I cannot say, but it has lingered with me into my late life as a lasting decision about what would make life worth while. I once tried to put this conviction into words by saying: to me geography is a calling. I might just as well have said the same about some other obligations and chores that life has presented to me.

Awareness of belonging to two worlds was fostered in several ways. We were not allowed to speak Chinese or to mix Chinese and Norwegian in our home. When we happened to blurt out in a mixture our parents pretended not to hear. They encouraged us to play with Chinese friends and a Chinese teacher came to give lessons in reading and calligraphy. We were instructed to follow the Chinese customs then prevailing of strict separation between teenage boys and girls. As girls we were not allowed to go out by ourselves. On our daily twenty minutes' walk to school we depended on our trusted washerman and water carrier. With him we could chat as freely as with a dear friend. He must have been young because thirty-three years later I met him again when I traveled in China in 1957 on an assignment by the Norwegian Broadcasting Corporation. I was having tea with a handful of people left in charge of the old mission buildings. He bumped into the party, radiantly reminding me of happy old days, apparently not aware of his pitifully emaciated face and tattered jacket, all the more ragged looking because it was a padded one, or what was left of such a piece of clothing. He must have been hungry to judge from the way he gulped down some food I offered him. My hosts, however, briskly sent him away in their generally futile effort to keep me from uncomfortable discoveries.

Some insight into social relations was gained while occasionally accompanying my mother on visits to the womenfolk in Chinese homes. While listening to their chatting and my mother's admonitions I learnt to balance a bowl and saucer with boiling hot tea on the cupped palm of my left hand and to cool the potage without stirring while holding the lid in my right hand. More important, however, was the education my father gave about the condition of the peasants. As an experienced gardener he could explain why and how they practiced various combinations of crops and sequences of vegetables. Being genuinely interested in the people he wanted to convert he had acquired some knowledge about the plight of the peasants, their lack of land, their indebtedness and dependence on landlords and moneylenders. His soft-spoken but firm condemnation of malpractices instilled in me sympathies and antipathies which, I believe, helped me to a better understanding of the upheavals of land reform with which I was confronted when I returned to China in 1957. The impact of these lessons probably also contributed to my interest in the Third World and its problems.

At about 1919–20 a wave of unrest and protest swept through the Chinese student body and even reached the otherwise secluded girls' school next door to my home. The air was full of rumors and excitement about boycott of Japanese goods and there was even looting in the shopping district uptown. Naturally, my parents sided with their missionary educationalist colleagues in disapproving of these activities. I secretly admired them and for my sympathy was awarded a paper and bamboo fan with a map on one side and the infamous twenty-one demands by Japan neatly written out on the other side. The fan is still in my possession but resentment toward the Japanese left me when a Japanese student whom I met at Harvard in 1951 let me read her diary with the account of her experiences immediately after VJ day in Manchuria where her husband had served as a doctor. To my surprise, I found that my hatred of the Germans, developed during their occupation of Norway in the Second World War, also had disappeared. I guess the disclosure of sentiments and events put down in the diary opened my eyes to looking for individual people instead of a brash and basically glib bunching together of whole nations.

My sense of dual loyalties got a shattering blow when I came to Norway in 1924 to enter high school. We settled down in an industrial town dominated by a big pulp and paper factory. Times were not very good, Social Democrats and Communists were fighting each other in the union. Reverberations of this penetrated classes and free-time activities although very few working-class youngsters went to high school. Into this world I stepped, full of expectations about the good life in Norway and eager to have Norwegian friends and be accepted by them. This had been my dream for the last few years in China, perhaps because a slight sense of boredom and distance had been creeping into my relationship with Chinese girls. Naturally, I cannot say what my classmates and townspeople at large thought about the newcomer. All I can do is recount my feelings at the confrontation. It was a profound disappointment to find that being of Christian missionary background was a drawback and that it obviously would take both time and effort to be included. I discovered that my apparel was strange and called for laughter. Comments from on-lookers were probably just an expression of curiosity and not disdain as I thought. Nevertheless, my urge was to get away and hide which I did through long lonely walks. For many years I tried hard to conceal my background and if that was not feasible to keep silent about it. I got a certain compensation from throwing myself into debates and other activities in the students' union. Not till I was married and had settled down to a lasting life in Norway did I completely relax about the circumstances of my childhood.

My father died when I was eighteen years old and about to enter university. The future did not look too bright and a university education seemed out of the question. However, missionary friends invited me to stay with them for the first year and a relative – an American Presbyterian missionary – gave me a lump sum of 1,000 Kroner, at that time no trifling amount. So, I started a conventional round of subjects for a degree in philology: philosophy, Latin, English language and literature, history and Norwegian language. But time and again I told myself that this could not go on. I lived frugally, saved on occasional dinners, and gave private lessons to backward schoolchildren but even so I felt it was too hard on my mother who meanwhile had taken on a job as teacher in a private school, poorly paid, but probably her only choice then. In the evenings and late at night she was continuously busy making and mending clothes for the seven of us. So, I suggested that I cease studying and take on a job, but my mother refused to consider such silly talk. She had decided that I should have my degree and nothing could deter her. It probably did not make my repeated proposals any more palatable that the kind of job I wished for was as a factory girl – a romantic but of course entirely unpractical proposition, considering the stringent times, so stringent that the young girl who came in to alleviate my mother's household duties was very happy occasionally to take home some bread or left-over food.

Meanwhile, I left philology and switched over to geography on the initiative of Stephan, my beloved life-companion. Rather trifling as this seemed at that time it proved to be of decisive importance to my life-course. It gave me a sense of direction which I had previously lacked during my university years, a new purpose in life because it linked my studies to my childhood years.

The fact that I was virtually pushed into the study of geography is significant, because it initiated a series of turns in my life, brought about by outside intervention or stimuli. My professional life is no exception to this. I cannot, like so many others, especially forceful personalities, boast of neat planning of my life, based on independent decisions. My professional career was a broken one, as I married before having quite completed my final Ph.D. degree. It would have been unconventional nowadays but at that time neither this nor the fact that I stayed at home, taking care of three children, seemed a real failure to stand my ground. Admittedly, I was not entirely happy with the situation in spite of our rich family life, not in the least less rich during the very difficult years of the German occupation.

Through sheer coincidence I was asked to step into a broadcasting program in 1948 to report on developments in China. This led to more than fifteen years of regular piece-work as a foreign news analyst with the Far East as my special field. It also led to a request by Professor F. Isachsen, my teacher before I married, to resume my studies which I did. I got my degree – with my two boys in the audience at the public oral examination – whereupon I was lucky enough to be able to choose between fellowships to study at Harvard or in Paris. I chose the former and had the privilege to get to know the outstanding sinologue, John K. Fairbank, who generously sponsored me in spite of my age. Taking courses in his China Program gave me a feeling of having 'come back'. A full turn of the circle occurred some years later when I was offered a tour of Japan and some places in South-East Asia by the broadcasting agency. The opportunity to revisit China in 1957 brought my two worlds together again – a mental wound had been healed. In this new holistic world of my consciousness I could comfortably switch my attention from one part to the other.

This confidence endured even when in 1960 I stumbled into a position as university lecturer and was assigned a tough schedule of teaching the geography of Norden. I was requested to give up broadcasting work but decided not to comply. Even when I later on applied for promotion and was turned down, largely because the evaluating body found my production too scattered, my feeling of security in the course I had charted was not shattered. The first outright recognition by someone in the field of geography came in the form of an unexpected request to take on a visiting professorship in Stockholm. In spite of the small-sized geographical institute I joined, the experience was highly stimulating. After half a year of commuting I took on a lectureship at the East Asian Institute in the University of Oslo, only to find that three years later I was welcomed back to the department of geography to a new position, somewhat higher in the academic hierarchy.

From now on I had the responsibility of developing the new discipline of development geography. It was a big challenge. Besides teaching I had to attempt to qualify myself in the diverse development theories. The years ahead proved to be most exhilarating. Engaged students made the work rewarding and research projects took me to countries in Africa and South Asia as inspector of field work, mainly financed by the Science Research Council but supported by our State Aid Agency which assigned to us areas for field work in the assessment of resources and development potentials. During these years I also toured China several times, in all making five visits to the People's Republic.

The last tour was unique. I was then accompanied by sixteen colleagues, mostly physical geographers, from different universities. The excursion was sponsored by the Chinese Academy of Sciences and proved to be very successful even though our gracious hosts did not in all respects comply with the professional standards on excursions that we are used to.

Just as turning to China studies in the early 1950s was like 'coming home' so the last tour of China and the positive response by my colleagues was a kind of professional full turn of the circle. Looking back from my retirement I cannot but admit that my interests have covered many fields, perhaps too many to give my production the character of having been directed toward a definite goal in the academic sense. There has, however, been an inner consistency.

My first major bit of research focused on rural settlement problems, specifically the depopulation of uphill country in south Norway associated with emigration to the United States. It was a resource-poor area and the labor and living conditions of the farmers on their tiny holdings made a profound impression on me. Insight into their problems made my research all the more worthwhile. Ever since, knowledge of life under different environmental conditions has been a sort of passion with me. I have somehow felt – perhaps naively – that added knowledge in itself might contribute to the improvement of an objectionable state of affairs. Research purely for its own sake has never appealed to me, probably because I had but slight qualifications for achieving methodological advancements. Once I quarreled with two of my departmental colleagues because I thought that in the results of our research during a field course we should include some recommendations to planners. They, however, argued that our job was merely to submit our findings.

Geomorphology, particularly the impacts of glaciation on the Norwegian landscape, was another exhilarating field of knowledge in my first period of study. The insight I acquired on excursions – however slight – proved to be an eyeopener to the importance of natural conditions for the activities of man. My interest in diverse aspects of nature has been a lasting one and the interaction between man and his natural environment a central theme of my thinking. Accordingly, I have been a strong adherent of the idea that geography should be a unitary discipline in the sense that as far as possible it should cover both physical and cultural aspects of our earthly habitat. Incidentally this is still done in our department. It is, of course, only practicable at the elementary level where teaching may give students an understanding of the necessity to view man and his activities in relation to the natural resources at his disposal.

My preoccupation with the problems of developing countries has fortified my conviction in this respect. I realize that determinism easily surfaces, e.g., in studies of man and land in tropical countries. However, what I have defended is an ecological approach: to study ramifications of over- or under-utilization of resources does not imply that environmental conditioning is of prime importance. Land use and rural settlement, whether in Norway, Slovenia, Tanzania, or China has been my chief concern. Integrated rural development, lately a slogan, to me means simultaneous advancement on all sectors; agricultural productivity, appropriate industry, health and nutrition and last but not least, the lot of women. This much too neglected aspect, was called to my attention for the first time when we had a Tanzanian student living in our home. He disclosed that he had never been to his mother's kitchen and while performing his share of dishwashing and other household chores vowed that he would tell his wife of the advantage to be had from such sharing of duties. He urged me never to forget the importance of the attitude of women to the general improvement of social and economic conditions in African countries.

My interest in land use and the geography of agriculture has brought me to a number of countries outside Norway as a corresponding member of the IGU Commission on Agricultural Typology. When my traveling caused sarcastic remarks from a colleague it left me completely unruffled because of my conviction that it served my purpose of getting to know as much as possible about the environments of different peoples. In this respect the conditions in some Central East European countries gave me invaluable new vistas.

My children were reared under the shadow of Hitler's Germany, a shadow which deepened during the occupation of Norway. When I resumed my studies a few years afterwards it was perhaps natural that I should take an interest in political geography. The German Geopolitical School and the Japanese philosophy behind their Co-prosperity Sphere, viewed as an equivalent of 'Lebensraum' did not make any lasting impression. However, H. Mackinder's ideas of the Euro-Asiatic heartland and the world dominance of whoever controlled it I found fascinating, as I did E. Huntington's climodeterminism as a possible explanation of movements of people across the Euro-Asiatic continent and of the difference between the physical and mental vitality of different peoples.

The brand of political geography preoccupied with border problems and the proper delimitation of boundaries seemed rather boring. The functional approach I used myself in analyzing Far Eastern policy problems which made up the bulk of my writings while I was working for the broadcasting up the bulk of my writings while I was working for the broadcasting corpora-corporation.

Mathematical and statistical analyses of geographical data appealed to me, each for its potentials. A few faltering attempts at grasping some of the initial elements, however, told me that I had neither the perseverance nor, more importantly, the qualifications necessary to be able to use computer-based analysis. So, I was happy to revert to work with land use and other rural-related problems along conventional lines, particularly mapping for analysis and interpretation. In a way, the problems lately posed by pollution and menacing depletion of resources came to my rescue as a good justification for working in the old manner – to subsume research under the rubric of inter-action between society and its environment, and man's stewardship of the resources at his disposal.

My study tours of the People's communes in China were both enjoyable and frustrating. Immensely enjoyable were the opportunity of meeting people, seeing different cultural landscapes, and watching what the diligent farmers were doing to improve their land. Frustrating, because it often was rather difficult to distill realities from verbose propaganda and, at times, exaggeration of achievements. To my great satisfaction discussions of such topics were several times interrupted by words like 'well, you know the old China', which was, of course, an acknowledgement of my ability to surmise certain things about their motivations.

Failing health unfortunately prevented me from making full use of material collected in China and about Chinese conditions. Partially this was due to the responsibility tied to my task of introducing development geography in our department. An inner urge, or perhaps ambition and vanity, made me give priority to teaching and research in this field which did not sufficiently over-lap with my Chinese studies. This preference of mine was stimulated by the overwhelming interest of students both at intermediate and advanced levels. These were truly rich years. Besides bringing me to some new countries outside Europe they made me learn a lot from the give-and-take relationships with students and young researchers. I do believe that this happy situation may be partly ascribed to the fact that I am a woman, incidentally the only permanent female faculty member in any Norwegian geography department. A woman, notably of my generation, will in general be less inclined than a man to assert herself *vis-à-vis* students, will be more accommodating, and more apt to recognize the points of superiority which will inevitably become manifest in bright advanced students.

Communication – dialogue – creativity

Reflections on the work of Aadel Brun Tschudi

Jan Hesselberg: To lecture about facts, models, and theories may be carried out in several ways. There is, however, a fundamental distinction between the mere transfer of facts and ideas and of shared personal and scholarly experiences of the 'world out there'. Aadel Brun Tschudi had the gift, rare nowadays, of stirring enthusiasm in her students at lectures. She was able to create a feeling of insight and reflectiveness among students. Not that she used many words to create such a feeling, a short remark or two sufficed to put the text of the day in a broader and meaningful perspective. Such remarks did not belong to a particular political ideology or to a fashion of the day. They went far beyond.

Not all of Tschudi's lectures were like this. In the last years her most fruitful lectures were those when she had relatively little time for preparation, and hence would advance *ad hoc* lines of argument which put the main text in a reflective light within a broader frame of reference. The atmosphere in the lecture hall then was very attentive and quite special. It was a unique experience to be part of such a journey into unknown territory. For once the answer was not given beforehand. One was confident, however, that the vast personal and intellectual experience of the lecturer would make for a logical sequence worth traveling along.

Not everybody got the same feeling of satisfaction after such a lecture. To communicate properly on this level needs an open and interested mind on the part of the listener. Opinions aired in the tearoom afterwards accordingly varied. The bits and pieces of reflection, which Tschudi offered, needed to be elaborated upon by the listeners themselves. This was, and still is, an uncommon situation for students of today, who are used to getting 'everything ready made for instant consumption'. A most interesting fact emerged during the tearoom discussions. There was usually little agreement among those expressing remarks about the major points and valuable insights derived from the lectures. Her reflective remarks were at a level of abstraction which had obviously met with individual experiences and individual compositions of knowledge and understanding. It has been said that lectures should not be written down and that what is written should not be directly lectured. There is an obvious difference between both the content and style of these two forms of communication. Tschudi observed this in her way of lecturing.

In my experience as a student, communication as Tschudi managed it was unique. Communication in the sense used here is, however, short of dialogue. The former implies a substantial inequality of knowledge and experience between the two partners, whereas the latter can only be rewarding if participants are on a fairly equal level regarding the topic of discussion. Accordingly, I did not at once come into a dialogue with Tschudi when I stepped from student to research fellow. She not only gave freely of her experiences (a rare practice at our university) but she actually listened and accepted viewpoints from us, the younger generation. I realize now in hindsight that I felt like being in a dialogue with Tschudi at that time. By creating that feeling in us, she undoubtedly put the stage right for us to be creative. She gave us responsibility for courses, and she acted on our opinions on different matters when we agreed among ourselves but happened to disagree with her. To be democractic one has to be strong. Tschudi gave us an example of this fact.

Selected readings

1934   Avfolkning av Vest-Agder og nedlegging av heigärdene saerlig i Sør Audnedal og Spangereid, *Norsk Geogr. Tidsskr.*, 5, 207–52. (Depopulation of West Agder and pasture-farm abandonment especially in Spangereid.)

1957   *Japan, soloppgangens land.* Aschehoug, Oslo. (*Japan, the Land of the Sunrise.*)

1963–64 Tomatøyene i Boknfjorden, *Norsk Geogr. Tidsskr.*, 19, 1–50. (Tomato-growing in Boknfjorden.)

1968   Land use problems on the urban fringe: The case of Sørkedalen, Oslo, *Norsk Geogr. Tidsskr.*, 22 252–63.

1969   (In collaboration with H. Myklebost), Hadeland jordbruk og pendling, *Ad Novas*, 7, 9–49. (Agriculture and commuting in Hadeland.)

1971   Asia, in H. W:son Ahlmann (ed.). *Asien, Sovjetunionen, Ostasien och Sydasien.* Natur-och Kultur, Stockholm, pp. 229–86.

1972   Ujamaa villages and rural development, *Norsk Geogr. Tidsskr.*, 26, 27–36.

1973   People's communes in China, *Norsk Geogr. Tidsskr.*, 27, 5–37.

1976   Rural development planning in China, *Norsk Geogr. Tidsskr.*, 30, 17–24.

1978   *Gjensyn med Kina.* Grøndahl, Oslo. (On Re-visiting China.)

Chapter 4

Fig. 4.1 William R. Mead, England

'It was about 11:00 p.m. around midsummer and we stood on a birch-clad tumulus overlooking Nordfjord. Our Norwegian company...broke spontaneously into "Ja, vi elsker". It was a perfect illustration of what George Steiner calls "a form of interanimation" between people and the land they occupy.'

A sense of temporality, historical depth, and esthetic sensitivity rings through the work of William Mead. With vignettes like this his spoken and written word has lured many to geographical awareness. Cradled in the Vale of Aylesbury, his own geographical sense developed early, and grew through excursions, diary-keeping, and adventures in literature. It was an accident of university curricular requirements that led him to economic history and geography rather than to English literature: *felix culpa* indeed, for to geography he has brought so much insight from poetry, saga, music, and regional novels.

Finland stands out as the cherished focus of emotion and reason: it remains for him 'in a continuous process of unfolding'. North America was once a close rival: it was by the summer shores of Lake Ontario that this essay was written. William describes himself as a 'promeneur solitaire', but I doubt if there is an aspiring humanist anywhere who would not find in his life and work the resonance of a kindred spirit.

Autobiographical reflections in a geographical context

The Vale of Aylesbury

Daniel Defoe's *Robinson Crusoe* began his narrative boldly and simply in the first person singular. It is a good example to follow and I emulate his opening paragraph. I was born in 1915 in a small market town in Buckinghamshire, some ten miles from the family farm from which my father (the fourth son of a typical Victorian family of seven) had been eased out after he left school. Church Farm, Stewkley, standing next to the sturdy Romanesque church in the centre of the village, had been in the family for several generations. The name of the family is scattered through the church records back to the seventeenth century; but, with an Anglo-Saxon name such as Mead in an Anglo-Saxon clearance settlement such as *styfic leah* (stump clearing), my ancestry may well precede the time of the Dane Law. My father, following the apprenticeship that was common in those days to an uncle who was a cornchandler and grocer, became a master grocer. My mother, who was also of farming stock, died when I was three and my brother was six months' old. My father always longed to return to the land. Indeed, together with an uncle he bought two farms – one with an eye to an early retirement which the war prevented. My roots are accordingly in the countryside.

The detail of the local landscape was interpreted to me as early as I can remember, though in an ecological rather than in a geographical way. I suppose that I learned the names of trees and plants as soon as I learned anything. I never remember not knowing the difference between oak and ash, elm and beech, wheat, barley, and oats. By the age of four, I was taking long walks with my father – sometimes following the escarpment of the Chiltern Hills, where chalk pits would be explored and where flint stones were gathered in order to strike sparks; sometimes, to the margins of the Corallian limestone country, where I never ceased to be excited by the foot-wide ammonites built into a parkland wall. There were also drives in a pony and trap. My favourite drive was on the north side of the heavy clay Vale of Aylesbury where, through gated fields, we would leave the gravelled track and take to the switchback grassland (later to be appreciated as the fossilized ridge-and-furrow of the open field system).[1]

On these occasions, I became aware of another feature of the countryside. Sometimes, the roads were wide, with broad grass verges: sometimes, they were narrow and entrenched, with overarching hedgerows on either side. Living in a market town, one was very conscious of the livestock sales and the droves of cattle and sheep that congested the streets on Wednesdays. The importance of the roadside verges for their drovers was clear. The weekly distribution of markets, mostly six to ten miles from each other and each on a different day of the week, made an early impact. (The pattern was immediately recalled when I first encountered R. E. Dickinson's work on East Anglian market towns about 1936.)[2]

Through these walks and drives, I gradually acquired a knowledge of an increasing number of localities. Some were highly attractive to me, some less so. A collection of favourite places began to emerge. In turn, these were inseparable from particular times of the year. As a result a kind of botanical calendar built itself up in my mind. I knew which fields offered the best cowslips, purple orchises or fritillaries (now all but destroyed by the plough and by chemicals) and which hedgerows offered the best blackberries, hazel nuts, crab apples, and sloes.[3] The search for each in due season was anticipated with zest. One favourite place was a double hedgerow (later to be speculated upon as manorial boundary which had never been incorporated in ploughland) where we found the earliest anemones, primroses, and bluebells. Beyond this, the difference between the chalkland and the clayland floras soon impressed itself. For one who lived in the Vale (a *champagne humide* as I came to appreciate it in a classical French context) there was already some-

Fig. 4.2 The Vale of Aylesbury

thing distinctive about the plant assemblages of the chalk country (or *cham-pagne pouilleuse*) – the scabious, knapweed, rock rose, bedstraw, toadflax, harebell, wild clematis, hornbeam, spindleberry (which came into my first poem). Above all, it was the wild thyme that evoked the chalk downs (its scent makes me sympathetic to the notion of an olfactory geography as conceived by the Finnish geographer J. G. Granö).[4]

Perhaps these plant associations help to explain the sensitivity with which I can respond to minor differences in the countryside and with which I can detect parallels between places. It was a revelation to discover uncannily familiar pockets of country when I first visited the Paris Basin. Some of its little valley-ways immediately suggested to me counterparts that issued from the lower chalk escarpment of the Chilterns. It made me think how very much at home some of the Norman French must have felt as they took up residence in and bequeathed their names to a scatter of springline settlements in my own home area.[5]

The Vale of Aylesbury is pleasing but not outstandingly attractive country, yet for me it has come to have those qualities that Georges Duhamel attributed to his Ile de France. 'To this narrow stretch of countryside I have bound myself through an intimate acquaintance with the trees, the seasons, the scents of the living earth, all sorts of tastes, joys and passions. From it, I have received my first and my most affecting impressions of the world . . . . This is my fief, my personal possession, my very earth.'[6] Much later, I encountered Restif de la Bretonne's *Monsieur Nicolas* and *la bonne vaux* to which he retreated.[7] My comparable green valley is a broad Vale. I can recall moments (recent as well as long ago) when time was suspended in it. It is a green vision, luminous in early June haze, best perceived floating downhill into the warmth of the Vale on a bicycle or riding along a bridle path with near-ripe grasses tapping the toes in the stirrup. It is just before the hay is mown and just before Matthew Arnold's 'high midsummer pomps' take over. Then, the sense of place is at its most personal.

I recall having the same response in Norway. It was about 11:00 p.m. around midsummer and we stood on a birch-clad tumulus overlooking Nordfjord. Our Norwegian company had clearly been moved by the same sensation and they broke spontaneously into *Ja, vi elsker*. It was the perfect illustration of what George Steiner calls 'a form of interanimation' between people and the land that they occupy.[8] For me, it strengthened the siren song that Norway had always sounded in summertime.[9]

A feeling for words

The feeling for words came much later than the feeling for places. Three points are probably relevant. First, I did not learn to read until I was seven, but I remember some of the earliest words that I was able to decipher. They were attached to a picture of Eskimos around an igloo and they were in Arthur Mee's *Children's Encyclopaedia*. They ran 'Will the Ice Age Come Again?' There was a second picture of a bright star approaching a dark world and it was captioned 'How will the world end?' (It all came back when I first saw Thornton Wilder's *Our Town*.) Secondly, dialect words made an early impact. A vocabulary assembled from mid-Bucks dialect became central to family anecdotes and private jokes. (Later in life I was fascinated by the generous employment of the glottal stop among the inhabitants of my father's home locality and wondered if it might be a legacy from the original settlers. Only in western Jutland do I recall a more vigorous employment of this throatal phenomenon.) Thirdly, foreign words made their first impression through French at the age of nine when we had some simple instruction at school. I still find French a peculiarly attractive language. Indeed, it seems to me to be the natural language of geography.

Although I was an avid reader by the age of fourteen (with Rider Haggard and Alexander Dumas swallowed whole), it was not until I was in my seventeenth year that I remember words dancing on a page and print becoming alive. Soon afterwards (and partly through sixth form teachers), I sometimes became so sick with the excitement generated by a page that the book had to be closed. On these occasions, a page could acquire a curious physical transparency and its contents might have several meanings at once. I imagine that a conductor reading a musical score can have the same kind of ecstatic experience.

At the local grammar school, I responded willingly to everything appertaining to words and places. There were botanical walks, when the Latin names of plants were matched to the English. We went to unlovely as well as agreeable places so that we might identify weeds and other noxious plants. Our first geography master, an ex-naval commander who had sailed the seven seas, gave lessons which reminded us of Puck who cast a mantle round the world in forty minutes. But he transmitted an enthusiasm for maps, which I began to draw for the sheer pleasure of it. (How I envy Alan Sillitoe his essay on maps.)[10] The commander's successor, who had come straight from the university, really brought us abreast of geography, putting over with great aplomb what I suspect were his undergraduate lecture notes. But the lessons in English literature affected me most, both because of the personality of the master and because of the eloquence of his exegesis.

Words now began to be committed to paper in a prodigal manner. I had kept diaries since I was about ten years' old. Next, came a phase of copious note-taking, to be followed by the inevitable adolescent verse and drama (the theatre had already captured me). Essays were always a pleasure and in a sixth form where there were precisely three students following the arts course they were treated tutorially. The habit of writing down impressions became well nigh automatic. Needless to say, there was total frustration during the early years of the war (especially in Iceland in 1941) when diaries and notebooks were strongly forbidden. Subsequently, the opportunity to write extended letters from Canada to England was a great relief. After the war, a series of notebooks, focusing principally on Scandinavian journeyings, was begun.[11] They incorporated (and continue to incorporate) extracts from books, manuscripts and conversations, soliloquies, home thoughts from abroad, and a series of rather more organized sections devoted to subjects of special interest. Even if their contents are not easily recoverable, such records prove invaluable for all sorts of purposes. A wise friend suggested that a notebook should never be so large that it cannot be kept in the pocket. There is much to be said for the principle of Thomas Gray that a word or two written on the spot is worth a cartload of recollections.[12]

Ends and scarce means

Having no advanced level school Latin, I could not read for a degree in English literature. The alternative was the London B.Sc. (Econ.) which at the time consisted of a miscellany of superficially unrelated subjects. Independently of the fact that I took two papers in geography and one in the strongly appealing economic history, two points of consequence may be recalled. First, translation from two modern European languages was required so that, perforce, I had to learn to read German. Secondly, I read and re-read several books with lasting profit. They included Graham Wallas's *Human Nature in Politics* (1908), R. H. Tawney's *Religion and the Rise of Capitalism* (1933), and J. B .S. Haldane's *Possible Worlds and other Essays* (1927). The first of these contained passages which were curiously apposite to the late 1930s: 'When a man dies for his country, what does he die for?' What is the personification of a country that emerges from his life's stream of sensations . . . 'the printed pages of the geography book, the sight of streets and fields . . . perhaps the row of pollarded elms behind his birthplace'?[13] Tawney did not deal with Scandinavia, but many of his ideas helped towards a fuller appreciation of that part of the world (so, too, at a later period did those of Daniel Boorstin). *Possible Worlds* was the first collection of essays that I read by a contemporary scientist and I have returned to them on a variety of occasions. (P. B. Medawar's *The Art of the Soluble* (1967) and A. Koestler's *Act of Creation* (1964) were to have a similar impact.)

Perhaps the most critical book at the undergraduate stage was Lionel Robbins's *Essay on the Nature and Significance of Economic Science* (1932). From it, I learned to look on behaviour as 'a relationship between ends and scarce means which have alternative uses'. The material relevance was immediate, but not until the age of twenty-four (see below p. 54) did it dawn on me that the most scarce means of all was time. Since then, I have been pathologically conscious that choosing one line of action means denying time to another. It was another way of looking at time from that engendered through phenology and the passage of the seasons.[14] In parallel, *Der Rosenkavalier* meant more because of Richard Strauss until I could appreciate the contribution of Hugo von Hofmannsthal[15]:

> *Die Zeit, die ist ein sondbares Ding.*
> *Wenn man so hinlebt, ist sie rein gar nichts.*
> *Aber dann auf einmal, da spürt man nichts als sie:*
> *sie ist um uns herum, sie ist auch in uns drinnen.*
> *In den Gesichtern rieselt sie, im Spiegel da rieselt sie,*
> *in meinen Schläfen flieszt sie.*
> *Und zwischen mir und dir da flieszt sie wieder.*
> *Lautlos, wie ein Sanduhr.*

Prelude to Finland

A kind uncle left me £500 in 1937 and that is how I came to settle down to a spell of postgraduate work on Finland. The origins of the Finnish interest defy explanation. All that can be said is that it had something to do with Sibelius, that it was inseparable from my first reading of *Kalevala* and that, in turn, this was related to an undergraduate country-by-country appraisal of the geography of Europe. L. D. Stamp had written an excellent paper on what he called 'amphibious Finland', but Kalevala was all that I could find in English translation from the literature of Finland.[16]

So it was that because of her interest in Europe, Hilda Ormsby, Halford Mackinder's first geographical assistant, supervised a master's thesis on the geographical background to Finland's foreign trade. There was no pressure to learn appropriate language, but in the company of a number of like-minded friends, I had lessons in Swedish. Swedish gave me access to official Finnish material as well as the possibility of rapidly acquiring a reading knowledge of Norwegian and Danish. I went to Finland for the first time in 1938. It cost £10 return from Hull to Helsinki – including three generous meals a day.

For two years, I read hungrily all that I could find about Finland, joined the Anglo-Finnish Society and sought out Finnish students. Then, as my supervisor pared down a far too wide-ranging dissertation, I recast the rejected material into papers which to my surprise found favour with editors. Thus, *Economic Geography* published a verbose paper (which I sometimes give to tutorial groups to criticize stylistically).[17] *Geography* gave space to 'Finland and the winter freeze' in a version much improved by H. J. Fleure's editorship.[18] *Baltic and Scandinavian Countries*, edited in remote Torun, immediately accepted a paper on Finland's foreign trade (which still reads quite well).[19] The *Geographical Review* found in 'Finland in the sixteenth century' a contribution which threw into relief the Finnish Winter War of 1939–40.[20]

By 1939, the Second World War was changing lifestyles. I had been short-listed for the post of Secretary to the Finnish–British Society in Helsinki in late August; but the vacancy was filled locally until the cessation of hostilities. Accordingly, I registered for a Ph.D., was evacuated to Cambridge with the London School of Economics, joined the RAF in November (though I was not required until the following June) and became deeply involved in a campaign to raise money for the Finnish Red Cross.

Canadian interlude

The war years were kind to me. They offered the opportunity to see something of Iceland, to acquire an intimate knowledge of the chalk downs of Wiltshire, and to make the acquaintance of Canada (with the chance of brief forays into the USA). It was in North America in 1941 that the meaning of time was borne in on me. Simultaneously, I became conscious of the release of physical energy. If Griffith Taylor (whom I had the pleasure of meeting in Toronto at the time)[21] and Frank Markham (whom I knew later)[22] conceived untenable theories about climate and the energy of nations, my personal experiences nevertheless suggest that the energy of individuals is by no means to be separated from climatic circumstance. In the New World, I never remember feeling tired.

I was not really interested in being sent to Canada where I was to play a small part in the establishment of an RAF school of navigation. As a result, for a while, I cut myself off from my surroundings, spent all of my spare time reading Scandinavian literature and sought contact with the nearest geographers. They were Jessie and Wreford Watson. Thanks to them and their circle of friends, my response to Canada began to change. Soon, I immersed myself in it, travelled as widely as I could, read all that was new about Canada and much more that was old. Indeed, I all but exchanged my Scandinavian skin for a Canadian skin. Wreford Watson invited me to join him at McMaster University the day on which I was demobilized in 1946.[23] If the first time that I let reason prevail over emotion in my academic career was in 1937 when I made a conscious decision to adopt Finland, the second time was in 1946 when I decided that I wanted to remain in Europe.

But a Canadian bond had been forged – perhaps more correctly a North American bond. The frequency of my visits to Canada has increased with the passage of time and the weekly correspondence has not flagged for thirty-five years. The correspondence began with a series of letters from Canada to English friends. On returning to England – to a tented site on the flanks of Old Sarum – I wrote similar letters to my Canadian friends. The Wiltshire letters are testimony to the affection that I rapidly found for the surrounding countryside – its prehistoric sites, its medieval churches, its unspoilt villages, all accessible by bicycle across the undulating chalk downland (which was now given new dimensions through the writings of the naturalists W. H. Hudson and Richard Jeffries).

Rooted in the war years is a geographical textbook on the USA and Canada.[24] My colleague Eric Brown, who had also been in Canada during the war, agreed to collaborate. We completed the book in about eight months in our spare time. I go to Canada most years, but no longer attempt to keep abreast of what is happening in that open-hearted and boundless land where, despite ample technology and limitless material potential, centrifugal forces and ethnographic tensions remain depressingly persistent.

The Liverpool experience

In November 1946, Clifford Darby appointed me to his department at Liverpool, thereby rescuing me from the civil service (in which I had been offered an appointment), from the temptation of Swedish and Finnish government scholarships, and from the Canadian dilemma. The three years in Liverpool – a city of great character despite its post-war demise – saw the resolution of a series of personal tensions. Divergent interests became convergent. To be encouraged to explore the English landscape, to be urged to deepen and extend the Finnish connection, to be expected to go to Scandinavia and to be asked to produce a lecture course on North America seemed a fate too good to be true. Simultaneously, I found myself in the company of colleagues who complemented the study of the landscape with the study of the land – an approach which, in the light of my childhood experiences, made splendid sense.

Above all, Clifford Darby encouraged specialization. I need no longer be inhibited by those who accused me of modifying Danton's dictum into

Finland, encore Finland, toujours Finland. I could enjoy the totality of experiences that the country offered, meet (or at least correspond) with everyone who was writing of Finland, and find out why those who had written and were no longer living had espoused the Finnish cause. I began to see something of relevance to Finland in almost everything that I read and accordingly to see the country from a continuously changing perspective.

The germs of other ideas also belong to the Liverpool years. Thus a lecture by W. G. Hoskins reminded me of the 'switchback' fields in the Vale of Aylesbury.[25] It became apparent that, although Maurice Beresford had undertaken sample studies of the coincidence between the strips on pre-enclosure maps and contemporary ridge-and-furrow, no one had conducted any extensive survey of the feature.[26] Buckinghamshire with its varied geology, soils, and relief seemed to me to be an ideal county in which theories could be tested. Accordingly, a ridge-and-furrow map for the county was completed in 1953 and other county surveys were added as research assistance became available.[27] Throughout a near decade of controversy, it was a continuous source of amazement to me that those who debated the origins and significance of ridge-and-furrow did so exclusively in an English context. Europe, with all its wealth of literature on field systems, was completely neglected. It was amusing in the summer of 1953 to have quoted to me on a croquet pitch in the heart of Finland Alice's comment to the Red Queen, 'why it's all ridges-and-furrows'.

From ridge-and-furrow, it was but a short step to the appraisal of field boundaries – both the hedgerow, with its childhood associations, and the stone walls of the Cotswolds, Dorset, and the north country. These and other features of the landscape became the object of student exercises during the field excursions that were such a regular and popular component of early post-war undergraduate activity. So, too, did the works of the regional novelists and early topographers. Among the novelists, Thomas Hardy claimed pride of place and entered into our departmental iconography.[28] Pilgrimages were made to Dorset, with novels in hand and declamations upon the spot where their action had taken place.

H. C. Darby's concern with the English topographers enabled me to re-assess the significance and stature of their Scandinavian counterparts. As a consequence, I returned to the work of Linnaeus and his school, and in particular to the travel diary of his favourite pupil Pehr Kalm.[29] Kalm rapidly became one of my great 'Integrators', because around him and through him my interests in Finland, Sweden, Norway, eighteenth-century England, Canada, and the eastern seaboard of the USA were drawn together.[30] In addition, here was a Finn who had spent three weeks at Little Gaddesden in the Chilterns.[31] The 150 pages of the diary that he kept there in the spring of 1748, gave a fuller description of the local scene than anything written by a British contemporary. For the better part of twenty years, I have led an annual undergraduate excursion in the footsteps of Kalm, read from his diary on the spot where it was written and, in the light of it, sought to contrast the mid-eighteenth-century scene with that of today.

It will be evident that H. C. Darby's approach to geography in general and to historical geography in particular has had a singular attraction for me.[32] The brief Liverpool interlude enabled me to see my home scene with new eyes and the Scandinavian world from new angles. If my *Historical Geography of Scandinavia* (1981) did not take shape unconsciously in Liverpool, it would never have been conceived without the Liverpool experience.

Finland – *ad infinitum*

I returned to Finland in the summer of 1947, firmly resolved to work in the field of agricultural geography. The return was by way of Stockholm and with introductions to Finnish geographers from William William-Olsson. Helmer Smeds was my initial mentor and he introduced me to the rural problems of his home province of Vasa, in a Finland sadly different from that of a decade earlier.[33] From Vasa, I was dispatched to Finnish-speaking Savo, where good fortune placed me in the hands of August Jantti, a practical academic with wide-ranging interests at the meeting ground of farming and forestry. Here the groundwork for *Farming in Finland* was laid.[34]

But I was soon to realize that my field of investigation was to thrust me into a socio-economic scene the causes, character, and consequences of which began to engage most of my attention. It was the problem of resettlement of displaced people and, in particularly, the farming communities that had been evacuated from the ceded territories. The resettlement programme affected everyone I encountered and was an emotional issue in which it was diplomatically difficult for any outsider to participate – let alone an Englishman whose country had been partly responsible for the Finnish predicament. The reverse of resettlement was the compulsory acquisition of farmland and forest to accommodate them. This problem was of national magnitude and solutions to it were needed as a matter of urgency. At the same time, Finland was required to pay a major indemnity to the USSR and was haunted by the threat of a Russian takeover.

Coincidence brought an American visitor to the University of Liverpool directly from Helsinki in 1948 and I had no idea that he was a representative of the Rockefeller Foundation. Since the Foundation had just provided substantial support for a sociological enquiry into Finland's resettlement problems my own concern immediately attracted attention – the more so in that my project and approach were complementary to those of Helsinki's sociologists. As a result of fellowship came my way.

Among the many aspects of rehabilitation none affected me more than that of the farmers who were being settled on pioneer holdings – 'cold farms' as they were called. In their activities there were overtones of the Anglo-Saxon clearances in England (not least *styfic leah*) and the colonization of wooded Ontario. I left Liverpool for a year in Finland in 1949. The problems that I was to face were very human and their roots were geographical in all senses of the term. I found a room in the home of the director of the Forestry Research Institute and the sociologists introduced me to a resettlement area in Lapinlahti, in the by-now-familiar province of Savo.

In this way, I came to 'adopt' a group of displaced farmers at different stages of resettlement and sought to find a common pattern in their experiences.[35] A questionnaire survey was conducted among some sixty of them and, against the background of the marked seasonal rhythm of agricultural life in Finland, I invited farmers and their wives to keep work diaries for a number of selected weeks throughout the year 1949–50. In order to gain the confidence of the families, it was necessary to live and work with them for at least a token period. It was even more necessary to keep a finger on the pulse of their activities throughout the year, so that regular visits had to be paid to them. As a result, I had some firsthand experience of primitive pioneering (for there was no capital to invest in machinery). And, since it was socially unwise to be seen in the cars of the many visiting officials, travel was on foot, on ski, and on bicycle.

During the year's residence in Finland, I also pursued a hobby which had been started in Liverpool.[36] It was the search for everything appertaining to Anglo-Finnish relations. The wide circle of Finns with whom it brought me into contact was in complete contrast to my friends in Savo. The interest in Anglo-Finnish relations sprang out of a paper on early British travellers to Finland that had been given to the Roscoe Society at the University of Liverpool. Among the members of the Society was C. N. Parkinson who, when he left for Singapore, bequeathed me his two volumes of the Naval Records Society covering the Baltic Campaigns of 1854–55 and besought me to follow up Captain Hornblower's Finnish connections. In Finland, I was soon to find myself tumbling over all sorts of material – archival and unused as well as printed and forgotten. I was especially thrilled by the manuscript diaries of Johan Jacob von Julin (in private hands and still unpublished), by the (then) unprinted and unedited diaries of Pehr Kalm's journey to England on his way to America, and by Zachris Topelius's journal of his mid-nineteenth-century visit to England (which led me on to his still unpublished lecture notes on the history and geography of Finland).[37] An unwritten book exists in the Anglo-Finnish theme.

The resettlement investigation could have merited a monograph, but it was reduced to a chapter in *Farming in Finland* and to a number of papers. The content of both the questionnaire and the work diaries provided material for the kind of diagrammatic representation that has always appealed to me. The results are embodied in activity graphs and in models of the process of pioneering. One of the models was repeated in a simple way following a visit to a number of the pioneer farmers twenty-five years later. It goes without saying that the reunion was a revelation – again in both human and geographical terms.[38]

The history of the representation of seasonal work rhythms is interesting in its own right. The miscellany of diagrammatic material of which it consists seems to owe most to the agricultural economists and the French regional geographers. Yet, for anyone living on Europe's oceanic fringe the seasonal round in northern Europe is expressed most strikingly in Finland's winter circumstance. The impact of winter on Finland as an independent political unit is unique in its effect on all socio-economic rhythms.

The experience of my first Finnish winter reminded me of Quebec and recalled for me the study of man and winter in Canada by Pierre Deffontaines.[39] For the book that winter prompted I wanted a collaborator to help gather material, to check the facts, and to monitor the manuscript. I found a sympathetic co-author in Helmer Smeds. He entered fully into the spirit of *Winter in Finland* (1967), which heralded new research fields in Finnish geography, provided a meeting ground for bringing the artistic and the scientific together, and offered a serious geographical presentation which was not without popular appeal.

All this activity began to draw me more closely into the circle of Finnish geographers. Stig Jaatinen spent a year at University College, London, during the course of which we investigated the representation of Finland in British geographical and cartographical literature – and incipient study of mental maps.[40] There were two principal results. First, our investigation led us to the library and atlas collections of the Royal Geographical Society (and, incidentally called my attention to the neglected and uncatalogued archive of the Society).[41] Secondly, it generated an interest in the history of Finnish geography and the geographies of Finland past. In this way, I was led to the library of Åbo Akademi and to the archive of the Finnish Economic Society.[42]

Other historico-geographical sources such as the Swedish military archive in Stockholm, with its detailed late eighteenth-century reconnaissance materials from eastern Finland[43] and the *af Schulten* hydrographic surveys, attracted my attention. Add to this an irresistible urge to dabble in the Public Records Office for papers relating to the British and French naval campaigns in the Baltic as well as the excitement of turning up a wide scatter of references in Finnish provincial archives and it will be clear that Finnish affairs were getting out of hand. They were brought into focus again by an invitation to write a general book on Finland.[44] A second general study on *The Åland Islands* followed in its wake.[45] Stig Jaatinen was a willing co-author of a book which will always recall visits to his summer home on the northern outskirts of the Vårdö archipelago.

There are three final points to be mentioned about a Finnish experience concerning which anything written can be no more than a silhouette. First, it may claim to have opened up source material and presented points of view which are new to the Finns as well as to others. Secondly, it has contained a strong interpretative element – not least in that it has deliberately attempted to diffuse the ideas and work of Finnish geographers and other scholars to a wider audience.[46] Thirdly, it has brought me a number of postgraduate students, two of whom – Michael Jones and Michael Branch – have become significant Finnish specialists in their own right. Michael Jones worked on the social and economic consequences of isostasy in the Vasa area of Ostrobothnia.[47] Michael Branch began his distinguished Finnish studies with a detailed appraisal of one of the East Karelian travel diaries of the nineteenth-century philologist and etnographer A. J. Sjögren, whose neglected archive had attracted my attention some decade earlier.[48]

The experience of Finland has been a continuous state of unfolding. It remains so. In general, I have not sought after new issues and problems to investigate. They seem to have sought after me. The perpetual dilemma is when to resist them or, if they are irresistible, where to check them.[49] The dilemma might have been even greater but for one major personal deficiency – my inadequate command of Finnish. Perhaps it was too late for a non-linguist to attempt it when over thirty. More likely, there have always been priorities to which it has yielded. It is a social shortcoming that, in a community where 93 per cent employ Finnish as their first language, I should slip more easily into the minority tongue.

Attachments are sensitive plants. Ultimately, I do not want to probe too deeply into the reasons that account for the Finnish connection. I am inclined to agree with Claude Levi-Strauss that 'When we make an effort to understand, we destroy the object of our attachment.'[50]

The Scandinavian connection

Side by side with the narrower and more intensive concern for Finland runs a broader Scandinavian connection. The return to Scandinavia in 1946 was an emotional experience of a high order. I was able to go to Norway to help at a British Council summer school for Norwegian teachers at a time when travel was still denied to British tourists. The pleasure of lecturing on English literature was repeated during five subsequent summers, though never again in the idyllic setting of Nordfjord. A series of educational excursions for geographers (which brought me to Mundal in Fjaerland) and the friendship of a Norwegian businessman (Harold Meltzer, who initiated me to fell-walking) strengthened the Norwegian association. Meanwhile relations with Swedish geographers in Stockholm and Uppsala had been firmly established.

It was not until 1953 that I allowed myself to be tempted by publishers to write a book on Scandinavia. (In the meantime, I had had the pleasure of a working visit to Fredrika Bremer's 'New Scandinavia' – Minnesota and Wisconsin – which broadened my appreciation of the Old World Scandinavians as well as introducing me to their New World kith and kin.) The book bore the title *An Economic Geography of the Scandinavian States and Finland* (1958). Its form and content were planned to complement Andrew O'Dell's *Scandinavian World*. Besides attempting to introduce to the English-speaking world the range of contributions made by Scandinavian geographers to the understanding of their lands, the book sought to display the diversity of methods of diagrammatic representation available to the economic geographer.

Naturally, the book generated argument – not least the title which, though acceptable to Finnish-speaking friends, was disapproved of by those whose mother tongue was Swedish. By way of compensation, a popular book (written jointly with Wendy Hall) automatically included Finland under the title *Scandinavia*.[51] Meanwhile, the members of the Nordic Council (i.e., the five countries of 'Scandinavia') set about urging the use of the collective Norden. So the argument is unresolved, though I am now sufficiently sure of my judgement that the outside world has to include Finland in the group of 'Scandinavian countries' despite ethnic and cultural differences. Contrastingly, I suffer an increasing measure of schizophrenia over the collective personality that I ascribe to Scandinavia, because as soon as I set foot in any constituent country, all the forces of national identity seem to assert themselves. All this leads me to reiterate continually that the most important fact in human geography is the nation state – a concept which geographers either appear to take for granted or else to reject as unworthy (or incapable?) of analysis.

Partly because I have become conscious of the ephemeral character of much economic geography, I have shifted increasingly into the field of historical geography. (George Steiner may be oversimplified in his theory that the scientist is forward-looking and the artist is backward-looking, but it may help to explain the conversion of a fair number of British geographers into students of the past.)[52] Accordingly, following the example of friends who have aspired to write their regional historical geographies, I have essayed an historical geography of Scandinavia. It was difficult to bring the task to an end. First, the volume of relevant material seemed to expand like the universe. Secondly, I was fearful of asking the advice and criticism of friends. To ask was to invite further debate about the balance between countries and topics, to raise matters of interpretation (on which no two might agree), and to expose omissions which could only delay the completion of the work without fundamentally modifying the content. Such problems must be common to all who embrace the study of other lands.[53] Thirdly, with the experience of advancing years, one is increasingly conscious of the form that one is imposing upon the object or area of study. Is the result nothing but a strange, often myopic produce of the mind's eye – albeit in the phrase of Paul Klee, a 'thinking eye'? Perhaps the best that one can hope to achieve is a little explanation and a little illumination.

Rêveries du promeneur solitaire

These autobiographical reflections suggest that in the earlier stages of a career one tends to walk alone. To engage in postgraduate work in pre-war Britain was to be different, to choose to work on Finland was to be doubly different. Even to select the field of geography was unusual. Forty years have changed the situation fundamentally. Geography has become a burgeoning research discipline. In addition, the generally gregarious character of those who practise it makes geography doubly acceptable to *le promeneur solitaire*. At the same time, there has emerged a company of scholars in Britain, well integrated with their North American colleagues, whose research centres on Finland (and they are all better equipped for the task than I am).[54] Geography is in a sense my public sector, but in one respect at least, Finland remains my private sector. Rockefeller grant apart, I have not sought after research money for Finnish purposes. I prefer to be independent and not to ask for public funds for my private interests. 'No academic is an island,' as R. J. Johnson writes, but in this respect I am almost one.[55]

Having taken an introspective line, a few further points may be added. These reflections have reminded me of the degree to which I am a visualizer. Past scenes are recaptured in great detail: the printed pages of books are recalled with striking clarity. Certainly, the training that I have picked up as a geographer has increased my capacity for observation and heightened my awareness of the forms of order in the landscape. Recall is inseparable from order, disorder has always disturbed me. There is pleasure in symmetry, though ironically the attraction of the romantic prevails over that of the classical (and that goes for my musical tastes, too). The remembrance of past scenes is also inseparable from colour – and disharmony in colour also distracts. It is perhaps natural that a visualizer should be drawn to open spaces and that confined places should be distasteful. There is always exhilaration in the high plains of the American West, on the wide fells of western Norway, and on the high seas. Contrastingly, caves and aircraft cabins are disagreeable and I am happier on a horse than in the pressed steel box of an automobile.

All this seeing and sensing has occurred within a system of options which has expanded sometimes logically, sometimes according to chance factors. There has been no conscious search for synthesis, though the kaleidoscope of interests has been brought intermittently into a state of repose. It is perhaps natural that the convergences should have been almost invariably in a Finnish context. Finland is a country of modest dimensions where paths intersect with greater frequency than in larger and more populous countries. Yet, while at one time an intersection may be anticipated, at another the long arm of coincidence can become alarmingly short. The superstitious might attribute it to the traditional Finnish necromancy, but it is equally reasonable to suppose that if one has a burning interest in a topic everything is drawn into its orbit in one way or another.

As a specialist working and playing with Finland, I find that almost everything that I read may have an idea, a point of view, a phrase or even a word that is applicable to that country. As a generalist in the field of geography, I find that I am frequently most conscious of the character of the subject when I am reading in other disciplines. It is partly for this reason that I find the division of universities into departments increasingly irksome. The free flow of ideas across the entire range of knowledge is of the essence of a place of learning. From the free flow, geography has gained much: to it, the subject has contributed not a little. The contribution has been integrative – and integrative in two ways, first through its synthetic approach and secondly through its human relations. In Britain, at any rate, geographers are generally less given to fratricidal rivalries than most academics. The clash of ideas may be healthy and inevitable, but it can impede progress as well as promote it. To reduce disharmony between those who further knowledge is to be able to increase the volume of pleasure springing from its pursuit. Is it entirely surprising that one of the few geographical autobiographies bears the title *Let Me Enjoy?*[56]

I am not a futurologist, but I do not foresee geographers losing their birthright. They have built up a very considerable tradition in a relatively short time and there are plenty who are competent to sustain it. I see nothing Spenglerian in the move from a springtime concern with the physical and rural to an autumn emphasis on the industrial and the urban. Nor do I anticipate a winter in which tyranny of theorists and technicians will drive out the poets.[57]

At least, that is how it appears by the summer shores of Lake Ontario with the wisdom of hindsight fitting the chain of events almost too neatly together. At another time, in another place, the explication might be much the same, but the implication would inevitably be different.

Notes

1. See Selected readings, 1954.
2. Dickenson, R. E. (1932) The distribution and function of the smaller urban settlements of East Anglia, *Geography*, **17**, 19–31.
3. Mead, W. R. (1966) The study of field boundaries, *Geographische Zeitschrift*, **54**, 101–17.
4. Granö, J. G. (1929) *Reine Geographie*, Helsinki. (An English translation is being prepared.)
5. For me, the Vale and the Downs are now identified in their French terminology, *champagne humide* and *champagne pouilleuse* respectively, cf. Mead, W. R. (1964) The Chilterns: an eighteenth-century appreciation of the cultural landscape, in Clayton, K. M. (ed.), *Guide to London Excursions*. L.G.U., London, pp. 110–13.
6. Duhamel, D. (1926) *La geographie cordiale de l'Europe*, Paris, a book which I originally read for purely Finnish reasons, cf. the beautifully written chapter 'Le chant du nord'.
7. Chadbourne, M. (1958) *Restif de la Bretonne*. Hachette, Paris, pp. 3–5.
8. Steiner, G. (1975) *After Babel, Aspects of Language and Translation*. Oxford U. P., London and New York, p. 453.
9. It is expressed in the introduction to Mead, W. R. (1959) *Norway* (How People Live Series), Word Lock Educational Company, London.
10. Sillitoe, A. (1975) *Mountains and Caverns, Selected Essays*. W. H. Allen, London.
11. I was particularly intrigued to encounter the notebooks of the Swedish geographer Kjellén, cf. Kjellén-Björkquist, R. (1970) *Rudolf Kjellén*, Stockholm. The notebooks are deposited in the University Library, Uppsala.
12. See Selected readings, 1975 and Kallas, H. and Nickels, S. (eds) (1968) *Finland, Creation and Construction*. Allen & Unwin, London.
13. cf. Qualter, T. (1980) *Graham Wallas and the Great Society*. Macmillan, London; and Lionel Robbin's testimonial, 'As a teacher, he surpassed anyone I have ever known.'
14. New vistas were also opened by Carlstein, T. *et al.* (eds) (1978) *Timing Space and Spacing Time*. Edward Arnold, London; reviewed *con amore* (1978) in *Geographical Journal* **145**, 299–301.
15. Hofmannsthal, H. von (1959) *Gesammelte Werke, Lustspiele, I*. Berlin.
16. *Kalevala, The Land of Heroes* (1961) (trans. Kirby, W. F.) Everyman Library, London, vols 259, 260 – and the consequential Mead, W. R. (1952) Kalevala Englanissa, *Kalevala Seura vuusikirja*, **33**, 119–31.
17. Mead, W. R. (1939) Agriculture in Finland, *Economic Geography*, **15**, 125–34.
18. Mead, W. R. (1940) Finland and the winter freeze, *Geography*, **24**, 221–9.
19. Mead, W. R. (1938) Anglo-Finnish commercial relations since 1918, *Baltic and Scandinavian Countries*, **2**, 117–25.
20. Mead, W. R. (1940) Finland in the sixteenth century, *Geographical Review*, **30** (3), 400–11.

21. e.g., Taylor, G. (1936) *Environment and Nation*, University of Chicago Press, Chicago.
22. Markham, F. (1942) *Climate and the Energy of Nations*. Oxford U. P., London.
23. We wrote a short paper together, Mead, W. R. and Watson, W. (1946) Canada in the American balance, *Culture, Quebec*, **4** (2), 85–92.
24. Brown, E. and Mead, W. R. (1962) *The United States and Canada*. Hutchinson Educational, London.
25. Hoskins, W. G. (1955) *The Making of the English Landscape*. Hodder and Stoughton, London.
26. Beresford, M. W. (1948) Ridge-and-furrow and the open fields, *Economic History Review*, **1**, 34–45.
27. Mead, W. R., Harrison, M. J., and Pannett, D. J. (1965) A Midland ridge-and-furrow map, *Geographical Journal*, **81**, 366–9; Mead, W. R. (1977) A ridge-and-furrow map of Leicestershire and Northamptonshire, *The East Midland Geographer*, **6**, 8, 48, 382–5; Mead, W. R. and Kain, R. (1977) Ridge-and-furrow in Cambridgeshire, *Proc. Cambs. Ant. Soc.*; Mead, W. R. and Kain, R. (1977) Ridge-and-furrow in Kent, *Archaeologia Cantina*, **42**, 165–9.
28. Darby, H. C. (1948) The regional geography of Thomas Hardy's Wessex, *Geographical Review*, **38**.
29. Mead, W. R. (1954) A northern naturalist, *The Norseman*, **12** (2), 98–106; (3), 182–8.
30. I read his travels for the first time in Liverpool. I have made my own translation of his journey to England from Martti Kerkkonen (1966, 1970) *Pehr Kalm Resejournal*, Helsingfors, but it is unpublished.
31. See selected readings, 1962.
32. Let alone his concern with stylistics, cf. Darby, H. C. (1962) The art of description, *Trans. I.B.G.*, **30**, 1–14.
33. cf. Smeds, H. (1935) *Malaxbygden, bebyggelse och hushållning: södra delen av österbottens svenskbygd*, Helsingfors.
34. Mead, W. R. (1953) *Farming in Finland*, Univ. of London, Athlone Press, London.
35. Mead, W. R. (1957) The margin of transference in Finland's rural resettlement, *Tijdschrift v. economische en sociale geografie*, 7/8/, 293–302; Mead, W. R. (1959) Frontier themes in Finland, *Geography*, **44**, 145–56; see also Selected readings, 1951 and 1958b.
36. Mead, W. R. (1949) The discovery of Finland by the British, *The Norseman*, **7** (4), 1–16, which was followed by a dozen papers exploring the same theme.
37. Mead, W. R. (1968) Zachris Topelius, *Norsk geografisk tidsskrift*, **22** (2), 89–100.
38. Mead, W. R. (1976) The changing geography of a Finnish lakeland parish, *Norsk geografisk tidsskrift*, **30**, 93–101; cf. also Selected readings, 1968.
39. Deffontaines, P. (1957) *L'homme et l'hiver au Canada*. Gallimard, Paris.
40. Mead, W. R. and Jaatinen, S. (1956) Finland in British maps, *Fennia*, **80** (1), 1–27; Mead, W. R. (1968) The delineation of Finland, *Fennia*, **97**, 1–18.
41. Subsequently, as Chairman of the Library and Maps Committee of the Royal Geographical Society, I have helped to put the Archive in excellent working order. Needless to say, I have looked at the Scandinavian materials in it. Cf. Mead, W. R. and C. Wadel (1961–2) Scandinavia and the Scandinavians in the Annals of the Royal Geographical Society, *Norsk geografisk tidsskrift*, 1–45, and Mead, W. R. Introduction to G. A. Wallin (1979) *Travels in Arabia, 1845 and 1848*, Cambridge.
42. Mead, W. R. (1953) Land use in early nineteenth-century Finland, *Annales universitatis Turkuensis*. A, **13** (2), 1–23, and Åström, S. E. (ed.) (1967) *Nakokulmia menneisyyteen*. Helsinki, pp. 87–100.
43. Mead, W. R. (1968) The eighteenth-century military reconnaissance of Finland, *Acta Geographica*, **20**, 18, and Geographie humaine de la Finlande au dix-huitième siècle, examen de quelques materiaux de base, *Norois*, **59**, 347–85.

44. See Selected readings, 1968.
45. See Selected readings, 1975.
46. Mead, W. R. (1977) Recent Developments in the Human Geography of Finland, *Progress in Human Geography*, 361–75.
47. Jones, M. (1977) *Finland, Daughter of the Sea*. Davison, Folkestone.
48. Branch, M. (1973) *A. J. Sjögren, Studies of the North*. Helsinki.
49. e.g. Mead, W. R. (1963) The image of the Finn in English and American literature, *Neuphilologische Mitteilungen*, 243–64, and Figures of fun, op. cit. (1976), 117–27.
50. Levi-Strauss, C. (1970) *Tristes Tropiques*. New York.
51. Hall, W. and W. R. Mead (1972) *Scandinavia*. Thames & Hudson, London.
52. Steiner, G. (1971) *Bluebeard's Castle*, some notes toward the redefinition of culture, New Haven, Yale University Press.
53. See Selected readings, 1963.
54. Sufficiently numerous to run a regular Finnish seminar in the University of London since 1968. Cf. Kirby, D. G. (1976) Historisk forskning rörande Finland i Storbritannien, *Historisk tidsskrift för Finland*, 2.
55. Johnson, R. J. (1979) *Geography and Geographers*. Arnold, London.
56. Spate, O. H. K. (1966) *Let Me Enjoy: Essays Partly Geographical*. Canberra, London.
57. Olsson, G. (1979) *Birds in Egg*, Pion, London.

Selected readings

1951 The cold farm in Finland, *Geographical Review* **41** (4), 529–43.
1954 Ridge and furrow in Buckinghamshire, *Geographical Journal* **120** (1), 33–42.
1958(a) *An Economic Geography of the Scandinavian States and Finland*. University of London Press, London.
   (b) The seasonal round: a study of adjustment on Finland's pioneer farms, *Tijdschrift v. economische en sociale geografie*, 7/8, 157–61.
1962 Pehr Kalm in the Chilterns, *Acta Geographica*, **17**, 1–33.
1963 The adoption of other lands, *Geography* **48**, 241–54.
1967 (with Helmer Smeds), *Winter in Finland: A Study in Human Geography*. F. A. Praeger, New York.
1968 *Finland* (How People Live Series). Praeger, New York.
1972 Luminaries of the north, *Transactions Institute of British Geographers* **57**, 1–13.
1975 (with Stig Jaatinen), *The Åland Islands*. University of London Press, London.
1981 *An Historical Geography of Scandinavia*. University of London Press, London.

Interlude One

Geography for a time was a 'poor relation' among academic fields on the western shores of the Atlantic. Throughout the nineteenth and early twentieth centuries a number of individuals displayed styles of thought and practice reminiscent of German, British, and even French Schools, but America was, after all, a new experience, inviting a radically different interpretation. Alternating waves of isolationism and internationalism have been observed during periods preceding the First World War (Warntz 1964: 159; Freeman 1977). An observer seeking a contextual perspective cannot fail to notice, however, the enormously different material object of enquiry, political and social climate, values and institutional structures, which attracted to the discipline quite another type of young recruit than those who followed the more established curricula of European Schools. From the 1920s on, geography in the United States begins to grow in self-confidence and manpower, developing graduate schools and field stations, and realizing success in selling geography to public interests as well as to designers of school curricula. A burgeoning profession harvests the fruits of early efforts by scholars whose definitions of geography are considered unpalatable, but whose proselytizing was evidently effective (Davis 1906, 1924; Wright 1952, 1966; James and Jones 1954; James and Martin 1979).

The following discussion (Ch. 5) focuses on selected departmental settings where university-level geography was nurtured in America. Marvin Mikesell chairs a discussion on academic milieux as experienced by students, researchers, professors, and chairmen: Leslie Hewes speaks of Berkeley and Nebraska, Preston E. James of Clark and Michigan, Clyde Kohn of Michigan and Iowa, and E. Cotton Mather of Wisconsin and Minnesota.[1] Time limits were to be those of experience – from the 1920s to the present – but the discussion tends to concentrate on the interwar period. Discussants also speculate on what an ideal department for the 1980s might be. Although emphasis rests on only a few departments, Chapter 5 may appear, to European readers, to be a characteristically 'American' story. They may nevertheless recognize some of those practical and philosophical issues which implicitly or explicitly stirred international interest during the interwar period.

From that vast agenda set by Humboldt's *Cosmos* (1845–62), Ritter's *Allgemeine Erdkunde* (1862), Vidal de la Blache's *Principes* (1922, trans. 1926), and even Davis's 'cycles' (1906), how was one to carve out a scientifically respectable, practically useful, and ideologically defensible way for geography? A priori, it seemed, such 'mystical' notions as oecumene, organism, regional gestalt and personality, were somewhat suspect.

Unlike many European colleagues, Americans seemed dubious about ecological questions. Tansley's notion of 'ecosystem' was evidently ignored or else considered someone's else's business (Tansley 1935); for only a few could geography be construed as the study of 'relationships between man and milieu' (Barrows 1923; see also Stoddart 1966; Mikesell 1968; Duncan 1980). Still there was a conviction that the discipline should provide something of a synthetic perspective; whether in regional or systematic endeavors, one should somehow gather the pieces into a coherent whole. How could one maintain this promise and at the same time be or become science? Could historical perspectives still be cultivated?

De Geer (1923), Hettner (1927), Sauer (1925), and Hartshorne (1939) proclaimed methodological principles for the practice of geography which were to gain, for a period at least, virtually universal appeal (Darby 1947; Lautensach 1952; de Jong 1955, 1962; Dickinson 1969; see also Schultz 1980: 95–217). Morphology, chorology, regional pattern elucidated by regional process and 'sequent occupance' (Whittlesey 1929), defined a geography which would be concerned with empirically grounded observations of differences and similarities on the surface of the earth. Global analogies, organic analogies, or any a priori world views could now be placed in parentheses, as it were; they could be regarded as irrelevant, ideologically offensive, or simply inscrutable. The geographer's task was to represent what he saw, to analyze it carefully, and to render his results articulate in verbal, graphic, or cartographic language which summarized and corresponded, allegedly as closely as possible, to the reality studied.

The map, as a key metaphor for geography, framed a focus on patterns – examining spatial distributions in comparative or systematic fashion. One could collate, superimpose, analyze correspondence among patterns – what other discipline would compete for such an agenda? European scholars were demonstrating the value of this approach in their impressive treatises about rural settlement, urban morphology, agrarian landscape, and cultural geography (Waibel 1933; Demangeon 1927, 1942; Darby 1936; Enequist 1941; see also later reviews, e.g., Schlüter 1952–58; Schwarz 1961, *Geografiska Annaler*, 1961; Dahl 1973). Generalizations could certainly be made about spatial form and process. Hypotheses could be tested (Krebs 1923; James 1926; Hartshorne 1939). Even if the field were to face some dispersion of effort, its methodological identity would be unmistakeable. One would not be seduced or pirated by apparently more sophisticated 'systematic sciences' because always one would ground cognitive claims on the mapping of phenomena in real situations.

'The Environment of Graduate School,' while touching on issues which were salient in other settings, may reveal quite as much about societal and academic circumstances in America of the 1920s and 1930s as it does about geography. The survey was obviously in vogue in sociology and other social sciences; micro-level detailed survey flourished (Coleman 1980). This apparently wholehearted thrust toward empirical work, with a problem orientation, did not explicitly espouse the philosophical foundations of pragmatism, a proto-typically American philosophy (James 1907, 1955); rather it sought to expunge what was perceived as the tyranny of imported a priori theories about society and environment. Methodologically this should have implied a clash of two world views: that of the formist (morphologist) and that of the organicist (Pepper 1942). In other disciplines, the clash of these two world views and their associated metaphors was also apparent (Theodorson 1958; Stoddart 1966). Morphology and chorology seemed to have won the day among most American geographers throughout the interwar period (Finch 1927, 1937; Whittlesey 1929, 1936; Hartshorne 1939).

Were there some empirical realities of interwar America which made philosophical abstraction repugnant, particularly those associated with absolutist or imperial politics? Did it not seem better to tackle concrete problems than to argue over theories? For the American business attitudes of the 1920s, as well as the New Deal policies of the 1930s, there could scarcely be a problem which could not eventually be solved. Perhaps also there was a pioneering attitude, undaunted by environmental constraints, which transposed the Frontier myth to the more tangible dream of People of Plenty cultivating Section-ed fields of golden grain? (Potter 1954). The 'contriving brain and skillful hand' could thus deny history and proceed with unquestioning faith in limitless resources and progress despite the devastating impact of Depression and Dust Bowl (Malin 1955; Zimmerman 1964). There it was – a vast arena of unexamined space for technical and practical activism to be explored 'with high boots and plaid shirts' by an energetic generation of map-makers and socially-concerned citizens (James and Mather 1977; Hart 1979).

From an outsider's perspective, one might suggest that the heated debates between Midwest and West Coast brands of morphology seem less significant, in retrospect, than they did to the admirers of Hartshorne and Sauer. Mentors of both schools were socialized in similar contexts, e.g., the Salisbury seminar at Chicago, the Michigan Land Economic Survey, and for some at least, the Tennessee Valley Authority project. Sauer and his students later sought other contexts to study, searching for the pristine and the 'primitive' in Middle America and elsewhere. Emphasis rested on field work, mapping, exploring formal associations among patterns. In the Midwest there was also, it seems, a desire to be relevant within the fact-finding survey-and-reconstruction ethos which was so generally supported by Federal moneys under the Roosevelt administration. For both schools, the earlier rhetoric over 'relationships' had intellectual and political implications which were offensive. North America, after all, unlike Western Europe, had virtually none of those *pays* or valley sections which had inspired and justified the 'home area', *Lebensraum*, or regional *gestalt* studies of organicist flavor elsewhere (Nelson 1913, 1916; Fleure 1918). Such themes, to transatlantic ears, may have been too easily associated with *Geopolitik* and were therefore ideologically repugnant to many. Was there something 'democratic', a sense of *e pluribus unum* (with or without history) about areal differentiation? The language of the map, however much it was accused of being an instrument of imperial domination elsewhere, was still a vernacular which could give voice to empirically observed reality in the fields. At least it offered a convenient grid system for an efficient inventory of resources. One's implicit world view was a dispersed one, each region and distributional pattern could be viewed in terms of its own morphological features.

Important differences, of course, developed within and between Midwest and West Coast traditions. Controversies might rage over genetic versus functionalist approaches, diachronic versus synchronic methods, between 'diffusion' versus 'independent invention' as potential explanations of cultural patterns. Intellectual debates over ideographic versus nomothetic goals for science would be readily imported from Germany. It seemed that Midwest practitioners, by and large, were content to explore the shape and dynamics of spatial and areal chorology, while the Berkeley people seemed eager to maintain a more explicitly historical and culturally relative perspective. And what a difference it was to make for the graduates of those milieux! Sauer's own work, and his periodic disclaimers over questions of methodology, seem oddly out of tune with what some of his own disciples later did. Sauer, according to Hewes, referred to his relationship with the historian Bolton and the anthropologist Kroeber in terms of the 'double play combination of Tinker to Evers to Chance' (Ch. 5, p. 73) but for his students such exposure to the thought and practice of other fields meant more than the competitive challenge of a Boston Braves game. A sense of history, of the rise and fall of civilizations, of cultural ecology, was to inspire a vastly different literature than the products of the Midwestern areal differentiation (see, for example, Thomas (ed.) 1956; Wagner and Mikesell 1962). Children of the Corn Belt would eulogize the 'now' landscape of optimistic spatial form and process, devise ingeniously refined codes, matrices, and cartograms to represent distributions and patterns, eventually to export an impressive technology, analogous to the International Combined Harvester, from the world's richest farmland to their agrarian colleagues abroad. Disciples of the Berkeley School would cultivate another way, winning a more selective clientele in the Angloworld, with less concern for foreign markets.

If there was an intellectual common denominator to match the key metaphor of the map, it was a Kantian vision of space and substance. It was a metaphor which would endure and re-emerge in the 1960s, after a short phase of enthusiasm for 'mechanism', with refurbished technology and again with Federal blessing, in maps of poverty, income, crime, minority status, and housing, in mental maps, atlases of economic development, and the factorial ecology of urban areas in never-never land. The halcyon days of mapping, and the contexts in which chorological awareness was fostered in America are brought to life in the discussion which comprises Chapter 5.

Note

1. Edited transcript of a video-recorded conversation at Lincoln, Nebraska, April 1978, included here with permission of NETCHE Studios.

Chapter 5

The environment of graduate school[1]

Fig. 5.1 (Left to right) Professors Clyde F. Kohn (Iowa), Leslie Hewes (Nebraska), Marvin W. Mikesell (Chicago, Chair), E. Cotton Mather (Minnesota), and Preston E. James (Florida) (Netche Studios, Lincoln, Nebraska)

Marvin Mikesell: Professors have always been inclined to talk about themselves and be interested in each other. Maybe there is a little more of that going on now because we all know that the opportunities for growth are very limited and will be for a number of decades. We are going to talk here about the influence of the environment of graduate school, try to remember our own experiences, and try to reflect a bit on what might be an ideal environment or at least a tolerable environment for intellectual creativity. There are several of us here, and we have a very interesting array of experiences. My own background is having been a graduate student in two big state universities in California. It began just after the Second World War, when most of the people were veterans on the GI Bill. I was a little too young to be in the Second World War, so I had the interesting experience of being in graduate school with veterans who were in a hurry to make up for lost time. I finished in 1959, and I find that one of the persons here whose experiences we can profit from is Jimmy James who entered graduate school nine years before I was born. Perhaps you could tell us a little bit about what it was like to be a graduate student back in that period.

Preston James: When I went to Clark University they had just set up the graduate school of geography in 1921. There was no program as such, nobody knew exactly what a proper training in geography ought to be. So in a sense I made my own program. I remember that one of the people at Clark told me that I ought to go to Chicago if I would ever amount to anything, because that was the best department of geography at that time. But there I would have had to take a lot of elementary courses before I could start the program. And at Clark I was ready to start my doctoral dissertation. So naturally I gave up the opportunity to go to Chicago.

Mikesell: Were there several faculties there at the time?

James: Well, it was an exciting time in this sense: people came from outside to offer lectures and seminars for, say, two or three weeks at a time. Most of the people that I was interested in working with had then been working on the Peace Conference in Paris. They were the people who drew the maps on which the new boundaries were drawn. This was a very important work and based on very accurate detailed mapping. I remember looking with great delight at the maps that Colonel Lawrence Martin presented to us, mostly boundaries of Hungary and Poland, the new countries of Europe. This was fascinating.

Mikesell: So you had the advantage of being in a department that was doing something of public concern and as a young student you could be drawn into that atmosphere. I wonder if others have had that experience of going to a department where there was a research program under way that they could be brought into as an apprentice; quite a different experience from having to find your own way.

Clyde Kohn: In the mid-1930s, Preston James was a member of the Michigan department of geography faculty where I was enrolled as a graduate student. Remember, at that time our country was experiencing one of its worst economic depressions ever. The federal government was making money available for field work, the TVA was being developed, and so on. As graduate students we were often involved in field studies that had very practical outcomes. We experienced the application of geography to public policies almost immediately.

James: Don't forget the Michigan Land Economic Survey.

Kohn: That was earlier, but when I came on the scene, there was a great deal of diversity within the department. McMurry, our departmental chairman, had been involved in that early Michigan study, and, I believe, Davis also when he was a graduate student along with Carl Sauer. A part of my professional training was by these fellows who had participated in the early Michigan study in the 1920s.

E. Cotton Mather: It seems to me, Marvin, that one of the big things between pre- and post-Second World War had to do with numbers, and the relationships in the graduate school between professors and students. I noticed this at the University of Wisconsin: there were really just six of us as graduate students before the war. We had more than one professor per graduate student. There was sort of an informal relationship as well as the academic formal relationship, and some professors had an input in both ways. After the war you had this mass influx, and there was a separation between graduate student and professor, even though we were supposed to be in a more informal age. Before the war too we usually had a foreign professor, mostly Europeans – for example, we had Leo Waibel – and that meant that often at the end of the seminar we went down to the Roman Inn where you had spaghetti and beer, and this seminar went on informally until they threw us out at 1:00 a.m.

Leslie Hewes: I think the informality at Berkeley around 1930 was one of its attractive features. Thornwaite, who helped steer me in the direction of Berkeley said: 'By all means stick around after seminars, because Sauer will talk.' And he did so as long as the students would stay around and listen. He talked off the cuff and quite frankly on subjects of interest in geography, and about the actors, the people in geography.

James: I think the tradition of talking that way informally came from the Salisbury seminar at Chicago. I went to Michigan after I had my degree, and I found that everybody else in the department had been to Chicago, and they were devoted to Salisbury and his seminar. So I picked up a lot of ideas. In fact, I got my training indirectly from Chicago and I have certainly benefited by the Chicago ideal. One of the ideas was this business of having an informal discussion of geographical problems, particular problems of how you proceed on a field study without any organization, a free-for-all discussion. Students and faculty participated in this without any separation. It was simply a tremendous experience for a young fellow.

Mather: Jimmy, was it true that at the time you were a graduate student, great importance was also given to physiography? That meant that you were in the field, and you had informal relationships. We have gone through vogues as every field has gone through like the aerial photo, and the computer. But there is not very much in personal relationships of the kind that you are talking about if you have the vogue of the computer. In the Salisbury days you went out and looked at the subject. And a stuffed shirt in the field is exposed very quickly.

Kohn: At Michigan, in the late 1930s, we were out in the field every Wednesday afternoon and all day on Saturday, and every summer. Many of our discussions didn't take place in a seminar room, but in the field. Jimmy was in part responsible for that because of his role in the annual spring conference. I recall how envious we graduate students were one year when Joe Russell was asked to go to that conference.

James: As an historian of geography, I am beginning to probe the archives and am finding out all kinds of things that I didn't experience directly. One of them is that there is a wholly new and different tradition in the development of geographical ideas which comes from economics rather than geology. Anybody trained in geology who became a geographer, was devoted to field study. But anybody trained in economics never did any. These fellows emphasized theoretical models, and used statistical procedures. Richard T. Ely is, I think, the person who was primarily responsible, maybe indirectly, for the development of geographical ideas among the economists. And, of course, the Horton School in the University of Pennsylvania was where economic geography really originated. These fellows never did any field work.

Kohn: Keep in mind that in training present graduate students to think in terms of models and deductive methods, we do require them to go out into the field to see if what they have deduced from theory can be applied in the so-called 'real' world.

Mikesell: Or at least you collect data from interviews.

Mather: Yes. Or more usually you don't collect data and you don't do field work. You write an article, for example, of research, and then you go out and take three photographs to prove that you were there. There is the difference. Before, you actually went out and did field research. Today most of the research by students and faculty alike has varied somewhat from that.

James: They get their data from the Census.

Kohn: Well, no, I am going to object to this a bit. Because with the swing . . . .

Mather: You come from a business school.

Kohn: No, no . . . with the swing to behavioral studies, I think we are spending more hours in the field with people than we did when traditional economic geography was more in the forefront in geographic research.

Hewes: The field course at Berkeley was a strong point. I think I took that course either formally or informally each time it was offered when I was there, and I think we all profited from it.

Mather: What did you do in the field? I am sort of curious. Did you hang around with soil augurs and act like scientists or . . . did you ever make maps?

Hewes: We did a bit of that. More commonly we went out to see what we could see. A few times we made maps, more often, I think, not.

Kohn: That's another way in which the Michigan program in the 1930s influenced me. The faculty at that time emphasized cartography, field mapping, and aerial photography. They demanded that we prepare field notes carefully so that the data could be mapped back in the laboratory.

James: That emphasis on making a map yourself began to fade out in the 1930s, because of the use of the air-photographs. The vertical air-photographs became the basis on which you actually plotted data.

Mikesell: In my field camps during the late 1950s, we went out with overlays and marked on vertical photographs.

Mather: I came in between, and I can tell you a good little story on that. Trewartha was my field instructor, and he made us go out with plane-tables. By that time plane-tables were out of date. But Trewartha knew what he was doing. He was using the air-photos. And I was out in the field with John Borchert and Wilbur Zelinsky in one team, and after about 30 minutes I thought, 'This is really baloney running up and down this ground moraine with a plane-table.' So I just ran up and sat under a tree.

James: That must have been quite a team that you were on.

Mather: That was! And pretty soon Borchert came up and said: 'What about this?' I said: 'You fellows can go where you want to. I'm going to do it the easy way.' Zelinsky was still going around over the ground moraine. Anyway, we made our maps for the whole term in about 1 hour with an aerial photo. Trewartha gave Borchert A and Zelinsky B. I got qualms of conscience for the first and last time, I guess, and I went in to see Trewartha, and said: 'Well, this is the way I checked it.' He said: 'That's the way it should be done.' But you see, we went through this period where you used methods that were unnecessary. A lot of us got sick of this, and we said that we didn't want anything to do with field work again.

Mikesell: I regret to say that we gave up our field camp just simply because we didn't have a single member of the faculty who wanted to do it. But in the last years, out in northern Indiana, we had an interesting experience. We sent the students out one day and said: 'You can only observe. You are forbidden to talk to anyone. You can't ask any questions ... you can't read anything.' And we had a discussion that evening on what was seen. It was fascinating because people saw different things. We had a foreign student who said that these little towns have such incredible abundance of shade trees. We had others who noticed the transportation facilities and so on. And we built up a discussion on the morphology of these towns based upon observation. That was the first day. The feeling was, well, description isn't that easy, but then questions came up about function, about trade hinterlands – all things that couldn't be determined by observation. So they went back a second day to interview, asking questions about nodal regions by finding out where the customers came from where people went on their Sunday trips, and so on, they came back for another discussion which was on interviewing. They still had additional questions, and so the third day they went back and looked into the historical record to get in the background. I don't know that anyone has attempted to break field work in this way, because normally when you go out in the field all this is happening at once: you are looking; you are interviewing; and you have probably read something.

Kohn: That does bring to mind discussions we used to have at Michigan at that time, whether or not you went out into the field with the problem or found the problem in the area in which you were interested. There were several of the faculty who believed strongly in the second part, that you went out to observe the area and you discovered the problems through observations.

Mikesell: This is another issue, too, in looking at departments – whether their primary purpose is to teach or encourage awareness of ideas or of methods. I think in my years at Berkeley it was very clear at least as far as Sauer was concerned that he was throwing out ideas, and he was giving people a sense of problem, and you were on your own to find the method to solve it, often going into other departments, taking courses outside. Many departments have had the reverse policy: they teach method and you are on your own to find the problem.

James: The Michigan group was strong on going out and find your problem. In 1924 I went to Trinidad to make a study of land use on the island. I came up with, I think, what would be reasonably called a theoretical model: a new crop like sugar-cane, for instance, would originally concentrate in an area right near the largest city, Port of Spain. Then the crop would spread all over the island, and then retreat eventually to the physically best suited places. Now that idea actually came from O. E. Baker, and only recently did I realize that O. E. Baker actually had it from the teaching of Richard T. Ely. This was the economic input into the geographical world.

Mather: Jimmy, what you are talking about is formal educational experiences. Forty years after you've been a graduate student you look back, and you find that some of the most important things in graduate school are the things that weren't in the formal structure. At Wisconsin, for example, when W. C. Finch wanted me to drive the automobile on weekends: going out, I was sitting in private with him, and listening to a seasoned geographer in the field. It was like, for example, with Leo Waibel, the German geographer, who had done settlement studies in Africa, South America, Costa Rica – it was like going out on a vacation with him. I had never noticed a snow fence, but he was so impressed with them, he photographed every snow fence in sight. He introduced me to the difference between Polish farmsteads and German farmsteads. He sensitized me, he taught me more than I'd ever learned in the classroom. The point I'm making is this: it was the classroom or the formal course that introduced me to the professor, and the professor then in informal ways introduced me to further opportunity. The Second World War brought us to assemble in Washington: that's when I met most of the people here at this table probably. But it was the fact that I had been introduced into the academic enterprise then, and I knew Dick Hartshorne and through him, you and... This is the way life rolls.

James: Well, if you have some kind of a practical problem that is thrust upon you, like a war situation, this is enormously beneficial for anybody who is not quite sure about what kind of a problem he wants to work on. I suppose really there are two professional fields that do in fact benefit from a war: one is medicine and the other is geography.

Mikesell: But also there is just the question of people getting together. You also had your national organization that could bring you together. I wonder if we could think a little bit about the question of change. We know that after the Second World War there was a tremendous growth: everything got

71

larger, more impersonal. We have to come to grips with the way things are now, or how they've evolved, say, over the last twenty years or so. Do we still have a good environment for creative activity, comparable to what we've just been recalling? And if we don't have it, then how can we get it again? What steps might be taken?

Kohn: To me one thing that makes the difference between a good and a not so good department is the openness of communication between faculty members working on research, and their willingness to talk about their research with their students – both graduate and undergraduate students. These students become stimulated when working along with these professors. Also, graduate students share what they are doing with their professors and with each other so that there is a complete openness in communication. I think that's one thing I found at Michigan when I was there. We had six professors, all with different areas of research. They worked together, or seemingly did, and they shared their research with us as graduate students. We were stimulated as a result. I see that going on in many institutions today. I thought you were getting at that, Cotton.

Mather: Yes. I think actually the apprenticeship relation is the most important in graduate study. Graduate study, as I see it, has changed, but it has changed less than any other level of education, and according to American valuation probably needs more changing. But I think it's the one part that has been saved the most from the standpoint of apprenticeship, learning. This is what graduate study is all about. You cannot learn to write just by reading, you cannot learn to swim by being lectured in it, eventually you have to jump in the water. What we have to do in geography is to have a working relationship with a master. Of course when we are talking about the old days, I'm trying to forget a lot of the old days, because some profs were not working geographers. I don't want to remember them. I don't think of this as good old days.

Mikesell: Unpleasant working relationships?

Mather: That's right. But there were old masters, and there are people today who are masters, but they belong to the '2 per cent club'.

Mikesell: I want to pick up something Clyde said that was really important: the idea of change of scale. We know that the geographical profession now has about 10,000 members in North America, and we have been talking about an enormously smaller enterprise. You talked about smaller departments, Clyde, and I believe you said, 'Well, one can still have that,' but your department is small . . .

Kohn: We have ten faculty members, and about forty to forty-five students.

Mikesell: In your case you have been able to maintain approximately the same kind of relationship I knew at Berkeley when I went there: six faculty and about fifteen students. What happens when it becomes a quantum leap and your department has forty people, and you have 200 majors, and you have a great big new building, and you are on the fifth floor, and so on? I'm thinking particularly of the post-Second World War kind of explosion, and to what extent it affects the intimate apprenticeship relationship that we talked about . . . what were the stresses on it?

Mather: Well, seminars are still listed as seminars. Previously, you had a discussion, you had maybe six people around the table, you went in a car out in the field – you had a discussion, you had personal relationships. Today you have a 'seminar'. If the professor really has something to say the students get the message, and you've got forty-seven people in there. Now what sort of a thing is this? It's listed as a seminar, and it's not a seminar. But I would like to make a positive note on this, and that is that when we were students, for example, in the Midwest, most of the students were from the Midwest. One of the big revolutions is the mobility of people. You sit today in a classroom, you have somebody from China there, somebody from the Balkans, you have Latin Americans. Another thing is you have Americans, Midwesterners who have been elsewhere, and when you have seminars they have Kodachrome slides. They show what they are talking about. They may not know, Jimmy, what we thought of their field work, but at least they were there, and they photographed it, and you can't get by with some of the stuff you old boys used in the old days.

Hewes: I never was a member of a big department, either as a graduate student or as an instructor. But at Berkeley under Sauer, we profited from some cross-fertilization. He was a great baseball fan, likened himself, Kroeber in anthropology, and Bolton in history to the double-play combination of Tinker to Evers to Chance. We took work elsewhere outside the department and, I think, profited from that.

Mather: Les, that brings up another point: in the old days you took minor work in other disciplines. As we have grown large today, we have minor work in the department. And so as our perspectives on the modern world supposedly get broader and broader in one sense, we become more ingrown at least in the sense that the discipline has been cloistered more today on to itself. Don't you think that's true?

Kohn: Well, not necessarily. In our department at Iowa faculty members are working closely with members of our geography faculty on urban and energy problems.

Mather: You are a small department at Iowa. I'm talking about the massive type of department like we have at Minnesota where the faculty is on three floors, and students are here and there and everywhere. We set up programs that we think are just marvelous; we have so much speculation within geography that we are cutting out the outside geography experience, and Clyde, you know very well that you've got an interesting argument only if you don't look at it too far. For example, in the old days we had two foreign languages, and Iowa is one of the traitors of this kind of thing – you don't even have one foreign language. In the old days at least you had some contact with the French or the Germans. We think that today we are getting broader and broader, and we are covering more physical space – three floors of professors at Minnesota. We are more ingrown in a way.

Mikesell: Well, with so many sub-departments I believe that often now geography seems not even to be a noun but rather some kind of suffix. So the great big departments that look so impressive with thirty or forty people, may in fact be about ten to fifteen sub-departments, each with only two people. The big department with only two people in its sub-fields is in fact smaller than a Berkeley of the 1930s or a Clark of the 1920s. I wonder if we could think about an issue that is obviously important, and that is appropriate administrative style. I am thinking particularly of the departments before the Second World War where you had the old style permanent head who probably had been in the job for a decade or more, or indeed for most of his mature years. As we look around the country today, we see rotating chairmanships, sometimes two- or three-year rotations. So there is obviously here a very different situation in the human environment. Would it be fair to say that most of us started in an authoritarian atmosphere, a kind of 'patriarchy' which we talk about in such a friendly way that we must have enjoyed it? And we are now in many cases close to something that could almost be called 'anarchy', or a kind of rotating egalitarian situation. How do we feel about that? As Americans we will naturally tend to think that the new situation is closer to our spirit than an almost feudalistic pattern. And yet, I sensed from the rather warm way that the past has been talked about, that there may be ambivalence on this issue.

James: I would certainly agree that the long-term chairmanship of a department is much better. These short-term jobs means simply . . . that the head of the department becomes simply a sort of office boy who sees that the chalk is put around, and that the rooms are assigned – jobs that ought to be done by an administrative assistant. But actually to keep a group of scholars happy together is a full-time job: it's a very important job, because scholars are peculiar people. It's not easy to keep these fellows talking to each other and avoid the kind of antagonism that sometimes develop and split a department right down the middle.

Kohn: One has to work closely with one's university colleagues, with administrators, and especially with chairman of other departments. I don't want to call it anarchy, but sometimes I think that is the pattern we are heading for in terms of the rotation of departmental chairmen.

Mather: And once, for the head who really administered the department, budgets were pretty much secret. He wasn't diverted from his professional motivation, but the minute democracy was introduced everybody started fighting about everybody else's $50 salary raise. Then on top of that, today you have the big bureaucracy above the department, and they are going to ask questions to the computer for all of us, and the most important thing is: how much in grants did you get this past year? And people get advancement by grantmanship, not by scholarship. Sometimes these overlap accidentally but sometimes they don't. What we really have, I think, with the change of chairmanship, is a whole complex of factors that made it very difficult for a person to go full time without being on seventy-seven committees, applying for grants, getting mixed up with the federal government, waving around a lot of equipment, worshipping science as a great deity.

Kohn: I think also it sets an unfortunate model for graduate students. They don't know what's going to happen before they finish their graduate work because of changing departmental policies.

Mikesell: It seems to me there is an answer to that. I've heard this question arise very often: what happens when you are in the department and you have a different atmosphere among students who are assistants on projects and those who are not. You can have a serious morale problem, particularly if the former group is well paid and the latter has a different standard of living. But I think one way of solving this problem is to be sure that your projects are related to your educational program, and that departments do not become in effect branches of the government. And I think a certain amount of real concern about taking on work that is intellectually compatible with the discipline and its orientation would resolve this problem. Certainly none of us would be at all offended if the work we are doing would be useful to someone. But there's another aspect of the problem. I suppose it's inconceivable that we will ever have a class of professional managers in geography, like hospital administrators. But we do have people who are full-time managers. I guess geography just isn't big enough to have a professional managerial class.

James: I'd like to know what happened in, say, departments of economics around the country. Do they have that problem?

Mikesell: Well, we can find three or four administrative assistants who were doing various things. I think this pattern was encouraged in the 1960s, which was the post-Sputnik panic period when a lot of money was spent on reform of science and engineering. A whole different style of life was started during that period, expecting grants, having a large infrastructure of administration, federal support, and so on. I think we are moving away from that now, and in many ways the atmosphere we are in the late 1970s going into the 1980s is like the 1930s. I have often had students come to me and say they want advice on employment prospects, and I have to say, 'Well, you know, I was a teaching assistant in a state university. I did my field work abroad on a grant from the Ford Foundation. If you go to the next office, perhaps you're going to find someone who had his education paid from veterans' benefits. If you want someone who can give you advice about financial problems you have to find someone who was a student before the Second World War. He will tell you what it was like to come in during vacations, pursuing graduate study part-time when you are a high school teacher, and so on.' So we have a difficulty, as I find it, in advising our own students, because of our personal experience, and in my case I feel guilty and uneasy about this: I feel that I was so spoiled, trying to deal with the people who are now in a tight market for jobs with nothing like the support we had.

Mather: Les, the student, the graduate student, usually has a propensity for eating and occasionally having a beer. What I was wondering, you knew the graduate school at the time of the depression, and you've known it all the time since. When students make applications to graduate schools, I suppose they still think about economics as we tend to think about the academic hall. I'd like to have your comment on what you think about this aspect of it. Jobs weren't really in surfeit in the 1930s.

Hewes: The economic aspect was certainly an important one, and no doubt still is for a good many of these students. I was a teaching fellow at Berkeley, and I got a modest salary but enough to live on. I think the chief benefit though was the work I did. We teaching fellows attended the lectures in beginning courses, and learned a lot from them. In fact, I have told some graduate students who complained about their stipends at Nebraska, 'Well, you ought to pay us for the experience you are getting.' But, of course, it wouldn't take care of the beans and the corn...

Mather: What do you think the situation is today, compared to the 1930s?

Hewes: The salaries are much higher.

Mather: Yes, but so is the cost of living.

Hewes: That's true. I suppose it's rather comparable really in those terms, and quite a few today do have teaching assistantships.

Mather: The change I see is that in the old days you got the stipend from the university. When you went as a graduate student to the department, there was a guaranteed sum from your local institution based on your scholarship application. Today when the graduate students come in, they get a guarantee like they did in the old days, but as soon as they get there, they find that certain professors have grants from foundations, and from the federal government. There is the possibility that their academic decisions might be tempered by the fact that there are rather fat little assistantships or additional remunerations, put into the system.

Mikesell: I read an editorial in a biology magazine where a very angry biology student said: 'Why don't you biology professors practice zero population growth for your own profession and quit enticing us into a field for which the opportunities are very limited!' It was a very angry letter. Then about a month later I saw a very thoughtful editorial by an astronomer who said: 'Give them the facts, be honest with them, and let them make the choice.' You know, they have a right to try to be astronomers even if there are only three or four openings a year. We are all caught in this. My own feeling is that we should give the best possible information of what the prospects are, and then let students make their choice and welcome them if they want to try to be geographers. But we also have to cultivate our placement office and be aware of the many things one can do with geography other than teach. And above all we have to be aware that geography is a good general foundation for almost anything you want to do. We can always sell it on that basis.

In the little time we have remaining I wish each of you would try to reflect a little bit on what you think would be an ideal environment. I know it's hard to speculate, because often it's the environment we had and remember that seems to be the good one. But what would be Camelot for us and what would be the size of it? Suppose you were chairmen of a brand new department with no constraints whatsoever. You could choose the best faculty and the best students, have all the money you want, and so on. What would be even one or two ideas that you think would go in the blueprint of making an ideal department?

Kohn: I think I'd start with that one you just mentioned: the best faculty you can get together. I don't think you are going to be successful unless you have men and women who are in the forefront in their profession, who are the themselves making contributions to and excited by the discipline and leading it.

Hewes: About how many?

Kohn: We have ten faculty members. I agree that we should have more diversity represented by our faculty, but there is an opposing argument. I think there is some kind of critical threshold in the number of faculty members in a given area of research. We are trying to think through that problem, for example in the area of regional development. We now have three faculty members working in that area, and there is a great deal of cross-fertilization taking place. I'd rather have depth in two or three areas of specialization within geography than attempt to cover them all with only one faculty member in each area. This does mean less diversity, and fewer areas for graduate students to explore, but it does give the students an advantage of having two or three professors at work in the same area of specialization.

Mikesell: That's the cluster idea, it's so fashionable now, having complementary relationships as against having a basic array of, say, the physical, the cultural/historical, the urban/economic as a nucleus. How do you feel about being in a general or pluralistic department or in one of the specialized ones?

Kohn: I'd like to have more clusters than we are now able to have at Iowa so that we could provide programs for students with different interests. We like our arrangement within the Big Ten where students can go to another university for a semester to study with faculty members in areas we cannot offer because of our small size.

Mather: I would react to that, Clyde, by asking the best faculty for what? I think that in our field we need technicians, and we need educated people. And most of our specialisms as they've evolved in the last two decades, are really concerned with vocational training. Things are changing very fast, even technologically, and almost every other way in the world today, it seems to me that while in the older tradition the aim was to be educated, the imperative today is much more for 'education' and training. I'm not against training, but I think if some departments want to have vocational training, we ought to label them as such. A Ph.D. in most of our geography graduate schools is not an educational experience, it's a training experience. I think it should be open to different departments to go different ways, because we need different people, and some people are not as easily educable as others.

Kohn: I think you are introducing the concept of an ideal department. What are the experiences you would provide graduate students given a competent faculty to offer them?

Mather: Well, one of the great things was going to national meetings by a railway train. You don't put this in a catalog or your course curriculum, but in the old days a few graduate students – and only a few that were going to become scholars anyway – got on the train, with their professors, they rode from a place like Minneapolis down to Madison, a few more professors and a few more graduate students got on, and the train rolled on to Chicago. This is the kind of thing that happens in life: the train rolls on and on, and people get on and off the train. Some professors retire on full salary aged thirty-two. Others continue like Jimmy approaching . . . What is it now? Fifty?

James: I'm going on ninety.

Mather: One year nearer every year. But you see, this is the way it is, and you have those experiences that are very difficult to lay out in a program. Marvin asked how do you do this. I think it is up to the individual professor, and this is the kind of thing that Sauer did: he created an atmosphere.

Mikesell: You see, that's the issue. We are dealing with a very delicate question that I think can never be answered, and it concerns morale. There is no doubt that Sauer's department at Berkeley had high morale. I think – this is my own view – the reason was that Sauer managed to convince most of the students that the work they were doing was extremely important. Not only that, but also that he was impatiently waiting for you to come in with your results, because he would just snatch them out of your typewriter the moment you had written them. And the effect of that was just sky-high morale because we knew he was a brilliant person, and he led us to believe that what we were doing was part of this intellectual enterprise. Literally, people went into the jungle and climbed mountains because they had this sense of spirit. That's something very personal. It worked very well there. There are certainly cases where people have gone to other institutions, and tried to act just like him, and couldn't carry it off.

James: That's a very important thing. If you have a good professor, somebody with imagination and with personality, you could see if you could get him supported in some way to undertake a particular kind of field study. I tell you, back in the 1930s I went out to Jackson Hall and made a study there with five or six graduate students, and later they became some of the leading people in the profession. These fellows combed Jackson Hall and made the maps of this area. The paper I wrote on this was republished for the Wyoming Legislature. The point is that there was a practical problem in which the students were involved, and in which faculty was involved, and there was an exciting experience. That's something you can't always plan in advance. You have to take advantage of what comes along.

Mikesell: That would be ideal if there would be this collective involvement with the effort. I think now we can imagine two terrible extremes. A graduate department that is in fact producing 'hired guns' who are ready to do what anybody tells them to do, and who get their intellectual initiative by orders. The opposite extreme would be very concerned, sensitive, intelligent people who would love to do something and they don't know how to do it, so they are moving around helplessly. And somewhere in between is that real world that we struggle with, and as we have been able to demonstrate, I think, in our discussion we've seen change from the 1920s to the present time as the discipline itself has grown. We have had a variety of circumstances, and we have created specialized departments and general departments of different size and character. In this process of trial and error, it seems to me that we haven't got the ideal situation and yet it has never been so miserable or unhappy that any of us wanted to leave it. So basically we enjoy a sense of dissatisfied contentment with our lot, and I think that is a pretty good reflection of how most professors feel about their lives.

Mather: But I think there's another thing that is involved, Jimmy, that you probably can answer better than anybody here: the most important decision for a graduate student in the field of geography is where on the map, if they are going to an American institution, are they going. Isn't it true, for example, that if you were in one corner, let's say California, that you might more likely get interested in things that were geographically pertinent there. If you look at this more broadly, isn't it true that certain academic institutions have more of an international than regional tradition? If the purpose is to get educated, it's not only the curriculum, the stature of the men, but where on the map you pick your graduate school.

Mikesell: Thank you very much for this interesting discussion of ideals and realities. Although geographers don't like to admit that human behavior can be determined by environment I think this review of our varied experiences has demonstrated that we have at least been strongly influenced by the environment of graduate school. Since that environment is always changing – with or without our consent – the conversation we have had here is only one event in an endless dialogue.

Note

1. Edited transcript of a videotaped discussion recorded at Lincoln, Nebraska, April 1978. Courtesy of NETCHE Studios.

Chapter 6

Fig. 6.1 John B. Leighly, USA

An eye-opener and a pleasure it has been to correspond with John Leighly who finds such 'immemorial and universal' interest in the practice of geography. The environment of his graduate schooling may have resembled in many ways those described in Interlude One but the environment of his professional career was special, across the continental divide, physically and socially.

From the Midwest USA, John Leighly worked his way through college at Mount Pleasant (Michigan) where his interests in landscape and life were evoked by one R. D. Calkins. This was geography with a strongly Davisian flavor, a style which he was later to reject emphatically. It was in 1922, on a field experimental station of the Michigan Land Use Survey, that he came under the spell of Carl Sauer who invited him to assume a position at Berkeley where most of his career has been spent.

Graduates of the Berkeley department will remember John Leighly for his scholarly standards, his devotion to teaching and field work. In fact, he has regarded teaching as his primary vocation to life, most of his research interests in climatology and physical geography arising from his experiences with courses and seminars. Fluent in German and French, and one of the earliest American geographers to cultivate a 'Scandinavian Connection', however, the horizons of his work have extended from toponomy to the history of ideas, from the Bay region to the Baltic.

Memory as mirror

An academic career of any kind was far from my expectations when I entered Central Michigan Normal School, now Central Michigan University, in September 1919. As befitted my extremely narrow circumstances, I aspired to nothing better than a high school teacher's certificate. I had returned from France and been discharged from the army during the summer, and had chosen the institution at Mount Pleasant as the nearest and least expensive place I knew of for continuing my schooling. A little money saved from my soldier's pay sufficed for the modest fees charged and for my living expenses until I could earn more. I was wholly dependent on my own resources, and was able to support myself through the academic year, mainly by part-time work as a common laborer on construction jobs.

I was required to take one or two courses in geography. These were taught by R. D. Calkins (1883–1955), who had studied geology and geography at the University of Chicago under T. C. Chamberlin (1843–1928) and R. D. Salisbury (1858–1922) at about the beginning of the century. Calkins's teaching was weighted in favor of physical geography, and at intervals he offered a few courses in geology. His courses interested me more than any others I took, though, having grown up in rural surroundings, my curiosity extended to all of the natural sciences. He emphasized observation of and contact with material objects more than printed words: his collection of teaching materials – maps, instruments, specimens of rocks, minerals, and useful natural products – was, for his purposes, the best I have seen in any department of geography. Perhaps the most stimulating of the courses I took from him was a field course, with Saturday excursions, in the spring term. An assignment to his field class on one of these excursions exemplifies his manner of teaching. Standing at a point overlooking the valley of the Chippewa River, which flows through Mount Pleasant, he pointed out to us an isolated hillock in the valley not far from where we stood, and instructed us to walk over and around it and then come back and tell him how it had been formed. (It was the remnant of a meander lobe, cut off when the river flowed, in the late Pleistocene, into one of the higher levels of what is now Lake Huron.)

He offered his students a definition of geography current at the time: it is concerned with the 'relations' that obtain between the physical environment and organisms, primarily the human species. This definition repelled me; the constituents of the physical earth were visible and tangible, but abstract 'relations' had no attraction for me. Undoubtedly some of my persistent biases are derived from my reaction against this definition of geography, which, I learned much later, was formulated by William Morris Davis (1850–1934) at about the time when Calkins was a graduate student at Chicago.

Calkins took a personal interest in me, and in the spring of 1920 urged me to go to the University of Michigan the following year. I had supported myself at Mount Pleasant with no great difficulty, but I had some trepidation about trying to do the same at Ann Arbor, where expenses were substantially higher. Calkins helped me out of that difficulty by recommending me for an assistantship, which I received, in the department of geology at the university.

The undergraduate in the college of liberal arts in a large American university is confronted by a daunting array of courses from which to choose. Requirements for the bachelor's degree at the University of Michigan were extremely flexible: a student needed only to accumulate the necessary number of 'points', of which a small number must fall within each of groups of subjects classified as 'humanities', 'social sciences', and 'natural sciences'. I devoted myself primarily to the many courses offered in geology and mineralogy, taking others to fulfill the loose requirements of 'breadth'. I neglected my opportunity to learn the mathematics and physics I have needed in my later years. My critical abilities would undoubtedly have been improved by some formal instruction in philosophy. I also regret that, already having a fair knowledge of Latin from high school, I did not learn some ancient Greek. I did improve my knowledge of French and German; the ability to read these languages easily has been the most valuable precipitate of my undergraduate years. When I received my bachelor's degree I could not attach much significance to it, since by that time I knew that I had acquired neither a mastery of any one field of knowledge nor a well-rounded literary education.

In the department of geology at Ann Arbor was a small sub-department of geography, in the charge of Carl Sauer (1889–1975), who had come to the university in 1915 as a newly-fledged Ph.D. from the University of Chicago. My adverse reaction to the definition of geography I had heard at Mount Pleasant prevented me from having any interest in it. But I became acquainted with Sauer in the summer of 1922, when he organized the first experimental season of field work of the Michigan Land Economic Survey, and offered me a position as field assistant for the summer. This summer's work was profitable, giving me experience in field mapping and familiarity with glacial and fluvioglacial land forms I had not seen earlier. My association with Sauer led me to take two courses with him the following academic year, and he then invited me to work in the field in the summer of 1923 on a geographical study of a part of Kentucky he had undertaken for the Kentucky Geological Survey.

Already in the spring of 1923, however, the event that determined my future career had occurred. Sauer accepted appointment as professor of geography in the University of California, at Berkeley, and offered me a minor teaching position there, with an opportunity to work toward the doctorate. My reservations concerning geography had been mitigated by my acquaintance with him, and he allayed them further by assuring me that I should be free to work on anything I chose. I was rather overpowered, however, by what he proposed that I should teach at Berkeley. He asked me to initiate instruction in climatology and cartography, of which I had little or no knowledge. But I gladly accepted his invitation, and already in the spring began to acquaint myself, by reading, with these unfamiliar matters.

Thus began my long association with Sauer, which ended only with his death in 1975. At Berkeley he was in a position to build a department to fit his own view of geography, which before he left Ann Arbor was diverging from what he had been taught at Chicago. He defined his new conception of geography in 'The morphology of landscape', which he began writing during his first year at Berkeley.[1] In this programmatic statement he abandoned the concern with 'relations' and 'influences' that had dominated academic geography in the United States for many years, and assigned to geography the task of interpreting the material content of the landscape, of both natural and cultural origins, by methods appropriate to each class of phenomena. I could willingly participate in such an enterprise, though I had reservations concerning Sauer's exclusion of physical processes from the purview of geography.

My first and most urgent task was to work up my courses in climatology and cartography, finding my way with no example to follow. I was dependent on what I read; and I read voluminously, especially in German and French literature. In addition to my part-time teaching, I shared in the instruction

Sauer commended to his tiny but closely-knit group of graduate students. We were isolated from the rest of American academic geography, both by distance and by Sauer's deliberate departure from the ideas and practices current among his contemporaries, so that reading, especially in the foreign literature, was our principal source of stimulation.

We read not only in contemporary geographical writings, but also in the literature of the past. Sauer's sense of historical depth, already prominent in his treatment of the cultural landscape in what he wrote before he left the Middle West, extended to his conception of geography in general. The names of Strabo, Varenius, Humboldt, Ritter, and Peschel were heard about the department more often than those of American contemporaries. These contemporaries, in the 1920s and earliest 1930s, included the two most influential exponents of geography in the United States in the first three decades of this century, W. M. Davis and Ellen Churchill Semple (1863–1932). I had read much of Davis, and participated in a seminar he offered as visiting professor (in the department of geology) at Berkeley while I was a graduate student. Sauer distrusted Davis's geomorphology as being built more on deduction than on observation in the field. I read none of Miss Semple's writings until later, and then only as a part of the history of American academic geography. Sauer had known her at Chicago, but already before coming to Berkeley he had discarded her mechanical interpretation of 'geographic influences'. He imparted his generally low opinion of his American colleagues to his students, and probably induced in us too much parochial smugness, of which we had to divest ourselves later.

Sauer's students were free to follow their curiosities in any direction they chose. Both from my early encounter with geography at Mount Pleasant and from Sauer's inclusive views I have retained a strong aversion to restrictive definitions of geography. Such definitions, which in America were derived mainly from W. M. Davis, seem to have been grounded neither in epistemological necessity nor in the existence of a body of original investigations that required a new designation, but in a craving for and assertion of a distinctive and undisputed place for geography in the structure of American universities.

Sauer described the intellectual atmosphere he aspired to in his department, and which he largely realized, in his ceremonial address to the Association of American Geographers in 1955.[2] Here he spoke, not in the first person but in general terms, of his own lifelong education and the practices he followed in guiding students into their intellectual paths. Though given full freedom, probably most of Sauer's students were led in some way to their choice of subjects for their doctoral dissertations by his suggestions. He did not make these suggestions overtly, but indirectly, so that the student arrived at his own decision. My dissertation was the result of such an indirect process. Sauer had been favorably impressed by some of the writings of Sten De Geer (1886–1933), then at Stockholm, and suggested to me that I apply for a fellowship of the American – Scandinavian Foundation to enable me to spend a year with de Geer. I received the fellowship, and went to Sweden in 1925–26. There, with much help from de Geer, I collected material for my dissertation, which I completed in 1927.

With my attainment of the doctorate in 1927 and of a regular appointment as instructor, I had to assume a full teaching 'load'. Until after the Second World War our department was small, and in order to provide a fairly wide range of instruction each of us had to teach various subjects rather than limit ourselves to individual specialties. I have no regret that I had no opportunity to specialize; I enjoyed shifting from one subject to another in the course of a week. I found, however, that I could not teach 'regional' courses to my own satisfaction, and after a few attempts avoided them. I never knew enough from my own observations to do justice to any region of appreciable size, and could not be comfortable in speaking of places I had not seen with my own eyes.

I always looked on teaching as my primary responsibility; original work and writing for publication were things to be done in free time. During the academic year I spent most of my time between class sessions in preparing material for teaching. I came to look on lecturing as an inefficient means of giving students an opportunity to learn, which is about all a teacher can do. Few students have enough skill in taking lecture notes to make these notes useful to them later. I preferred to give the students reproduced material – notes, maps, diagrams, exercises of various sorts – and to use class sessions for discussion and answering the students' questions. In spite of all the effort I expended, I doubt that members of my classes remember me for my teaching as such. One of our former students, now retired from a career of university teaching, once remarked to me that a good teacher must be something of an actor. I do not have enough histrionic or oratorical ability to fulfill that requirement. Moreover, I never had enough time to prepare for classes as well as I wished. I think I have done my best teaching since my retirement, at Berkeley or elsewhere as visiting professor, when I have had only one course, or at most two, and so have had sufficient time for preparation.

Convinced that students learn from their own activity rather than from that of the instructor, I directed my efforts toward eliciting actions on their part that yielded visible results: numerical or graphic exercises or written papers. In this respect I found my courses in cartography, in which the important work was done in the laboratory, especially satisfying. Even students who had little skill in drawing to begin with always improved greatly in the course of a semester.

The best situation for fruitful interaction between students and instructor is in classes conducted in the seminar format, in which a small group sits about a table on the same plane. At Berkeley we were free to choose themes for our seminars. I used mine principally for enlarging the range of material presented to students beyond what they encountered in their more formal courses. In one of my early efforts I introduced students to Walther Penck's *Die morphologische Analyse* (1924), long before its publication in English translation. I translated orally the essential parts of the book, paragraph by paragraph, and let the students discuss Penck's ideas. Several times I devoted a seminar to the oceans, giving special attention to their effects on climate, so conspicuous on our West Coast. On a few occasions I directed the students' attention to the study of place names. Introduced to them by Isaac Taylor's old *Words and Places* and to American names by George Stewart's *Names on the Land*, students always found place names interesting. I knew the European literature on place names well enough to interpret representative names on topographic maps from England, Sweden, Denmark, Germany, and France, which I gave the students to examine.

Sauer used his seminars for a different purpose. The material he dealt with was usually what he was working on at the time: his seminars gave him the opportunity to elaborate his ideas orally before committing them to writing. They served the students by enabling them to see a mature scholar in action, and often suggested themes for their dissertations. His reliance on the ability of students to find their own way led him to overlook their shortcomings in knowledge unrelated to their dissertations. I was more attentive to their general education, conscious as I was of my own deficiencies. In advising students I habitually told them, 'First become an educated person, and then think of specialization.'

In time I took over from Sauer instruction in the history of geography, which I found congenial and retained until I retired. The field is large, and some selection was necessary, especially in the one-semester course, conducted in seminar form, that we offered to undergraduates. After trying to cover too much ground within a limited time, I gradually restricted myself to academic geography in the United States, to which Sauer had never given any attention. I made this restriction for two reasons: because this part of the history of geography was closest to the students; and because they could not be depended on to be able to read any language other than English.

Much remains to be done on a critical history of American academic geography. William Warntz has provided an excellent account of what he called 'the first cycle' of its history, but left a great deal unwritten on his 'second cycle', which is continuous with the present.[3] In most writings about the rise of our academic geography attention is fixed too closely on what was called 'geography' in courses of instruction and in the titles of books. One part of the early phase of Warntz's second cycle is certainly the presentation given in the textbooks in physical geography of the 1870s, by Arnold Guyot, M. F. Maury, and E. J. Houston. But a more prominent part is what W. M. Davis (in defense of etymology) called 'physiography', which dominated our textbooks from the middle 1890s to the 1920s. The roots of Davisian physical geography were mainly in the scientific exploration of the western territories of the United States, first carried out by special expeditions but later embodied in a permanent Geological Survey. The stupendous land forms of the dry West, their structures unobscured by forests or a deep cover of soil, enabled J. W. Powell (1834–1902), G. K. Gilbert (1845–1918), and others to formulate principles of land sculpture that remade geomorphology. One needs only to compare the textbooks written or inspired by Davis and published in the 1890s[4] with their predecessors of the 1870s to see the profound transformation American physical geography experienced under the impact of new observations and generalizations derived from the 'surveys' of the American West. The forms of the land, now presented in terms of processes rather than descriptively, are assigned more space than other aspects of the earth. (In the two books cited, this imbalance is more conspicuous in Tarr than in Davis.) The publications of the Geological Survey and its predecessors make up a large fraction of the literature cited in them. It was this kind of geography, with its emphasis on the forms of the land, that found a place in our universities, usually within departments of geology.

In my teaching of the history of geography I emphasized this native background. I directed the students' attention to the vast literature in which scientific work in the West was recorded, and to the nurturing of our geography in departments of geology. I commended to their attention, in particular, Nathaniel Southgate Shaler (1841–1906), W. M. Davis's teacher, whose range of interests and sympathies was immensely broader than Davis's. He was the most prominent intermediary between the field investigations of the second half of the nineteenth century, nominally geological but in fact inclusively geographical, and academic geography.

The immense store of new information about the surface forms of the earth was scarcely incorporated into our textbooks before an ultimately more influential tendency appeared: emphasis on the earth, in words still encountered in our literature, as 'the home (or world) of man'. W. M. Davis seems to have been responsible for this new emphasis, but I have no doubt that he derived it from Arnold Guyot (1807–84). Guyot had studied under Carl Ritter at Berlin, and his eloquent lectures published as *The Earth and Man* (1849) – kept in print in successive printings and editions into the early years of the present century – were the most conspicuous exposition of Ritter's ideas published in the United States. Guyot's examples of putative relations between earth and man appear frequently in Davis's textbook of 1898. In his preface Davis defines geography as 'the study of the earth in relation to man' and physical geography as 'the study of man's physical environment'. The earth's physical features 'must not be presented apart from the manner in which they affect man's way of living'. His text contains the full range of verbs dear to the environmental determinists of the following decades, from the mild 'affect' through 'influence' to the stern vise of 'control' and 'determine'. Davis's exposition of the physical qualities of the earth is admirable, written in his usual clear and logical manner and illustrated by his exquisite drawings. Against this background his examples of the effects of the earth's physical features on 'man's way of living' give the impression of being dragged forcibly into juxtaposition rather than associated by logical necessity.

Davis's redefinition of the ancient and inclusive term 'geography' has dominated American academic geography throughout my career. As I have encountered it from time to time I have often felt that I was being defined out of geography, since my interest in the earth does not depend on the presence of its human inhabitants. That definition, it seems to me, betrays an enormous arrogance; it would be equally appropriate to view the earth as the home (or world) of oak trees or lizards. Its roots are, I suppose, in the theological notion that the earth was created for the use of mankind, prominent in Ritter's writings but given a secular garb borrowed from the post-Darwinian ethnology of the late nineteenth century. Its purpose was apparently the same as that of other restrictive definitions.

If I have been dismayed by the long-accepted Davisian definition of geography, I am even more saddened when I find the discipline referred to in contemporary writings as a 'social science'. The earth is assuredly not a social phenomenon. Few who were present when Carl Sauer's former students assembled at San Diego in the summer of 1973 to celebrate the fiftieth anniversary of his coming to Berkeley can have forgotten his parting injunction to them: 'Remember, geography is not a social science, it is not a behavioral science; it is an *earth* science.' Some features of the earth's surface have a cultural origin, and are to be accounted for by reference to human activity. But it would seem supererogatory to ask a discipline concerned by name with the earth to interpret the modification of physical geography by human action, in the words of the subtitle of George Perkins Marsh's *Man and Nature* (1864), by an appeal to the doctrines of human psychology. This modification belongs with the other changes wrought by organisms during the long history of the earth: the reduction of carbon dioxide that provided the oxygen of the earth's atmosphere; the accumulation of rocks of organic origin; the action of plants in the formation and retention of soils; and the changes in vegetation effected by herbivorous animals.

Economic geography, which has been prominent throughout my career, has had no interest for me, and I have read little of its literature. The basic necessity of gaining a livelihood from what the earth provides, for humanity as for other animals, is obvious. But the means by which mankind obtains that livelihood are given by culture, not by the earth. I have preferred to look on cultures and their artefacts from the viewpoint of the humanities, taking that term in a broad sense, rather than from that of 'social science', which seems still to cherish the hope of finding mechanical explanations of human activities. By a humane view I mean acceptance of and wonder at the apparently unlimited capacity of human beings to diversify and elaborate, far beyond immediate necessity, the elementary processes of living and thinking – through spontaneous invention, diffusion of innovations, and gradual, unconscious modification. These changes, like all processes of human history, are contingent, not determined unpredictable, but comprehensible after the fact.

A humanistic view of mankind and its role in the transformation of matter and energy at the earth's surface could have been founded, in the formative years of our academic geography, on N. S. Shaler's humane ideas, as expressed, for example, in his *Man and the Earth* (1906). But his disciple Davis preferred the mechanistic concepts of 'control' by the physical environment and 'response' by organisms. Still later, the demise of environmental determinism made a place for a humanistic cultural geography, but the hope of salvaging some remnant of determinism has persisted. Even in this unfavorable atmosphere, however, such a cultural geography has achieved a modest growth among us, exemplified in morphological studies of settlements, both dispersed and agglomerated, and of such constituents of the cultural landscape as houses, barns, fences, and cadastral patterns.

Though my own writings in cultural geography are few and far in the past, I have never lost interest in it, and have occasionally included it in my teaching. Rather than economic interpretations, I have preferred to consider subjective, esthetic motives. The branches of scholarship most congenial to cultural geography have seemed to me to be the histories of art and technology. My thinking was influenced not only by observation of European cultural landscapes but also by Lewis Mumford's early writings on American culture history, notably *Sticks and Stones* (1924) and *The Brown Decades* (1931), in which Mumford revived the memory of G. P. Marsh and recognized N. S. Shaler as an exemplary figure.

I have taken only a small part in the organizations that represent American academic geography. As a member of the Association of American Geographers I occasionally attended its annual meetings, and for a time participated in its official activities. But when at the end of the 1940s its membership was opened to all-comers, and in the following decades grew into the thousands, its meetings, now held in large metropolitan hotels rather than in academic surroundings, became much less attractive. I found more satisfaction in working with our regional organization, the Association of Pacific Coast Geographers. Its limited regional basis has preserved it from the gigantism of the national Association and from the intrusion of non-academic, sometimes anti-intellectual, tendencies that gigantism has fostered.

Because of the open and inclusive view of geography cultivated at Berkeley, successive waves of fashion passed over us with little effect. Indeed, it might be argued that in the long view these fashions, however loudly trumpeted as they appeared, have only temporarily obscured the 'immemorial and universal' interest that Sauer never wearied of invoking. I may be deceiving myself, but I have the impression that at present our academic geography is less hampered by restrictive definitions than it was in my earlier years. In particular, I am happy to see some attention paid to features of the physical earth, once excluded from the content of geography though never wholly suppressed.

Speaking to his colleagues in the address 'The Education of a Geographer' I have cited earlier, Sauer remarked: 'It is, I think, in our nature to be a heterozygous population.' I am sure that here, as elsewhere in that discourse, Sauer was thinking of his department at Berkeley: of himself, of those he chose to work with him, and of the students who were attracted to his department. Only such a hospitable attitude toward 'different temperaments and diverse interests', in words with which Sauer continued, made a place for me, with my defective formal education, in the academic world.

Notes

1. Sauer, Carl O. (1925) The morphology of landscape, *University of California Publications in Geography*, **3**, 19–54; Sauer, Carl O. (1963) in *Land and Life: A Selection from the Writings of Carl Ortwin Sauer. University of California Press, Berkeley and Los Angeles, pp. 315–50*.
2. Sauer, Carl O. (1956) The education of a geographer. *Annals of the Association of American Geographers*, **45**, 287–99; Sauer, Carl O. (1963) in *Land and Life*, op. cit., 389–404.

3. Warntz, William (1954) *Geography Now and Then: Some Notes on the History of Academic Geography in the United States*. American Geographical Society, Research Series No. 25.
4. Tarr Ralph S. (1896) *Elementary Physical Geography*. The Macmillan Company, New York. William Morris Davis, assisted by William Henry Snyder, *Physical Geography*. Ginn and Company, Boston, 1898.

Selected readings

1928   The towns of Mälardalen in Sweden: A study in urban morphology, *Univ. Calif. Publs. in Geogr.*, **3**, 1–134.

1937   Some comments on contemporary geographic method. *Annals, Assoc. Am. Geogr.*, **27**, 125–41.

1938   The extremes of the annual temperature march with particular reference to California. *Univ. Calif. Publs. in Geogr.*, **6**, 191–234.

1939   The towns of medieval Livonia. *Univ. Calif. Publs. in Geogr.*, **6**, 235–313.

1941   Effects of the Great Lakes on the annual march of air temperature in their vicinity. *Papers Mich. Acad. Science, Arts, and Letters*, **27**, 377–414.

1954   Climatology. In Preston E. James and Clarence F. Jones (eds). *American Geography, Inventory and Prospect*. Syracuse U. P., New York, pp. 334–61.

1956   Extended uses of polyconic projection tables, *Annals, Assoc. Am. Geogr.*, **46**, 150–173.

1967   *Land and Life: A Selection from the Writings of Carl Ortwin Sauer*. Univ. Calif. Press., Berkeley and Los Angeles.

1972   *California as an Island*. Book Club of California, San Francisco.

1978   Town names of colonial New England in the West. *Annals Assoc. Am. Georgr.*, **68**, 233–48.

Chapter 7

Fig. 7.1 Walter Freeman, England

Rumor has it that Walter, just after he was born, took a withering look around the room as if to say: 'I don't think much of this place.'

To be son of a Methodist preacher meant an 'itinerant' experience of many places, all of which aroused a keen curiosity about people, landscapes, history, and eventually the prospects for international peace or war. What a challenge it must have been to write a regional text on my native land! A tireless researcher, generous correspondent, and eager hiker in the countryside, Walter Freeman embodies the virtues of the liberal and compassionate scholar, cherishing the values of his own style, but never failing to support initiative and a pluralistic, ecumenical attitude toward thought and practice.

Walter has served as Secretary of the International Geographical Union's Commission on the History of Geographical Thought since 1968 and edits its bio-bibliographical series, *Geographers*. At our Edinburgh meetings in 1977 I wondered if my autobiographical suggestion was to win another 'withering look'. Wiser in the ways of academe than I, and cognizant no doubt of many an inside story from the Hundred Years, he offered some words of caution. Typically a good sport, he later embarked on the writing of his own story and since then has become for me an invaluable source of inspiration and encouragement.

A geographer's way

Interest in places was natural in my childhood. Our family moved every three or four years as Father was a Methodist minister and the tradition of John Wesley's itinerant preachers still survived. As a family we loved these moves as exciting events, though doubtless they were a source of strain to our parents. My later crucial schooldays (1922–26) were spent at Bridgend, South Wales, where geography was a favoured subject. Despite a warning from the headmaster that geography was a 'very difficult subject', I chose it as one of the three required for the Higher Certificate examination (now called Advanced level) two years later. The other subjects were English and history, from which I acquired a lifelong devotion to English literature and an unending curiosity about the past. South Wales has so wide a variety of landscapes that the idea of regional units seems logical, even obvious. They are sharply defined, with a geological basis and limited by lines (in fact geological outcrops) rather than by transitional zones. In lowland Glamorgan, history is clear on the ground for there are Norman castles, prehistoric 'dolmens', ancient churches, and remains of early Christian or medieval monasteries, all accessible by foot or bicycle. Upland Glamorgan is a different world, with farms and villages on the lower hill slopes and moorlands above them and congested mining villages and towns in the valleys. Much was explained in the National Museum at Cardiff where, at a time when most museums seemed to be filled with stuffed birds, exhibits in bright galleries unfolded the story of the area's geological evolution, vegetation development, and prehistory.

Our next move was to an industrial area of Leeds where I became a student at the university with its small but enthusiastic department of geography, presided over by A. V. Williamson. His student days had been spent at Liverpool under the guidance of Professor P. M. Roxby, whose articles on China were then eagerly read. Roxby thought that all prehistoric and historical time should be considered in any treatment of a landscape. Even on a local scale, say for the 50 square miles that might be considered in an undergraduate dissertation, this was a daunting assignment. (This view was shared by H. J. Fleure, whose *Corridors of Time* books, written with H. J. E. Peake opened fascinating problems of the early settlement of Europe and its margins as they appeared from 1927 onwards.)[1] Much loved by Roxby, as by many of his disciplies, was Vidal de la Blache's *Atlas d'histoire et de géographie*, first published in 1894 and re-issued with new data several times since then.[2] There was the human story in an inspiring outline, with its varied civilizations and empires evolved in recognizable environments, diffused by enterprising rulers over land and sea, meeting barriers where the momentum of advance might or might not cease so that the Greeks never went effectively beyond the areas of Mediterranean climate, though the Romans carried their military organization to the inhospitable uplands of Scotland and through the forests of the Rhine–Danube frontier. The Mongols spread from the interior of Asia to the marginal lands: so the story went on, perhaps the most fascinating of all stories of human endeavour, explicable not only in terms of the ability and imagination of rulers but also of the varied challenge of environment. Few British geographers gave allegiance to the views of Ellsworth Huntington but none could ignore his work for in it he raised the problem of man and environment, anywhere, any time, always with the recognition that man was a living soul of body, mind, and spirit.

To some, all this conspectus seemed attractively superficial and vaguely inspirational. Far more promising, it would appear, was an approach through detailed local work, a concentration of town and countryside, studied in detail, indeed on foot, as far as possible. This was in accord with French practice, for their regional monographs were widely respected and their methods spread through the Oxford School of Geography by Roxby and a few others into the newer universities. Regional geography was revered and thought by many geographers to be the crown and summit of the subject. This did not preclude historical study and therefore it seemed natural to choose as a subject for the undergraduate dissertation the early historical geography of

Glamorgan, to the Norman Conquest. Where did the earliest settlers live and why just there? Why were the upland valleys virtually uninhabited before the industrial age with farms and villages on the lower hillslopes and the ancient trackways on the ridgeways? And why were the Romans and the Normans successful settlers on the lowland but able only to hold the upland with military power? It was a subject of more problems than one realized at the time, for in it there was the concept of the gradual modification of the landscape by successive settlers according to their capabilities and circumstances.

Looking back to one's years as a student with the hindsight of forty-three years of university teaching, my conviction is that all one can do is to help people to educate themselves. For the M.A. degree, a growing interest in international affairs led me to write on 'Modern Asiatic Migrations' (from India, China, and Japan), inevitably compiled largely in libraries. This seems a brave, even foolhardy choice now and my idea that in time I could write a political geography to supersede Bowman's *New World* seems a comic example of the arrogance of a new graduate still glistening with the dew of a first class degree.[3] But in those days we thought in terms of the world, its major areas and problems. International affairs were constantly discussed and in 1931 I was fortunate enough to attend the two-month course at the School of International Studies in Geneva, where Sir Alfred Zimmern, then Professor of International Affairs at Oxford, gathered together some 200 students from forty nations for lectures, seminars, and informal discussion. There one learned the human reality of international problems in a way that even Isaiah Bowman's fascinating book *The New World* could not achieve; one saw the tension between the French and the Germans, the depression of a charming Spaniard over the approach of civil war, the deep regret of many Americans that they were outside the League of Nations, even the aggression of a picked group of Fascist Italian students.

In so divided and torn a world, with diminishing hopes of a peace under the guidance of the League of Nations, many students became negative and cynical, living for the day while it lasted. That to me seemed an evasion of responsibility and purpose, for the earlier teaching in geography combined with the evolutionary approach of geology gave a vision of one human world that could lapse into barbarism or advance towards universal peace. 'Make peace not war' meant as much to us as to later generations of students and some followed the lead of persuasive prophets in the Peace Pledge Union. The shadow of the First World War hung over our childhood and by the mid-1930s few doubted that another war must follow, sparked off by some real or manufactured crisis. Where then was there some faith that could provide stability and hope for the individual? The old rationalism with its unswerving faith in science as the explanation of all human and natural phenomena no longer seemed satisfying and there were warnings from prominent scientists that it was futile to suppose that science could save the world. Most Christians no longer believed that the world was made in seven days (however long) for the Darwinian revolutionary theories, developed and modified by later workers, showed that the story of the world's history was far longer and greater than even the gloriously imaginative treatment in the Book of Genesis. But just as there was a unity in the physical world (as Vidal de la Blache had splendidly shown in the first chapter of his *Principles*)[4] so in the human world failure to grasp the necessity of universal brotherhood, or at least universal tolerance, clearly opened the door to global tragedy. Whether this was based on religion or on sheer pragmatism, the immediate prospects for civilization seemed bleak in the 1930s.

Obvious dangers lay in the world depression of trade and widespread unemployment, especially among the industrial nations, for in adversity the natural human tendency is to look for a scapegoat, such as Jews in Germany or other ethnic minorities. The new governments in Eastern Europe were particularly vulnerable in such adverse economic circumstances. Even if the British government appeared to think that out of all the turmoil in Europe, finally in Spain, peace could emerge, it was hard to share their optimism. For the young the outlook seemed bleak but after a year of unemployment (which gave a permanent sympathy with the workless) I became an assistant in the Department of Geography at Edinburgh University (much to my astonishment) where the outlook of Professor Alan Ogilvie owed much to his experience of continental European and American scholars. Also in the Department was Arthur Geddes, son of the famous Sir Patrick, with his liberal humanistic approach, inspirational but at times emotional and romantic, never dull. In him there were elements of the radicalism of Kropotkin and the Réclus family, oddly combined with a somewhat sentimental Scottish nationalism, but however much one might disagree with Geddes one could never ignore his views.

Ogilvie and Geddes became lifelong friends for they, with David Linton, were quite different from one another in outlook and also quite different from any geographers I had met before. Previously much of what inspiration I had received was drawn from the views of Roxby and Fleure and my tendency was to look beyond the economic approach of Chisholm and of the rising school of economic geographers in London. Naturally I was not at that time sufficiently acquainted with the admirable breadth of Chisholm's mind, never adequately investigated to this day, but economic problems appeared to be part of a wider human malaise, for the poverty of multitudes of the world's population seemed strange at a time when its resources and the scientific ability to use them were greater than at any earlier historical time. Youth is always eager to reform the world, preferably by next Tuesday, and therefore it seemed reasonable to consider some vocation rooted in eternal values, such as the Christian ministry. Much the same outlook has led people into a variety of occupations, including politics (the comment that people become politicians 'only for what they can get out of it' is cheap), social service, medicine and many more – not least education.

Any possible change of profession seemed irrelevant when, in January 1936, a surprise move to Dublin came my way. Then it seemed clear that an academic career was the one for me. In fact there was no time to turn one's soul inside out in self-examination as there was far too much to do. In Trinity College (Dublin University) there was no Honours course and the first need was to establish one, for which students appeared immediately with the first graduation in 1940. The Far East now seemed remote though as it happened I became a part-time worker in Cambridge under Professor Roxby on the China Handbooks prepared for the Admiralty from 1941–44.[5] In Dublin the main and obvious need was to develop geographical work on Ireland about which little had been written. An earlier interest in migration and population problems seemed as relevant as ever but in a different way for Ireland which, of all countries in the world, had experienced a continuous decline in population from 1845. I found some admirable material in the Reports of the Congested Districts Board for 1891. These reports dealt with poverty and overpopulation in the poorest areas of a country of surprising variety in natural resources from one area to another. Through the Geographical Society of Ireland (founded in 1934) and the much older Statistical and Social Inquiry Society of Ireland (founded in 1847), it was possible to make friendly and helpful contacts (never a problem in Dublin anyway) with others interested in the land and its people.

In the years to 1939 it was possible to travel cheaply in continental Europe, partly by helping to run tours of the Le Play Society or attending some conference or congress. Of these, the 1938 International Geographical Congress in Amsterdam was an inspiration. It was good to realize that there were interested colleagues all over the world, and to get to know some of them as friends. Academically it was a strong congress but one which made one question the statement of the British government that there was no German or Italian intervention in the Spanish Civil War. One had no illusions about the near future of Europe and it was perhaps the last chance to travel before the lights went out. The charm and courtesy of the Dutch hosts, the splendidly organized tours of their own fascinating land, the receptions in their famous art galleries, lifted the spirit. It would be impossible to exaggerate the stimulus I received from the Amsterdam Congress; I cared little that the British delegation was small and not fully representative for it was people from other countries that I had gone to meet.

Nearer home there were the British Association meetings in September and various meetings in London after Christmas, including those of the Institute of British Geographers (a small body before the Second World War). After the war, conferences developed to such an extent that attending them became almost a profession for some people. The value of conferences is great and yet in the end scholarship depends on isolation, on contemplation, on silence as well as on debate and discussion. Yet some of us are individualists – never eager to jump on a bandwagon, still less to drive one – eager to induce research students to make their own interpretation of their material unimpeded by the domination or adulation of others. By 1948 my book on the geography of Ireland was complete, though it did not appear until 1950. It now seems to be an old-fashioned regional geography, aiming as was then usual at showing the totality of the environment and its people. (Were I beginning the work now I would probably focus the whole study on population.) The regional geographer can only describe and analyse the land as he sees it and in later editions I made an effort to bring the material up to date, for Ireland has changed markedly since 1945. The first edition of the book bears the mark of the difficult adjustments made in Ireland to the exigencies of war and of tentative peace. Both in Northern Ireland and in what became the Republic of Ireland in 1949 the effects of war were profound (though different), for emigration rose to a high level, shortages of imported fertilizers made problems for farmers of the twenty-six counties while those in Northern Ireland shared in the intensification of British agriculture, and industrial growth was hampered in the impending Republic while in the northeast there was a boom in textiles and shipbuilding, as in the First World War.

The main difficulty lay in the lack of material for so little had been written on it by geographers, though a long-vanished Ireland had been imaginatively treated by Grenville Cole and others in the *Oxford Survey of the British Empire*, a much-neglected classic of 1914, edited by A. J. Herbertson and O. J. R. Howarth.[6] Happily there was excellent statistical material available, particularly for the Republic, and this made possible a close study of population trends from the Famine period (1845–51) onwards. Many counties, predominantly rural in character, had only one-third of their population a century earlier, recorded in the 1841 Census. But strange anomalies emerged, for there appeared to be no general correlation between population decline and poverty of environments as decreases in rich lowland areas like County Meath were as great as, or even greater than, those in the western congested districts. The highest densities of rural population occurred in the 'congested west' (if one excluded the uninhabited areas) and the lowest in County Meath. The correlation one sought was in the prevailing size of farm and clearly there was an enormous difference in the standard of living between the farming families of the west and those of the east. True, there were governmental aids to the poorest families, of whom many also derived support from a wide variety of ancillary occupations such as road repairing, intermittent labouring, emigrants' remittances or government doles. But the contrast remains. By now it has been alleviated and any geographer ready to analyse closely some western area would find a very different world from the one revealed to me thirty and more years ago.

Actual methods of dividing Ireland into subregions presented an interesting problem. For guidance on a possibly physical regionalization, the highly successful work by J. G. Granö in the *Atlas of Finland* (1925–29 edition) was closely studied, but it soon became clear that the landscapes of Finland were far more homogeneous than those of Ireland, and any attempt to emulate the crisp formulae of Granö would be impossible.[7] The only possible course appeared to be actual study of the landscape (a comment that may evoke pitying horror from some 'modern' geographers), and this was in accord with the outlook of Ogilvie, from whom I learned so much on the art of observation, and others. It was easy and enjoyable to cycle 40 or 50 miles a day, pausing here and there to observe anything of special interest, perhaps to talk to the people as opportunity arose. And in the evenings one would learn something from the commercial travellers and government officials who were likely to be staying at the small town hotels; when they were not looking the crucial points were written down in field notebooks. Had not the great Demangeon walked along every road in Picardy? Later, field tours with the

Geographical Society of Ireland (from 1945) and with student parties from Manchester (from 1950) gave further examples of making sample studies.

Two elements in this approach must in fairness be mentioned. First, one used all possible statistical sources (and incidentally spent many hours slide rule in hand); and second, one cared for the people. It is a little difficult to be patient when younger geographers suggest that their predecessors were neither aware of statistical sources nor concerned with the way people lived; quite simply, they were. One knew very well what poverty meant even though conditions had been greatly improved from the year 1891 when the Congested Districts Board began its work. Since 1945 certain areas of Europe have become 'problem areas' receiving aid from EEC funds as well as from national govenments and it is good to have had some infinitesimal share in drawing attention to such areas. To cast a romantic aura over peasant life never seemed to me realistic. Nor was it helpful to ascribe all the woes to Ireland's unfortunate political history for the real problem was to see what alleviation of conditions was possible. A later work, *Pre-Famine Ireland* (1957) was based on material available for the ten years before the Famine began in 1845, using government reports, maps, Census and other statistical data.[8] The fashion of the time was to give a period study in historical geography, a cross-section based on one definite date, perhaps because the airy sweeps through the centuries of some geographers had seemed superficial and unconvincing. In Ireland there was an endless opportunity for detailed study and when the Geographical Society of Ireland began to publish a journal, first as a humble 'Bulletin' and then – greatly daring – as *Irish Geography* it was good to help with the editorial work. Wisely this journal has kept its original policy of concentrating on Irish material and under a succession of devoted editors has earned widespread respect and a worldwide circulation.

After fourteen years in Ireland I still wished to live in England (partly for family reasons) and in 1950 I returned with my wife and family (all born in Dublin) to Manchester. Through the work of C. B. Fawcett, L. Dudley Stamp, and Eva G. R. Taylor and others, geographers had become increasingly concerned with planning problems and this led me to various studies on British conurbations and on the geographical basis of planning problems. It is not for geographers, or at least for this geographer, to give a directive to planners. Nor indeed could I accept the view that geographers should be the supreme coordinators because their training had been broad and, at least until the Second World War, normally designed to give some insight on physical and human aspects of the landscape in town and countryside (perhaps more the latter for the teaching on towns was somewhat sketchy before the war). When, in Manchester, I taught geography to some undergraduate planners I found that many of them were so dazzled by their hopes of making everything new that they cared little for the past. Indeed contact with such students, good and delightful people as they were, made me wonder if planning should be taught to undergraduates, for to them it provided too heady a wine for their age. (Was it one of P. G. Wodehouse's butlers who said that 'the young lordship is not ready for our brown sherry'?) The graduates were different, for they saw that the rebuilding must fuse the new and the old and so retain the mixture of architectural aspirations that makes British towns of such interest to those who know them.

Planning is an activity of wide scope in which virtually every part of human experience is involved. What the geographer can do is to induce people to see the life of people in terms of environment, in terms of home and workplace and the journey between them, in terms of recreation, land conservation, and those amenities providing adequate living conditions for an increasing population. That these aspirations are not always achieved by carefully considered plans is abundantly shown in many inner city areas of Britain, now much reduced in both population and employment. Can it be that as a reaction to the baleful congestion of people and workplace central city areas are becoming human deserts? Each solution to a problem appears to raise new difficulties unforeseen in the earlier flush of hope. Whatever the answer, it would be arrogant to assume that the geographer alone can provide it. The point is that many of us through the years have been interested, indeed constantly concerned, with human welfare, with how people live, only to discover that Utopia is not built in a day. We never were, as some young modern geographers appear to think, flag-flapping imperialists, intolerant of racial and national differences, nationally chauvinistic, ignorant of the deprivation of millions through poverty and disease.

University teachers develop many interests, and a devotion to the history of geographical thought may seem to be a retreat to the 'peaceful groves of Academia'. To me the first action came in 1947 when a British Association committee was formed to collect material on people connected with the growth of modern British geography from 1887 when Mackinder became Reader at Oxford. There was a limited but useful response to a circular sent out asking for information, but practically no response to appeals printed in various journals. Nevertheless interest grew and one fine contribution was the presidential address of O. J. R. Howarth to Section E (Geography) at the 1951 Edinburgh meeting of the British Association, reviewing its work over a century. An entirely unexpected invitation came to write *A Hundred Years of Geography*[9] which appeared in 1961 and was followed in 1968 by *The Geographer's Craft*,[10] which dealt with individual geographers. At the New Delhi International Congress of 1968, I became a member of the newly-formed Commission on the History of Geographical Thought, which later held meetings in London, Manchester, Edinburgh, Paris, Warsaw, Leningrad, and Montreal. Finally in 1977 the first issue of *Geographers: Biobibliographical Studies* appeared, edited by the Chairman of the Commission (Philippe Pinchemel), the Research Associate (Marguerita Oughton), and myself. An earlier publication of the Commission was *Geography Through a Century of Congresses*, issued at the 1972 meeting in Montreal.[11] The history of geographical thought, so broad a subject, is now treated in numerous papers in various journals, some on individual geographers and others of a more general character in France, America, Great Britain, and elsewhere. The denigration of all past geographers and their work as irrelevant had fortunately been followed by a new consideration of them, much of it among young people.

It is not possible to divide a life into sections and to say that in certain years one's work was on a particular topic, followed by another at a later stage. My interest in Far East problems was revived during the war years when life was divided sharply into two parts, one spent at Cambridge preparing the *Admiralty Handbook* and the other in Ireland when any time left over after essential teaching and administrative work was given to my Irish research. That interest abides, partly through family circumstances, while the work on China, though a source of fascination through many years, belongs to the past, dealing with a China markedly different from that of the present day. Others may well be able to develop competence in this area in time, for just as Russia remained an enigma for several decades so China may be opened to geographers from Europe and America. Indeed there are signs that this is beginning.

The wish of any researcher is surely that others may begin where he left off, that they too may become 'committed' (to use a word much in vogue but hard to define). But the causes to which a researcher becomes committed depend largely on circumstances. From early youth one carried forward a love of the countryside, from later experience a concern for housing especially if one has seen something of conditions as they were forty or fifty years ago. In any life great changes are seen, some of them fortunate such as the removal of square miles of urban slums or the increasing prosperity of rural areas, but social problems remain and those whose concern lies in the rehabilitation of the environment are led to ask what has gone wrong. All academics, however desiccated by experience, hope that the reforming zeal of students will be channelled into practical activity through some profession helpful to society. Happily, since 1945 geography graduates have entered many other professions besides education. There was little alternative available in the 1930s, when most of the students were destined to be school teachers and many were recruited into the modern universities through Departments of Education. (It is the work of these people which has made possible the vast expansion of geography in the schools and universities.) But the various administrative, planning, commercial, and industrial posts now open to students did not exist for geographers. With many other subjects, geography may now be regarded as a liberal education, preparatory to some form of specialized study or employment.

To have a vision of a better world is common among geographers. Roxby saw it in harmonious race relations, in acceptance of universal brotherhood as a fact of life. Fleure envisioned it in a planned economy as the next stage in a long evolutionary struggle from the beginning of time, with liberal education devoid of prejudice as a stage of further enlightenment. Chisholm looked to world economic advance with the intelligent use of capital as a solution to poverty and frustration. Patrick Geddes with his followers thought of it in terms of planning a world in which all the people were of significance. Mackinder sought a political solution, with a world in which no great power could arise in Germany or Russia to threaten world peace. Stamp was pragmatic, determined to persuade others that land use mattered, at one stage pessimistic about feeding the world's population but later, as shown in *Our Developing World* (1960) cautiously optimistic;[12] his work was based on factual enquiry as well as on a kindly attitude to all the world. But can one really believe, as one geographer (still alive, so no name is given) has said, that in the end the economic motive outweighs all others? Perhaps Stamp went too far in his pragmatism, for he loved to say that he never had time to read Hartshorne's *Nature of Geography* and therefore did not know what geography really was. But he was a generous and good man eager to show what could be done to develop the world's resources through survey and action.

Many visions of geographers appear to be incapable of realization. Hartshorne's *Nature of Geography* (1939) shared the widespread veneration for regional geography of its time and covered aspects of environment and life.[13] The revolt against regional geography was not actuated solely by the crudity of many supposed 'correlations' but also by the sheer impossibility of meeting the challenge of completeness. And when, after the Second World War, some geographers put forward the idea that the geographer was the supreme correlator of knowledge acquired from other specialists, and therefore the planner by the very nature of his training, that too failed to carry conviction even among many geographers. With notable exceptions, much that was published as regional geography appeared to be trite and unconvincing and this in part impelled the growth of the systematic approach, which some would describe as a retreat into specialism, with themes of human and physical geography steadily subdivided as research continued to show new complexities and uncertainties where supposed veracity had existed earlier. That a regional approach was also necessary, however, was shown by the growth of regional science, by the growing interest of economists, sociologists, political scientists, planners, and many more in a geography whose concern lay in a more equitable society, nationally and internationally, including a search for understanding of how people have lived in the world's varied environments. People do not exist solely in national groups, cohesive entities such as city populations or industrial communities. Always and everywhere they are individuals, inevitably and inescapably united into social, economic or industrial communities. To some extent regional study provided a way of considering people, in areas of small as of large extent, in a multitude of differing circumstances; such as enquiry inevitably reveals the inequalities of human society.

Mackinder of his *Britain and the British Seas* (1902) drew attention to metropolitan and industrial England and by 1980 this was more apparent than ever, for in general the higher rates of unemployment are found almost wholly within his 'industrial Britain', in its economic heyday when Mackinder wrote but now beset with problems. It is in no way remarkable that, excluding some artisan city areas, almost all 'metropolitan' England returns Conservatives as Members of Parliament while Labour Members come from industrial England (excluding the richer areas), Scotland, and Wales.

Valuable as many statistical enquiries have been, the need remains to consider carefully the accuracy of the data. Much has been said of the outward spread of towns, and indeed during the interwar period many towns were virtually doubled in extent, without any marked increases and in some cases even a decrease of population. A significant, but often neglected, figure is the number of households, for families are smaller and more childless couples or even single persons have their own dwelling. Slum clearance has been followed by some disastrous experiments in re-housing; in the 1950s the status symbol of an enterprising municipal authority was the multistorey block of flats, but how many people want them now? Well-meaning councillors have been defeated by the most elementary of all human reactions – perception. People see the environment they want and within the constraints of financial resources do everything possible to find it. Any geographer dealing with human aspects is confronted with a constantly dynamic situation and may find comfort in a model or a paradigm. Some indeed appear to fear that they may be one paradigm too late.

Any geographer lives through a time of social, economic and political change and of academic development. Reviewing a life's work, he may see that it has all been very different from what he expected, perhaps due to new opportunities presented by a change of circumstances, such as a move to another country or university. Assuming that he is eager to retain research interests, so easily submerged under the pressure of teaching and administrative duties, there will never be any dearth of themes worthy of attention. Some of these will arise from teaching and research will strengthen teaching. He may be content with specialization, or he may want something more. Some controversies, such as those between a regional and a systematic approach, seem ridiculous to those who can shed the view that nothing matters except what is done by themselves and a few favoured friends, possibly their own research students.

To me geographical study remains concerned with place and people, on three scales. Of these the first is the world scale, for the aspiration of our Victorian forebears to know the world still seems relevant. The second scale is national or possibly concerned with a group of countries, for example Scandinavia, the Indian subcontinent or even Great Britain and Ireland. In such grouping it is possible to deal with regional and systematic problems, constantly interpenetrating. The third scale is local and geography has always advanced in strength when this has been recognized. Facile generalizations on urban geography, some of them little more than an historical survey with a few references to the site, situation, and morphology of a town, have been superseded by increasingly sophisticated studies of distributions within a town, of its particular quarters, of town centres. Similarly in rural areas, detailed mapping, investigation of individual farms, of cropping and physical conditions has made possible generalizations based on hard data rather than on inspired guesswork.

If in long experience I were asked to define what has been crucial my response would be field work. My schoolboy experience was of abiding value, because one learned to ask vital questions and to explore the local area for oneself, asking why roads should have such apparently erratic courses, why the older settlements of upland Glamorgan were on the hillsides and not in the valleys, why even in highly industrialized areas some steep hillsides still retained natural woodlands self-regenerated through time and never cleared, why so many railways competed for the limited amount of flat land in the valleys of the coalfield where the areas of scarred and derelict land had once been the scene of industrial prosperity. This was both human and regional geography, concerned with landscape and with the people who had remade it. That one found some of the answers in economic history was no more surprising than that the military historian may find some of the answers to his questions in geography. The unnecessary railways seen in some valleys had competed for the coal trade when the rich deposits had first been exploited; the mounds of industrial waste had been accumulated, even on valley floors, when land was cheap and capital abundant, when it seemed that coal mining could go on for ever and the depression, bound to come in the end, seemed as improbable as landing a man on the moon.

Inevitably as one looked beyond the home or any other personally known environment there was an acute dependence on the map, to a degree that was excessive. Harm as well as good lies in the trite comment that 'the map is the tool of the geographer'. A map is only a document, showing some things and ignoring others. There was a tendency to lose sight of the land and the people, to infer more than the map showed, to generalize with a plentiful dash of imagination, even guesswork. For example, as I worked in Ireland on population distribution and migration movements, the essential geographical question still remained: How is this reflected in the landscape? What do people see? What matters is not how far one travels or how frequently, but how one travels and what one observes. In the lecture room it is possible to get away from reality – indeed it is all too easy to do so – but in field work it is difficult to do so and therefore other teachers may share my view that taking people on tours, of any duration from half a day to several weeks, is time well spent, however, myopic some of the students may be. To those whose work began fifty years ago the historical approach seemed to be crucial, with the historical and regional work of Vidal de la Blache and others as a shining example of the view that the present landscape is an end product, evolved through centuries of human effort and experience.

Not for a moment would one argue that all geographers should be interested in history, for some vill find other studies – economics, sociology, geology, botany, zoology or even psychology or philosophy – relevant to their own studies. Certainly one would wish to avoid demarcation disputes between subjects, for divisions between them are arbitrary. Geography has its own distinct field of enquiry and is not merely correlative of other specialisms. It deals with space and man in time, knowing that in man there is the most complex of all studies. It can give much to other studies and receive much in return. It recognizes the changes made currently as well as in past time on the face of the earth and studies the physical mechanisms as well as the human decisions through which they occur. And if such a study leads at least some geographers to hope that they may in some small way contribute to the making of a better world, at least through giving others a better understanding of it, their work will be more than a dream and a shadow.

Notes

1. Peake, H. J. E. and Fleure, H. J., *Corridors of Time* series. *Apes and Men; Hunters and Artists; Peasants and Potters; Priests and Kings* (all 1927); *The Steppe and the Sown* (1928); *The Way of the Sea* (1929); *Merchants Venturers in Bronze* (1931); *The Horse and the Sword* (1933); *The Law and the Prophets* (1936); *Times and Places* (1956); Clarendon Press, Oxford.
2. *Histoire et Géographie. Atlas général Vidal de la Blache.* Paris, 1894; later editions, 1909, 1918, 1922, 1938, 1951.
3. Bowman, I., *The New World*, World Book Club, Yonkers-on-Hudson, 1921 1923, 1924, 1928.
4. Vidal de la Blache, *Principes de géographie humaine.* E. de Martonne (ed.), Paris, 1921, translated by M. T. Bingham as *Principles of Human Geography.* Constable, London and New York, 1926.
5. Naval Intelligence Division, Geographical Handbook, *China Proper*, 3 vols, 1944, 1945. P. M. Roxby (ed.) (work by present author in vols 1 and 2).
6. Herbertson, A. J. and Howarth, O. J. R., *The Oxford Survey of the British Empire*, 6 vols. Clarendon Press, Oxford, 1914.
7. There have been four editions of the *Atlas of Finland*. 1899, 1910, 1925, 1961.
8. See Selected readings, 1957.
9. See Selected readings, 1961.
10. Freeman, T. W., *The Geographer's Craft.* Manchester University Press, 1967.
11. International Geographical Union, Commission on the History of Geographical Thought, *Geography through a Century of International Congresses.* Montreal, 1972.
12. Stamp, L. D., first published as *Our Undeveloped World* 1953 and then as *Our Developing World* 1960 and 1961, Faber, London.
13. Hartshorne, R., *The Nature of Geography.* First published 1939 as *Annals of the Association of American Geographers*, parts 3 and 4, pp. 171–658.

Selected readings

1950 (reprinted 1965, 1969, 1971)	*Ireland: Its Physical, Historical, Social and Economic Geography.* Methuen, London and Dutton, New York.
1957	*Pre-famine Ireland. A Study in Historical Geography.* University of Manchester Press, Manchester.
1958 (reprinted 1967, 1974)	*Geography and Planning.* Hutchinson, London.
1959 (1966)	*The Conurbations of Great Britain.* Manchester University Press, Manchester.
1961 (1965, 1971)	*A Hundred Years of Geography.* Duckworth, London.
1967	*The Geographer's Craft.* Manchester University Press, Manchester.
1968	*Geography and Regional Administration.* Hutchinson, London.
1971	*The Writing of Geography.* Manchester University Press, Manchester.
1977 (continuing)	Editorship of *Geographers: Biobibliographical Studies.* Mansell Publications, London. See, for example, 'Edward William Gilbert' *ser. cit.* (1979), **3**, 63–71.
1980	*A History of Modern British Geography.* Longman, London.

Chapter 8

Fig. 8.1 Ilmari Hustich, Finland

Born in a slum environment in Helsinki and as a child witness to poverty, injustice, and war, Ilmari Hustich became a leading figure in political life and an expert on regional economic development in his native Finland.

From Lapland to Labrador span the horizons of his research on landscapes, life, and climate; from Romania to Estonia his curiosities about political boundaries and his sensitivity to issues of social justice.

Ilmari Hustich is the only essayist in this volume whom I have not met personally. Through correspondence over this essay, however, I have caught glimpses of his fascinating personality, thoroughly enjoying his openness and realism, his occasional irony and understatement. Delightful indeed it is to hear this saga from the deep forests of Kovero, Finland, where he resided until his death in 1982.

An autobiographical sketch of the 'life-path' of a geographer

My life-path has been both psychologically brushy and scientifically curved, stamped by simple opportunism, enforced by various external factors and different internal tensions. The following is a rendering of some facts and circumstances concerning my career, without any claims on a deeper analysis.

Beside the primary biologic–genetic programming, one's life-path is obviously to a high degree shaped by the material and psychological environment of childhood. Child psychologists quarrel over the relevance of the various factors – is for example the impact of environment 40 per cent or 60 per cent of the whole? The older I become, the more weight I attach to milieu and events of childhood in the shaping of both personality and life-path.

I was born in 1911 in Helsinki and lived as a child and a boy in the slum quarters of the city. I also stayed for some periods at a 'home for children' while my mother slaved in a factory. A background of that kind strongly influences personality. I witnessed social injustices at close range, and as a seven-year-old boy I experienced the civil war in 1918 as the poor people of Finland experienced it.

Astrid Olin, one of my school teachers, became a decisive influence in 1924 when she convinced my mother to let me continue in secondary school as a

103

so-called 'free pupil', despite all the sacrifices it must have meant for herself at that time. It was probably a mere accident that I came to attend secondary school; it was quite exceptional for that milieu. The normal course for a healthy boy after primary school was to become an apprentice in one of the nearby big mechanical industries. I remember how I used to wrap my schoolbooks in a newspaper so that nobody would notice that a boy who was already twelve to thirteen years of age still went to school. Perhaps my mother regretted her step later. Perhaps she dreamed that schooling would enable her boy to become a 'social reformer' who worked for the poorest in society, for her equals? I never realized her hopes; I marched the straight road into a bourgeois existence.

I became a rather ambitious pupil in the Lyceum. Two enthusiastic biology and geography teachers (Börje Olsoni and Ole Eklund, both well-known botanists) influenced the direction of my future studies. They encouraged in school that interest in nature which I had already shown. Later I have seen in this interest in nature an unconscious reaction against the shabby slum environment in which I grew up (perhaps it is only wisdom after the event). Interest in geography woke early, I assume as a 'longing for the far away', common among most boys, something which even school geography stimulated at that time. Later on school geography unfortunately became much else.

The generally strong political feelings in our poor quarters immediately after the civil war awakened a serious desire to find out what was happening in the world; to follow consciously political events. This early-founded political interest has never slackened. Later it was invaluably helpful in my work as a teacher in economic geography; the economic sciences frequently suffer because their practitioners often do not understand the impact of politics.

Summer holidays during my school years invariably meant work to earn money, first as a messenger boy or as a builder's workman in various parts of the city. After that came forest work, something which later gained a certain importance for the future plant geographer.

I qualified as a university student in 1931. But what was I going to study? It had to be something dealing with nature. At first hand the profession as a forester, sane and secure and providing both nature and money, attracted me. The two final summers before my student exam I had been in forest work in eastern Finland. During the years 1930–31 depression ravaged Finland as other countries and older foresters warned me not to go into the profession. Instead I began studies in natural history and geography in order to become a teacher of these disciplines. In this effort I was clearly stimulated by the memory of my teachers in these fields.

Already during the first student years – simply in order to finance my studies – I wrote newspaper articles on a variety of topics, predominantly biology and geography, but also on various societal matters. At that time it was not considered proper for a prospective scientist to write 'popularly' in the daily press. Perhaps this writing also strengthened the trait of superficiality that colleagues like to find in my personality.

The first two summers during my university studies I spent as a forest worker and a surveyor's assistant in Lapland, subsequent summers were spent in the mountains, supported by small stipends, to carry out the field work for a doctoral dissertation about flora on the isolated mountains of western Lapland. At the end of 1934 I became B.A. (candidate of philosophy) with honors in botany, geography and zoology and a pass in geology and chemistry. I studied quickly, feeling obliged to do so first of all for my mother's sake. To pass this B.A. exam, however superficial it was, marked a starting point for the future. Should I proceed along the academic path?

As I gradually won a certain reputation for knowledge about Lapland, the way opened for a salaried position with the tourist organization of Finland in the autumn of 1935. By that time I definitely gave up plans of becoming a school teacher; besides, my teacher training courses, for various reasons, were rather unsuccessful.

In 1935 I received a large stipend for a geographic study tour to the Carpathians. I wandered over the Tatra mountains from Czechoslovakia to Poland and back again, in the same year visiting Austria, Hungary, and Roumania. In addition to the field studies of nature I had the opportunity to make a close acquaintance with a still quite exotic Carpathian mountain people and their settlement, as well as to travel in the politically rather unsettled Donau area.

In the 1930s Lapland was becoming a popular goal for excursion people and tourists. I myself had wandered and gone skiing quite frequently in the north in connection with studies and later on behalf of the tourist organization. To begin with, I preferred to make detours around the farms in Lapland. People were not my concern, only the untouched nature was of interest, or so I thought at that time. But later my interest in the people there awakened, and I tried to understand how they lived and worked. At the end of the 1930s, I even wrote simple articles on colonization and settlement in the north, not only on plants and animals.[1]

I once had chosen the flora and vegetation of the low mountains in western Lapland as the topic for the 'pro gradu' dissertation. After certain hesitations I decided to try to widen it into a doctor's thesis. Although my professor in botanics, Alvar Palmgren, did not like the choice of subject, he gave me his full support. I defended my thesis in spring 1937, when I also married. I became a typical 'bourgeois', far away from the social reformer that my mother probably had wanted – an inferiority complex developed which in various connections appeared in my later life.

The theme of the thesis was the changing distributions of flora and vegetation on the mountains of western Lapland, against the background of recent changes in climate.[2] It was not conventional plant geography but rather an experiment in noting the dynamics in nature. In hindsight the thesis regrettably has to be described as a hasty piece of work, even if it contained a certain ecologically fruitful idea – the dynamic relations of vegetation and climate which I later dealt with in many contexts.

In the summer of 1937 I had the opportunity to participate in an expedition to Newfoundland and Labrador together with the geomorphologist and anthropologist Väinö Tanner and the geologist Håkan Kranck. I served as the plant geographer and as a general assistant in the expedition. Väinö Tanner, a learned man with wide interests, had for a short time been my teacher in geography. The expedition on the whole realized a hot childhood dream – to take part in an exploration, seeing foreign coasts and experiencing native people (Eskimos and Indians) in their own milieux. At that time, of course, Labrador was still an unknown country with regard to its natural history. This expedition initiated my activity in plant geography in Canada with the flora and forests of Quebec and the Labrador Peninsula as a central theme.[3]

After the war (1947) I took part in another expedition, financed by the National Museum of Canada and the Arctic Institute, along the eastern coast of Hudson Bay, still rather unknown. I had visited the east Canadian subarctic already in 1937 and 1946. Some publications on forest botany, on tree and forest limits and on general plant geography were the outcome of this pleasant effort.[4] Nowadays I like to believe that this Canadian part of my geographic publication activity is relatively tenable; it forms to a certain extent a completed whole and has been overestimated as a 'pioneer input' by friendly Canadian colleagues. It was not proper 'geography' according to the pattern of the 1950s and 1960s, of course, but a rather simple, descriptive ecology without methodological edge. I believe, however, that such a 'primitive' field geography is still needed to check and balance the often sterile model-building of today.

By the end of the 1930s, just before the Second World War, my situation was unclear. Since 1935 I had a relatively good position at the tourist organization as editorial secretary and specialist on Lapland, and I had defended my thesis in 1937. But a scientific path was by no means clearly staked out. Was I going to have 'science' only as an interesting side activity or should I seriously continue along the academic line? And if so, should I be a plant geographer or deal with 'human ecology', with people? Geomorphology, pollen stratigraphy, and the so-called 'pure geography' which were popular in Finland in the 1930s and 1940s, did not attract me, though the professors who then conducted these topics, among others Väinö Auer, researcher of Patagonia, in themselves were inspiring personalities. One thing was clear: I had always felt a positive interest in 'the geography of man' and in societal problems.

But how to combine plant geography with the 'geography of man'? I probably did not wonder over the problem very much at that time, yet it must have lurked continually in the background. I did not have the qualifications for a chair in botany in the ordinary sense. Besides, by the end of the 1930s plant geography had become a rather old-fashioned discipline. There were better botanists in the country.

Then came the war, the problem was postponed for some years. The war itself had little importance for my geographical life-path. With or without the war, I probably would have entered or been pushed into the same academic road.[5]

During the winter war (1939) I spent a short while as a military home front official in Lapland. In March 1940 my category moved into ordinary military service which lasted approximately to the end of the interwar peace period. I was later mobilized as an ordinary soldier in the summer of 1941 when our war began anew, first at the Hangö frontier and then at a regiment in east Karelia at Svir. I belonged to the older age groups (born 1911 and earlier) and had a light military service.

In spring 1942 I was given a leave of absence, among other things, for a short journey to Siebenbürgen which at that time belonged to Hungary; it had been Romanian in 1935 when I had been there. The tremendous alteration which had taken place in Siebenbürgen after the rulers had been interchanged was a circumstance which almost changed my 'geographical' life-path. Political geography, and in particular the prostitution of geopolitics in the service of power, was a question that I for a moment wanted to study seriously and perhaps make into a lifetime task. Afterwards the whole thing became limited to an essay on the Siebenbürgen problem[6] and a few meddle-some newspaper articles in the middle of the war on the task of science, directed against our own geographers' and other scholars' attempts to show 'scientifically' that Far (i.e., Russian) Karelia belonged to Finland. This was one of the few non-opportunistic elements in my life-path which has otherwise been characterized by opportunistic steps into the academic hierarchy, mainly according to the law of least effort.

Just before the war I had been appointed associate professor (i.e., docent in botany, specifically plant geography). Shortly after the war I received a three-year stipend for associate professors. Combined with a position as research assistant, this provided enough income for me to leave the position at the tourist organization. Thus plant geography remained my main scientific interest into the 1940s. During the long 'calm' periods of the war I had rather keenly carried on floristic studies in Far Karelia, areas which were relatively unknown from a botanical point of view.[7]

During summer 1939 I became captivated by a problem that was indirectly connected to the problematic of my thesis and which was to occupy me for forty years, i.e., the growth of the Lapland pine in relation to climate, elucidated, *inter alia*, by the variations in the growth of the annual rings.[8] I had to revise some of my earlier opinions concerning the relation between growth and climate. Gradually the whole question of the tree and forest limits became a central part of my interest.

Geography in Finland immediately after the war, was still to a large extent physical geography.[9] Yet in the 1940s and 1950s even a plant geographer could with relative ease become a 'real' geographer if he wrote some additional papers with a human geography content. Our older geographers had all gone along this perilous and difficult road of 'double qualifications' to the still undivided chairs in geography which included both physical and human geography. It was therefore not so remarkable that a young plant geographer became an economic geographer.

In 1942, when the professorship in economic geography at the Swedish School of Economics in Helsinki became available for application, I applied for the position, more for 'fun', because only with goodwill could some of my publications be classified as economic geography.[10] Another and better qual-ified candidate, Helmer Smeds, was appointed to the position, quite correctly. Then in 1949 the so-called Swedish professorate in geography at the University of Helsinki (a chair which still embraces both physical and human geography) was announced for application. This position could be viewed as lying somewhat closer to my training and scientific production. Among other things I had written a larger work in 1945 on the variations of the harvests in Lapland, on 'the climatic hazard coefficient', etc. My colleague Helmer Smeds was given the first place, of course. With few complications, in 1950 I was appointed to his professorate in economic geography at the Swedish School of Economics.

In the hard light of after-rationalization, I now see that my 1950 appointment as a professor in economic geography – a field for which I had not sufficient knowledge or talent – was an outcome of opportunism: I wanted to reach quickly a position that gave a good income and a certain amount of security. This story illustrates in many ways the characteristic uncertainty about the identity of geography in the 1940/50 period; it also indirectly indicates the lack of competent geographers in Finland. It began to dawn on me earlier than 'my geography' should probably elucidate the larger question of man and his environment rather than the narrower kind of plant geography I dealt with in the 1930s and 1940s. But I had no clear picture myself of what I really wanted. The new position partly solved that problem.

Economic geography in the strict sense as it was understood at a School of Economics remained for me, I think, more a 'bread-and-butter job' than a real life-task. Now after my retirement from this office, for which I was responsible for twenty-five years, I still feel foreign to the basic market-economic philosophy of the economic science. Perhaps my intellect fell short of following the nooks and crannies of economic theory.

I have wondered many times why I nevertheless persevered for such a long time in this job. It may be a question of intellectual inertia or simply pure convenience. In 1950, when I came to the office, I was already thirty-nine years old. And after that age one does not easily change either job or place to live, even though I had a few opportunities to do so, both at home and abroad. To deal with my professorate in economic geography in an approximately satisfactory manner resulted in the 'duality of personality' which perhaps unfortunately marked my scientific production afterwards. For I continued, obstinately and indeed with rather a lot of enthusiasm, two main themes in plant geography from earlier years:

1. Canadian ecology and plant geography. For this purpose I made several small expeditions and study tours to Canada (in the years 1937, 1946–48, 1952, 1956, 1959, 1963, and 1967); the journeys resulted in several publications, the latest ones from the 1970s.[11]
2. Studies concerning the growth of pine, its rejuvenation and dependence on the climatic variations in the north. These were published in the period 1940–78.[12]

Disregarding a couple of compendia and textbooks in economic geography[13] and in political economic geography I have not produced any truly scientific work in economic geography, apart from the more descriptive summary work on the skerry guard of Finland (1964).[14] On the other hand, I was interested in lecturing in general economic and political geography, and I probably treated that part of the job relatively well. But to the students of a School of Economics my kind of geography unfortunately remains a side topic (just now when it is more than ever needed in a world that despite everything is becoming more and more international)! I have not had any advanced pupils (apart from Sigvard Lindståhl, an associate professor who died at the age of forty in 1978). There is one sector of economic geography however, to which I have devoted myself more seriously: regional economic variations. In 1952 Professor Lars Wahlbeck and I wrote a longer essay on the geography of income in Finland and its regional variations, a theme which interested me and in which I had some competence.[15]

In the years 1961–62 I served as Minister of Trade and Industry in the Finnish Government, another result of the play of accidental circumstances. In that connection I also became chairman of the national delegation for the marginal island areas in 1961–65, one outcome of which was the publication mentioned before on the skerry guard of Finland. For about a decade I was a member of state committees and delegations for the development areas. These positions gave me the opportunity to contribute to decisions concerning the development of marginal areas. After many detours, I finally had arrived at human geography. During the 1960s and 1970s I wrote some smaller papers on this problem.[16] The regional policy of Finland and her marginal areas is the only field in which I have at least for a short time had a certain position as an expert.

In this circumstance one can see a reflex of earlier experiences and milieux. The reason for my interest in our marginal areas was probably laid down during my summers as a forest worker in 1930 and 1931 at the Russian border, where I still vividly remember the poor people and meager farms. Indirectly I also saw the same fundamental problems in Russian Karelia during the war years of 1941–44 at Svir. Perhaps the fact that development areas have come to play a main role in my later activity as a geographer is not just a play of circumstances; perhaps it is a question of an unconscious will and a line which has forced itself into the open belatedly.

An important factor in the shaping of a personal scientific life-path is the intellectual milieu in which someone carries out his daily work. It makes a considerable difference if a geographer works in a so-called one-man department or within a larger university or research institution which nowadays could offer joint research projects, collegial togetherness, and scientific atmosphere. As indeed is the case for many humanistic scholars, the milieu of a geography professor in a small country is frequently a one-man department (within the best case perhaps one professor, one lecturer, one assistant, one secretary, and one map-drawer). This was my working environment for twenty-five years. I have in many ways tried to break out temporarily from that milieu, for example, through active work in the Geographical Society and through research in plant geography outside the frame of my department. I also have been engaged periodically in societal matters in different ways. It seems natural for a human or economic geographer to be interested however superficially – as in my case – in the society around him. For someone in economic geography to remain insensitive to global political and social issues surely would be difficult.

A one-man department does not have to be a depressing milieu; I do not believe that my own scientific contribution would have been any better inside a large department. Perhaps the small format even suited me better, since I have been more of a 'lone wolf', more of an individualist than a collaborative personality. I have always chosen the scientific topic or those ideas I dealt with on my own, and I have had the privilege to continue in this way to the end of my life. In fact, a professor in a small department is one of the few civil servants in today's society who still has a certain freedom in his work.

The leader of a one-man department is forced by circumstances to become a jack-of-all-trades. He cannot cultivate just one sector of, for example, economic geography. He cannot deal only with his special field of science and fail to provide himself with a broader base of knowledge for the sake of his students. For my own part I had already at an early stage been programmed for a cross-disciplinary perspective on many subjects. This is satisfactory for oneself, but most frequently not appreciated in academic circles. The question of the one-man department is probably an essential problem, and also has relevance for the whole academic system of promotion. There are examples of geographers who have, so to speak, dried up in the isolated, provincial atmosphere within a too narrow department at an unimportant college.

My life-path as a geographer has been an opportunistic sailing tour in small waters, but at least in a home-made boat in so far as my world of ideas is concerned.

My publication activity illustrates quite well the transformations of geography's spheres of interest in Finland during the last decades, but it does not reflect any particular 'schools' and methods. My publications are for the most part simple, descriptive accounts of facts and relations, perhaps even sometimes a useful new idea, but seldom a deeper analysis. My scientific contribution has no doubt suffered from my dealing with many things, to a large extent a question of temperament. Yet, I have had the advantage of finding it rather pleasant in the home-made and wide geo-ecological niche in which I have been working.

Contrary to most colleagues I cannot, thus, look back upon a consummate, straight path as a geographer. My contribution to geography has been distributed mainly on three, each separately rather broad, fields:

1. Studies in plant geography and forest botany in Canada;
2. Studies concerning forest and tree limits and the importance of climatic variations for the growth of trees and the annual amount of organic production;
3. The development areas in Finland and the problems of economic geography related to them.

My work in geography reflects unusually little of the schools and trends that have dominated international geography during the period. Perhaps this is due partly to a certain intellectual independence, but still more to a partly enforced provincialism. Only after retirement did I participate for the first time in a large international geographical congress (1976). Within ecology I have been somewhat more mobile, internationally.

I must admit that, unfortunately, I have been rather uninterested in geographic theory and method. These discussions have frequently seemed to me somewhat overstated or even unimportant in the light of concrete work in the field. But I realize it is very important that different types of researchers are free to emerge. It is of particular importance that geographic theory is developed at a time such as the 1970s and 1980s when the discipline is experiencing a period of unprecedented change.

On the whole I represent what one could call 'grocery-store geography', an older generation of geographers who think that they can deal with everything between heaven and earth. During my active time in the Geographical Society our council members were both biologists and geologists; 'cultural' and human geographers were for a long time a minority. In Finland physical geography dominated for an exceptionally long period, until the end of the 1950s. The dominance of human geography really begins only with the so-called quantitative revolution which changed the form, content and, indeed, the task of academic geography with exceptional speed during the 1960s. Perhaps it has been a positive development, I do not know.

Against this background my activity as a geographer after the 1960s undoubtedly looks like an anachronism. I have also usually felt myself rather isolated and perhaps have been treated as an 'outsider' by colleagues in Scandinavian geography. I have for too long been sitting inconveniently on two chairs – as a geographer among botanists, as a botanist among geographers. However, I have not suffered from any greater negative after-effects of this, particularly because a sound cross-disciplinary movement, still unknown in the 1950s is beginning to gain ground in geography. 'Human ecology' is such a bridge field, one which I have tried to represent in later years, regardless of whether it is to be called geography or not.

The sky's the limit for geography as long as this term, *geography*, is still used. It seems as if development leads continuously toward an ever greater degree of specialization and to a continuous separation of different research directions and tasks, once seen as geographic. It will become more important to retain an ecological and synthetic grasp of the many important problems and research tasks in departments which, for a period ahead, still will be called physical and human geography.

Notes

1. Ajatuksia Lapin viljelysmahdollisuuksista (Summary: Thoughts on the possibilities of agriculture in Lapland), *Terra*, **52** (1940), 86–98.
2. Pflanzengeographische Studien im Gebiet dev niederen Fjelden im westlischen

finnischen Lappland, I, *Acta Botanica Fennica*, **19** (1937), 1–156; Id. II. *Acta Botanica Fennica*, **27** (1940), 1–80.

3. Notes on the coniferous forest and tree limit on the coast of Newfoundland–Labrador, including a comparison between the coniferous forest limit on Labrador and in Northern Europe, *Acta Geographica*, **7** (1939), 1–77.
4. Notes on the forests on the east coast of Hudson Bay and James Bay, *Acta Geographica*, **11** (1950), 1–83.
5. It is clearly foolish to touch upon the war, this huge catastrophe, as if it were only a five-year parenthesis in the life of a single human being. To so many, the war meant death, tragedy, and sorrow.
6. Siebenbürgen-problemet (The Transylvanian problem), *Terra*, **55** (1943), 73–93.
7. Eine pflanzengeographische Übersicht über das Gebiet Kuuttilahti am Sywäri-Swir (Fern-Karelien), *Acta Societatis pro Fauna et Flora Fennica*, **63** (1943), 1–53; Pflanzengeographische Übersicht über das Kuujärvigebiet am mittleren Swir in Fern-Karelien, *Memoranda Societatis pro Fauna et Flora Fennica*, **20** (1945), 46–77.
8. The radial growth of the pine at the forest limit and its dependence on the climate, *Commentationes Biologicae Societas Scientiarum Fennica*, **9** (1945), 1–30.
9. Reflexioner om geografiens förändrade inriktning i Finland (Reflections on the changed direction of geography in Finland), *Terra*, **80** (1968), 90–94. See also, Ajatuksia maantieteemme suuntauksista (Geography in Finland and its direction), *Terra*, **91** (1979), 31–41.
10. See Selected readings, 1945; see also On variations in climate, in crop of cereals and in growth of pine in northern Finland 1890–1939, *Fennia*, **70** (1947), 1–24; Yields of cereals in Finland and the recent climatic fluctuation, *Fennia*, **73** (1950), 1–32.
11. On the forest geography of the Labrador peninsula. A preliminary synthesis, *Acta Geographica*, **10** (1949), 1–63; The lichen woodlands in Labrador and their importance as winter pastures, *Acta Geographica*, **12** (1951), 1–48; Forest-botanical notes from the Moose River area, Ontario, Canada, *Acta Geographica*, **13** (1955), 1–50; On the phytogeography of the subarctic Hudson Bay lowland, *Acta Geographica*, **16** (1957), 1–45; see also Selected readings, 1954.
12. The growth of Scotch Pine in northern Lapland 1928–1977, *Annales Botanici Fennici*, **15** (1978), 241–52; see also Selected readings, 1948.
13. *Finlands råvarutillgångar. (The Raw Material Resources of Finland.)* Söderströms förlag, Stockholm 1953; *Finland förvandlas. (Finland in Transition.)* Schildts förlag, Helsinki, 3rd edn 1972; *Världen av i dag. (The World of Today.)* Schildts förlag, Helsinki, 6th edn 1971; *Världens klyftor och broar. (The World's Cleavages and Bridges.)* Schildts förlag, Helsinki, 2nd edn 1977.
14. See Selected readings, 1964.
15. Wahlbeck, L. and I. Hustich (1952) Inkomstnivåns geografi i Finland 1948 (The geography of the income level in Finland), *Meddelanden från geografiska institutionen vid Svenska handelshögskolan i Helsingfors*, **1**, 1–46.
16. Suomen kehitysaleuista ja niiden kehittämisestä (The development areas in Finland and how to develop them), *Kansantaloustieteellinen aikakauskirja*, 4 (1948), 275–90; Hajainsia mietteitä syrjäseutujen ongelmista (Thoughts about the problems of our peripheral areas), *Terra*, **88**, 1965, 185–95; see also Selected readings, 1968.

Selected readings

1945   Om det nordfinska jordbrukets utveckling och årliga produktionsvariationer (On the development of agriculture in northern Finland), *Fennia*, **69** (2), 1–102.

1948   The Scotch pine in northernmost Finland and its dependence on the climate in the last decades, *Acta Botanica Fennica*, **42**, 1–75 + 7 t.

1952   (Ed.) The recent climatic fluctuation in Finland and its consequences, *Fennia*, **75**; A comment 25 years later: A change in attitudes regarding the importance of climatic fluctuations, *Fennia*, **150**, 1978, 59–65.

1954   On forests and tree growth in the Knob Lake (Schefferville) area, Quebec–Labrador Peninsula, *Acta Geographica*, **13**, 1.

1956   Correlation of tree-ring chronologies of Alaska, Labrador and Northern Europe, *Acta geographica*, **15**, 3.

1964   Finlands skärgård (The archipelagos of Finland) *Svenska Handelshögskolans skriftserie Ekonomi och Samhälle*, **10**.

1966   On the forest-tundra and the northern tree-lines, *Report of the Kevo Subarctic Station*, **3**, 7–47.

1968   Finland, a developed and an undeveloped country, *Acta Geographica*, **20** (12), 155–73.

1973   Den arktiska, subarktiska och boreala regionen – definitioner, folkmängd och framtidsperspektiv (The Arctc, subarctic and boreal region – concepts, population and future), *Societas Scientiarum Fennica. Årsbok LB*, **8**; partly also in *Intern. Counc. of Scient. Unions Bull.*, **31**, 5–19.

1979   Ecological concepts and biogeographical zonation in the North: the need for a generally accepted terminology, *Holarctic Ecology*, **2**, 208–17.

1981   Fogelberg, P. "The written work of Ilmari Hustich", pp. 5–13 in Geographical Society of Finland, *Ilmari Hustich: Seventy Years, Fennia*, Special Issue.

Interlude Two

Thus far we have heard from authors of Anglo-American and Nordic background. Even for many of these scholars, however, the richest source of ideas during the first half of the twentieth century was France. William Mead speaks for many when he claims that 'French is the natural language of geography'. A number of social, ideological, and linguistic barriers have impeded much interchange between Europe's Latin and Anglo-Nordic civilizations, barriers which this dialogue effort of ours had some difficulties also to transcend. Thanks to the graciousness of Claude Bataillon (Ch. 9) and Jacqueline Beaujeu-Garnier (Ch. 10), however, some glimpses may now be offered on the thought and practice of the French School. Piecing together extracts from individual interviews, Claude Bataillon (Université de Toulouse) has designed an 'imaginary round-table discussion' among key witnesses of university geography during the 1930s and 1940s: Pierre Birot (Université de Paris IV), Jean Dresch (Université de Paris VII), Henri Enjalbert (Université de Bordeaux III), Pierre George (Université de Paris I), André Meynier (Université de Rennes), Pierre Monbeig (Centre National de la Recherche Scientifique, Université de Paris), and Louis Papy (Université de Bordeaux). An abridged translation of this discussion follows in Chapter 9. Bataillon adds some comment and query about this period and its consequences; let it suffice here to offer some personal interpretations.

Often indeed have I wondered: wherein lies the charm and appeal of *la géographie humaine*? Its mystique has certainly been acclaimed in countries of Latin civilization; it has earned pious genuflection and occasional appreciation from individuals within Anglo and other worlds. For me, it was something, like Mead's attraction to Finland, which I have never sought to rationalize. 'When we make an effort to understand,' Mead recalls from Levi-Strauss, 'we destroy the object of our attachment.' The writings of Vidal, Sion, Sorre, and others offered me a breath of fresh air in the often heady atmosphere of regional science models with which we worked at the University of Washington in the 1960s. 'Geography is a meditation of life,' Sorre used to say. Often he referred to Vidal's hunch that one could gain more from an evening spent with people on a village square than from tomes of statistical data. These scholars certainly spoke to the realities of my own childhood milieu in Ireland. Their metaphors were those to which I could easily attune; their art of poetic description and perceptiveness toward the human condition made me feel that geography could indeed become a vocation to life (Buttimer 1971).

Unlettered then in the nuances of Structuralist, Materialist, or Idealist theories of history, I did not probe carefully the institutional structure of French geography. Now, thanks to the work of Vincent Berdoulay and others, one can appreciate better some of the social and political features of the academic setting in which this school developed (Claval 1972; Clark, 1973; Berdoulay 1974; Broc 1974). And institutional structures, after all, play an enormously important role in shaping the lines of research curiosity, the rubrics for teaching and the prospects for thesis topics, in any university field. *Droit de cité* for geography at the Sorbonne was, Berdoulay suggests, to a large extent due to Vidal's superb diplomacy, justifying the field on ideological as well as scientific grounds in a society at once regathering its morale after the humiliating experience with Bismarckian Prussia and cultivating the vast panorama of the *pays d'Outre Mer*. Part of his platform was to anchor geography solidly in collaboration with geology and history, and every student would have to pass these two distinct thresholds of competence before becoming a practitioner of the art (McDonald 1964). The elegance and perennial value of French theses, regional and systematic, can be appreciated better when this educational prerequisite is considered.

Unity was the cardinal principle, and for this Vidal supplied both intellectual defense and structural guarantees. Intellectually speaking, the disciplinary unity of geography should reflect the idea of terrestrial unity. Within such a broad horizon a greatly diverse agenda could be framed, be it to capture the 'personality' and individuality of particular *pays* or regions, or to seek generalizations, principles, even 'causes' underlying the physiognomy of the earth's surface (Vidal de la Blache 1913, 1922). In studies of the biophysical realm, a general, systematic, and law-seeking perspective could be envisaged, whereas in history one would inevitably find the element of chance, of contingency, in human events, values, practices. *La géographie humaine* needed both perspectives, in order to elucidate the *paysages humanisés*, to seek general principles, and also to be aware of how relative such notions as 'region' and 'boundary' were (Vidal de la Blache 1904, 1911). Ludicrous it was to argue for or against environmental determinism: at best one could speak of probabilities, each culture (*civilisation*) construing its own biophysical milieu through the filters of experience, values, and traditional practices (Vidal de la Blache, 1903a, 1922). Institutionally speaking, unity could be monitored via that highly centralized and hierarchical arrangement whereby the Sorbonne could screen candidates and appoint professors throughout the entire French system.

The second generation of Vidalians did not see much need to query this general *ex cathedra* legacy. Instead, they could direct energies toward implementing it in teaching and research. Right up to the Second World War most graduates had the Vidalian double qualification in geography and history, and many had become deeply immersed in the life and landscapes of the regions about which they wrote. In the late 1930s and early 1940s, however, geography claimed its autonomy from history. This round-table discussion (Ch. 9) should illustrate well some of the ambiguities surrounding institutional change, and raise questions about the validity of either dogmatically 'internalist' or 'externalist' interpretations of disciplinary thought and practice. Did mentalities change to fit the new structures? Did the institutional autonomy of geography facilitate or impede communication with other fields? Were there important personal and regional differences in the impact of this reform?

The institutional context is, of course, only one aspect of milieu. Material realities of life and landscape, vicissitudes of political events at home and abroad, and culturally defined values and *zeitgeist* must be borne in mind. Protracted debates over the 'scientific' value of regional monographs, for example, only poorly disguised profound ideological differences among participants (see Berdoulay 1974). Overtly the questions of methodology were aired: individuality versus generality (Simiand 1906–09; Febvre 1922; Brunhes, 1913); positivist versus idealist approaches (Durkheim 1897–98, 1898–99; Vidal de la Blache 1898; Berr, 1911); and the relationship between morphological and ecological approaches to milieu (Claval 1964).

Probings into the record of French thought and practice which focus on structures, ideologies or epistemology can, of course, shed partial light. But they tend to reveal as much about the lenses worn by the interpreter as they do about the texts. Nor do they exhaust the issue of why French seems such a 'natural' language for geography, or why, to my knowledge, nobody has ever asked 'What's French about French geography?' (L'Information Géographique 1957; Journaux *et al.*, 1966). The question of language opens up a whole realm of curiosity about symbols, and the modes of symbolic transformations of reality perceived in landscape and life to the prose of *la géographie humaine* (de Dainville 1941, 1964; Langer, 1957). Could one not suggest that, in fact, it was in the wealth and variety of their key metaphors that the French have shone (Schlanger 1971; Ricoeur 1975; Berdoulay 1980)? Organic analogies had their proponents and opponents. 'Organism,' in Vidal's view could be an appropriate metaphor to describe natural regions, *pays* and plant communities, and he was obviously so impressed with Michelet's *Tableau* that he himself portrayed the 'personality' of France (Michelet 1833; Vidal de la Blache 1903b). To apply the notion of organism to artificial structures like the State, however, as Ratzel had done, was offensive and inappropriate (Ratzel 1898–99; Vidal 1898). The value of this metaphor was to spark imagination in evoking a sense of holism, and an awareness of processes relating parts to the whole, especially within 'la grande civilisation'. It thus could afford a synthetic perspective within which a number of different analytical agenda could be framed, and an edifying pedagogy for developing a sense of patriotism.

The Vidalians, at the same time, were evidently conversant with a philosophy known as conventionalism, often associated with names like Poincaré and Boutroux, and this would have encouraged a relativist, contextual approach to their work (Poincaré 1913; Boutroux 1874, 1908; Berdoulay 1978). There was nothing incompatible in this: the world could still be regarded as an organic whole, but the drama of human history was *frappé de contingence* (Lukermann 1965). Brunhes's 'Val d'Anniviers' (1910), Sorre's 'Andorra' (1913), several essays in the Gallimard series of the 1930s (e.g., Deffontainnes 1933; Blache 1934; de la Rué, 1935), all illustrate the possibilities of a contextual approach to particular subjects or regions, while never denying that the bio-ecological milieu could be regarded in organicist terms. Some of the finest literature in social history, as well as geography theses influenced by the *Annales* school, display this combination of 'organism' and 'arena' as key metaphors (Deffontainnes 1933, 1948; Le Lannou 1948, 1949; Braudel 1949; Rochefort 1961). What they share, among other features, is a basically 'ideational' orientation to landscape and life, and as such had to confront opposing metaphors which seemed more attuned to the 'artefactual' and the tangibly observable (Buttimer 1971).

Challenges to these two modes of symbolization stemmed from the growing concern for analytical rigor and the need to align geographic research with that of other sciences. Vocal counterpoints to the 'organism' were, of course, the 'map' and the metaphors of spatial form, epitomized in landscape study and the topographic map, those indispensable tools in the pedagogy of both physical and human geography (Vallaux 1925; Demangeon 1927; Cholley 1942). Here was the analytical style *par excellence* for the geographer, demanding no a priori assumptions about the organicity of landscapes, regions, or the world. From the vantage point of cognitive claims, 'map', like 'arena' implies a dispersed vision of reality, the former resting its cognitive claims on a correspondence theory of truth, the latter on an instrumental theory (Pepper 1942). 'Organism' implies an integrated view of reality and rests its claims on a coherence theory of truth (ibid.). This juxtaposition or combination of styles may baffle the epistemologist; it may, as one participant in this discussion suggests, be explained in terms of uncontested ideological assumptions. I like to think, however, that especially in the regional monographs, there was actually an attunement of *genres de pensée* to *genres de vie*. Did the prose of these theses actually not reflect some of the latent contradictions in the lived horizons of *pays* themselves – elements of organicity in the *structure mentales*, traditional practices and values on the one hand, and the formal realities of physiography, economy, and settlement on the other? To represent life and landscape as they appeared – combining observations of relief, drainage, and soils with those of human occupance, habitat, and livelihood – seemed to be quite as important as to make those representations consistent with scientific prescriptions (Blanchard 1925, 1958; Sion 1934).

But how, as Bataillon asks in this discussion, could landscape study become really scientific? In mid-twentieth century one needed to reach out toward dialogue with colleagues abroad, but also to make geography at once more scientific and practicably useful at home (Cholley 1942; Birot 1950; Birot and Dresch 1953, 1956). The 'map' was certainly an analytically oriented metaphor, atlases could provide comprehensive summations of information, but not necessarily explanations or scientifically based integration. Here I suspect that the topographic map, and its relationship to the geological map, may have stirred scholars to seek an understanding of process, and eventually may have opened the way for another metaphor, viz., 'mechanism'. If forms (patterns) could be explained in terms of process, and process construed within a general systems framework, then one could have both analytical rigor and integration. French enthusiasm for Davisian geomorphology and climatology, at a time when some of Davis's own compatriots were disowning him, suggests a curiosity about mechanisms, synthetic and deductive reasoning, long before human geographers elsewhere adopted a 'systems' approach

(Brunhes 1902, 1910; de Martonne 1925; Sorre 1933, 1943; Baulig 1950). Of course, geomorphologists and bio-geographers would find this metaphor pleasing; Sorre especially showed the continuity between 'organism' and 'mechanism' in his medical geography (Sorre 1933, 1961), Birot showing its value in elaborating relationships between climate and morphology (Birot 1959, 1960), and eventually Chabot, Beaujeu-Garnier, and Bastié, demonstrating its usefulness in an urban context (Bastié 1958; Chabot and Beaujeu-Garnier 1963). The prospects for developing a 'systems' approach to human geography were delayed for a long time, partly because of the vociferous denunciation by Vallaux and Febvre of mechanistic approaches to the study of society proposed by Durkheim and his school (Febvre 1922; Vallaux 1925). After the Second World War, however, the 'systems' approach gained increasing support, its power appealing to scholars of radically different ideological orientation (Gottman 1947; Cholley 1948; Sorre 1961; George 1951; Labasse 1955, 1966).

To look at *la géographie humaine* through the lenses of metaphor does not, of course, exhaust the issue of language. Neither does it 'destroy the object of my attachment' because it opens up the possibility of a richer perspective on thought, structures, and ideology. Each metaphor had its own range of possibilities, its strengths and limitations, its tolerance and/or flexibility to different structures, research problems, and teaching settings. Scholars sometimes moved freely among and between these different metaphorical styles, and during the classical period at least, thereby dramatized the tension between the imperatives of 'science' and 'humanities', between general knowledge and particular understanding. Unity amid diversity could well be a characterization of this period.

The discussion comprising Chapter 9 should shed light on some of the diversity of personal meaning which particular authors have found in their practice of geography. Emphasis rests on Paris primarily, and the picture portrayed does not claim to be a comprehensive one, even for this period. The text should help dispel, however, any monolithic images of the French School which some of us outsiders may have projected, and should leave little doubt about the significance of key individuals in shaping the horizons for the practice of geography in that country.

Chapter 9

French geography in the 1940s

André Meynier
(Rennes)

Jean Dresch (Paris
VII)

Louis Papy
(Bordeaux)

Claude Bataillon
(Toulouse, organizer)

Fig. 9.1

The context

Claude Bataillon: Prior to 1940, geography and history were combined in the French University System. Between 1940 and 1945, however, largely through the influence of Emmanuel de Martonne, a fundamental change occurred: degrees of *licence* and *agrégation*[1] could be obtained in geography independently. The field grew rapidly in numbers of students, teachers, and new institutes of geography were created. Most of the present-day geographers who are engaged in graduate teaching have been influenced by that system which was established during the early 1940s. The system has, of course, been radically changed since the events of 1966–68, but this has happened in

119

a piecemeal fashion, and some of the dominant features forged during the founding years have by no means been erased.

The persons interviewed in this 'round-table discussion' were all between thirty and forty years old during that turning point of the 1940s. They were students or young researchers in the 'old' system shepherded by de Martonne, A. Demangeon, and their colleagues. For an entire generation, they have made the 'new' system work and they are the best qualified to understand and evaluate it. This discussion is also an opportunity to learn how one became a geographer in France during the 1930s and to re-create the atmosphere of an era. Let me take the opportunity here to thank them for having approved the publication of large excerpts of their answers in the form of a cut-and-paste document for which I alone take responsibility.

Round-table discussion

Bataillon: All of you were students during the era when graduate-level geography teaching was considered an annex of history. It would be nice if you would reminisce about what made you choose geography, a choice which at that time was somewhat special.

André Meynier: My father was professor of history and geography and ever since primary school I have been interested in these fields. I spent much time looking at maps and found it exciting to read an atlas. As a child, traveling with my parents, I was fascinated by landscapes, villages, and the rivers we crossed. So when I entered the university my choice was already made: I chose geography right away. I knew nothing about modern geography, of course, just notions about place names and cartography. I was attracted by the idea of a *licence* in cartography. I was yet to discover that there was more to this than designing maps: Lucien Gallois swept us along in drafting and also in interpretation. The *licence* program at that time allowed me to have two written geography tests out of the four, the other two being in history and Latin. It was essentially a *licence ès lettres*. For my future work as a teacher, however, I felt that this education was really inadequate. I tried to increase the amount of geography in my classes. But I often noticed that in the third and fourth grades the teaching of history gave better results than the teaching of geography.

The idea of a research project in geography only occurred to me at the university level. Work for the Diplôme d'Etudes Supérieures (D. E. S.)[2] involved practical tests, explication of texts, history, and auxiliary sciences. Like everyone else at the time I chose to work on a particular *pays*, the Brive Basin, where I used to spend my vacations with my parents. Monographs were the rule of the day, and E. de Martonne would not accept a dissertation unless a substantial part of it addressed questions of human geography. Demangeon was my director, because he had already written about the Limousin.[3] The workload as a teacher at the Aurillac lycée was light and I would have liked to do a thesis on the mountains of Cantal, but this subject was already taken. Demangeon advised me to study Ségala.

I was allowed to define the content of my work as long a documentation was adequate. I would report regularly to Demangeon and de Martonne, but they never pressured me on any aspects of my work. During that era they were the only two 'masters' apart from Augustin Bernard whose courses I had followed and thanks to whom I received a scholarship in 1922 to visit Morocco, more or less as a tourist. To work outside of France was still the exception during my generation, although Ruellan worked in Japan. The dominant impression was that there was still much to be learned in France itself.

The idea of specialization hardly appeared during this era of monographs on France. Deffontaines was reprimanded for not having done any physical geography, and Birot around 1937 was the first person to do nothing but physical geography. Besides, up until quite recently a 'complementary thesis' on the other branch of the subject was still a requirement – a desirable idea indeed because the essential subject was regional geography which required the double education. History was an important part of monographs, and in my study I had to cover the period which began in the eighteenth century.

Pierre Birot: During my secondary education I was interested in all subjects except metaphysics. My father was professor of history and geography at a Paris lyceum and thus my readings in this domain were more precocious. Before entering college and when taking philosophy, I hesitated between archeology and physics. Thus, during my two years of work on a *licence* of history and geography, I passed the baccalaureate of elementary mathematics, followed by a course in general mathematics. At that point I found in Emmauel de Martonne's physical geography a discipline which allowed me to combine my two fundamental aspirations. I always search for the possibility of formulating a law and to quantify. Field experience is also for me a hygienic and esthetic need. I could never have been a historian, seeking my pasture in the archives; I would, however, have liked to be an archeologist because they also go in the field. The rural landscape appeals as a complement to the physical landscape in this respect. I have researched the historical roots of rural landscapes, and people's motivations interested me, especially during my investigation of Portugal. There is an obviously esthetic dimension in my taste for the Mediterranean countries: the pleasure of the eye is tied to curiosity about the landscape. By contrast I hardly feel comfortable among the old pedestals, and even less on the great plains.

During my entire career – for the *agrégation* at Saint-Cloud, and afterwards future *licenciate* at Paris-IV – I have taught the regional geography of foreign countries. I consider this regional instruction fundamental in the education of the citizen if it is coupled with an historical instruction which promotes understanding of the current world. I see therein an antidote to ideologies.

Jean Dresch: I developed a taste for history and geography rather early in life without there being any particular reasons for it, either because of family ties or otherwise, since my father was a German scholar. Family tradition oriented me toward teaching and also toward the *Ecole normale supérieure*.[4] During my preparation for university I was certainly interested in history, but there was no geography at all required for the entrance exam. Toward the end of my secondary school studies I had developed so great a taste for philosophy, that I entered the university with the label of philosopher.

However, I soon became restless with cultivating discourse for the sake of discourse; I thought it was better to stop while there was still time, and I remembered my taste for history and geography. While engaging in geography for the *licence*, I discovered that this field offered in the concrete exactly what I had earlier searched for in the abstract. I chose the discipline which appeared to be the most universal in scope, but on the condition that I could keep both feet on the ground.

This taste was reinforced by de Martonne and Demangeon, professors at the young Geography Institute which opened its gates in 1926. My friends at the time, J. Weulersse, P. Vilar, J. Bruhat, and I, all had a great admiration for Demangeon. He presented the formal university tradition whereas de Martonne, however typically a university man, was rough, and made no effort at all to bring smiles to the lips of young graduate students. These young intellectuals were rather horrified by the excesses of physical geography, but my reaction was the reverse. It is true that at the time when I chose physical geography, I engaged in sports a lot, and as a Toulousian, I developed a taste for the mountains. For three years I undertook all the difficult climbs one could make in the Pyrenees with a group of young Pyreneans. During these frequently solitary trips in the Pyrenees I wondered why I felt a love which was not associated with sport; it was tied to the contact with the rock and a good muscular coordination needed to make a ledge or a rock wall: this is what led me directly to geomorphology and geology at that moment I chose a career. I thus established a better relationship with de Martonne than with Demangeon, since I was a bit disappointed to find in him that same art of the word which had disappointed me at the university. My relations with de Martonne were deep and trusting and they have certainly influenced me. Neither of us was very communicative; our spontaneity was dampened by a natural shyness and this no doubt attracted us to one another. I could have had equally good relations with Blanchard – we were good friends – but later on we belonged to two different worlds, I being a Parisian while he was from Grenoble. With Cholley I was polite.

Geomorphology for me provided the most complete explanation possible of the landscape. That is why I studied geology and natural sciences during my first two years at the university. I had even thought of working on the geomorphology of the Pyrenees for the preparation of my D.E.S.

Pierre Monbeig: I very seriously hesitated between the study of English or of history and geography. My mother pushed me toward getting a *licence* in English, she told me that the history and geography professors did not have special classes, but that the Anglophiles did. My parents were former high school teachers and their life had not always been easy; the financial aspect of a career choice really counted. My father, whose influence was more formidable and more subtle, adored history; he had earned a diploma of university-level history studies while he was a fourth form teacher in the Beauvais high school.

What attracted me to geography, as well as to history, was a certain taste for political, economic, and social problems. Contemporary history held only a little attraction because of the way it used to be taught then: it was still diplomatic history. There was one historian, however, Henry Hauser, who was moving toward geography without realizing it. During this period when class sizes were small, professors had a greater degree of direct influence on their students and Demangeon recommended those whom he advised to take Henry Hauser's course on economic history.

As with most geographers of my generation the question of a choice between history or geography did not arise until it was time for the D.E.S. By the time I arrived at the Institute of Geography, I had been totally seduced by the field trips. I liked the contact with the landscape, with Demangeon and de Martonne – probably also the satisfaction to be among friends, to always find the same crew of people who got along so well together. One realizes later on that this was a very fertile situation. When the moment of choice came, I took human geography essentially because Demangeon had such an attractive clarity in his teaching and showed a certain friendliness toward the students, the lack of which qualities failed to attract me to M. de Martonne. I was not particularly excited about physical geography because I was a horrible draftsman; to work with de Martonne it was necessary to know how to draft.

Pierre George: I was led to history and geography by a certain taste for concrete reality. I was quite disillusioned by so-called 'classical studies' which were oriented more toward grammar than to the understanding of culture or civilizations. History and geography touched concrete realities; they were virtually synonymous with living the life of my era. I hesitated for a while between ancient history and geography, because for me those two formed a unity. Ancient history consists of everything that is simultaneously concrete and imaginative – the archeology and reconstruction of the past.

I was educated in the old school where history and geography were closely associated. I chose to do my first research project in geography for the D.E.S. because I simultaneously was attracted by historical culture and by the 'objective', physical sciences.

Geography allowed me to cultivate these complementary interests: on the one hand it was necessary to use the archives and do the work of an historian, and on the other, it was necessary to associate those long-term phenomena of which natural realities consist with the short-term phenomena of social realities. It was quite natural for me to become a geographer without turning my back on history. Later on, I felt the need to round out my education in the fields of natural and objective sciences.

My science degree brought me the anticipated satisfaction in the area of geology where the teachings of Ch. Jacob or Piveteau in paleontology provided me with the basis for regional geology which was necessary for the study of regional physical geography. In botany, however, what I recall is endless gazing through the microscope at sections of roots, stems, or reproductive organs; during that era phytogeography and ecology were minor disciplines for the botanists. Contacts between geographers and botanists were rather uneasy.

Physical geography appealed to me for its promise of 'objective science'. I was also fascinated by the physical activity which is involved in field work – certain efforts to literally conquer space. In so far as it has become primarily a laboratory science, physical geography has made me feel remote. I am not really opposed to laboratories, of course – during the war I worked with my friend and colleague Rivière on the problem of the millstone quarry in the Parisian Basin, but the loss of contact with current reality discouraged me.

After the war I moved more toward human aspects of geography because they appeared more urgent and at the same time more compatible with my general preoccupations. I maintain, however, that physical geography can yield some explanations of questions raised in human and economic geography.

One's choice of a research subject can come like a bolt of lightning, especially if one feels comfortable in the work environment; a man like Blanchard experienced this in the Alps as well as in Quebec. To work in a setting where one was not in congenial contact with the inhabitants turned out to be rather disappointing; the best guarantee for the pursuit of a chosen work is the fond attachment one feels for the study object.

The lower Rhône is an area with which I was quite familiar before I knew that I would be studying there; my mother had lived there while she was young and still had friends in the area. However, it was above all an area where I felt more comfortable than anywhere else: I experienced this same attraction to the Mediterranean environment when I had the opportunity to work in North Africa.

Bataillon: The lower Rhône was to become the base for numerous theses of regional geography, especially on southern France: the lower Rhône is not an old traditional peasant country.

George: One minimal argument for studying there was that this area was still free whereas studies on the Alps were in progress, and Faucher worked on the middle part of the Rhône. The lower Rhône was a very fascinating physical environment where Fontannes had already done some geological research. It was an environment in the process of mutation for about thirty years or so – a poor country, disinherited, where, in Daladier's words, one sold land by stone's throw (one measured the field by the length of a stone's throw without bothering with a survey for a sale). Suddenly, through irrigation and the railroad, the area entered into an economy based on speculation. There were, at the same time, longer-term historical phenomena – the history of Comtat and Avignon, the penetration of Jewish influence following the popes – quite a collection of exceptionally fascinating human facts.

The remarkable archives on the lower Rhône also added to its interest: one could trace agrarian history as far back as the Middle Ages. Grafted on to this traditional heritage one found an entirely speculative economy undergoing extremely rapid mutation-like development: exceptional flows of money occurred with all possible frauds and price speculations. Everything was easily accessible too – it all took place at the microeconomic level.

Two types of influences deserve to be recognized in one's fundamental professional choices; on the one hand the intellectual admiration that one can have for very great minds, the mandarins, without important personal influence because they were held in awe. One considered oneself privileged to be received by them once a year. The other type of influence came from the research director, always available, who read even the smallest bit of writing one sent him; that is incontestably the role André Cholley played for me. He helped me the most even if he has inspired me less than the super greats, Demangeon and de Martonne. Blanchard, too, inspired me, but it was awkward to see one professor when I was another professor's student.

Louis Papy: I certainly was not carried toward geography by an irresistible vocation. Geography lessons in secondary school, which comprised a heavy list of place names, did not fascinate me. I was very interested in classical literature and in history. I chose history and was attracted to the humanistic faculty in Bordeaux because of the fame of some excellent teachers who taught there in the beginning of the 1920s, Augustin Renaudet, Albert Dufourcq, Louis Halphen, and Georges Radet. I owe much to them, and my D.E.S. treated an historical subject. In geography raged the awesome Camena d'Almeida, a man of great culture, but no student ever dared address him. Let us simply say that the flame which puts life into the works of Vidal de La Blache (they published scholastic manuals together) was not to be found in the lessons of my good teacher from Bordeaux. Then there was Henri Cavaillès, an historian who was a latecomer to geography. He was an unpretentious man with a winning personality, curious about everything, and he had contact with the landscape and the people. His teaching was the kind that could inspire vocations. It is he who taught me to work 'in the field'. We would hike for a long time, climb mountains, and he would ask questions of the Pyrenean shepherds and make them tell him about their everyday life. That delighted me. Once I had my *agrégation* (1926) and was appointed to La Rochelle, I wanted to do just as he used to do: to travel along the coast, throughout the islands, to make the ordinary people from Charente talk – they were so different from a Gasconian like myself – that would fascinate me far more than to stay locked up in a library or to go through archives. That is how I became a geographer. And then my friend Sermet who taught at Rochefort introduced me to Daniel Faucher and Raoul Blanchard – men who also inspired enthusiasm. I participated in some of their excursions. One day Cavaillès advised me to see Demangeon: I had vaguely thought of writing a thesis. In Rue Saint Jacques,[5] I saw Demangeon who encouraged me lavishly; then I saw de Martonne whose muteness chilled me. It was not until later that I had a chance to get to know de Martonne better and came to appreciate the quality of his feelings: once he adopted you, he never abandoned you. In 1935, I traveled for several days from the Vendée to the Pyrenees with

Albert Demangeon. He was marvelous. He wanted to know everything. 'To travel as a geographer,' he used to say, 'means to know how to analyze landscapes, to learn what man had added to them – not an easy task.' Indeed! Geography has since then widened its horizons, refined its methods and techniques. Yet if one were to re-read those theses from the beginning of the century – Demangeon's, Blanchard's, de Martonne's – one could discover much in them.

Henri Enjalbert: I am the son of a small farmer who owned all of 6 hectares (about 15 acres) in Ségala.[6] I was the third child, and so it was quite normal that I looked for a position, or rather that my teacher looked for a direction for me to take. In that era it might have been a profession in the postal services, a tax collector, or a school teacher. I chose to become a teacher because, for one thing, it would give me a chance to travel, even by bicycle, during vacation times and to see other countries. The Director at the teachers' college at Rodez helped prepare me for entrance to Saint-Cloud. I chose history and geography because that seemed to offer me a vocation, and the hoped-for opportunity to travel. The prestige enjoyed by certain professors, especially de Martonne, also played a role: I followed his practical courses in Paris, also Demangeon's and later Cholley's. Then, quite fortuitously, A. Meynier's thesis on Ségala was published in 1931–32. That was about my country! I read it with the perspective of a farmer's son and there were certain problems which I wanted to explore at greater depth. After the war, A. Meynier very kindly came to spend some days with me in the area after my first article on physical geography was published, and this was an opportunity to juxtapose and discuss our viewpoints.

After graduation from Saint-Cloud I was appointed to a teaching post at the Ecole normale[7] of Angoulême and had to teach 'local geography' to the third year class. At the time there was almost no material available on Charente. I bought a small car and criss-crossed the whole *département*, while finishing work on my *licence* and *agrégation* in 1938.

When I was appointed professor at the Bordeaux lycée I right away found that the pupils in large classes were often dissatisfied with a rather traditional and trite teaching style. This led me to the study of general geography, that is, physical *and* human geography, conceived on a large scale and without worry about specialization.

The war and captivity took me to Pomerania. At first I gave lectures to my friends almost from memory, and then I began to use this forced vacation to take up the physical geography course again. In Leipzig I bought the books of the 'pre-Davisians' – Penck, Passarge, Hettner, Johannes Walter, and I spent my time reviewing these works and forming my own opinion which, to begin with, was improvised and incomplete for lack of knowledge of the landscapes in various parts of the world. To read the twenty volumes of the *Géographie universelle* one after the other only gives a bookish view of the world.[8]

When I returned from captivity I wanted to compare these thoughts with the countries themselves; that meant entering university teaching in Bordeaux where I lived, and where one had the opportunity to preside on juries for baccalaureate exams either in Africa or on the Antilles and thus visit numerous countries. In this manner I have made contact with the tropics and also with the Andes and the Rockies. In a few years I acquired a great deal of field experience, which – combined with the theoretical base that I acquired while in captivity – enabled me to publish papers tracing relief and surface formations on a world scale.

The unity of geography has always been fundamental for me; the concept of space had universal validity for both human and physical geography. I do not trust schemes, theories, principles, and I remain a naturalist both as far as the physical environment and the works of men are concerned. This allows me to keep in contact with things while at the same time submitting them to comparisons, yet without becoming too abstract.

Rural geography has always been dear to me because of my origins. I know how to use the sickle to harvest and also how to adjust a combine-harvester. I have worked with hand plough as well as with gang plough tractors. During my trips to Mexico I have always been able to chat with the farmers because I knew what maize was and the soil where it can grow.

Traveling in Latin American towns, one gets a feeling for the historical dimension, one finds a past relatively unaffected by change, and this provides elements for comparison. In physical geography, too, one finds a heritage, a past history. I have undertaken to follow this complex history from the end of the Tertiary and the Quarternary periods of climatic change. For me, being a geographer means also being an historian: in physical as well as in human geography the present is explained on the basis of the past. One does not study the past for its own sake, but rather to understand the present and prepare for the future.

Meynier: In my undergraduate education there was a definite emphasis on physical geography. Geomorphology was of prime importance when it came to map analysis, human analysis following in sort of a semi-deterministic view. Each professor led several excursions during the year, and there also was the inter-university excursion in which most of the students participated. These excursions were more didactic than participatory, i.e., the professor showed what he saw in the landscape, posing a few questions to add some color. These formal descriptions with sketches were not based on analysis of superficial deposits until later, during the 1940s, under A. Cholley's influence. Visits to farms and factories were added. In Paris de Martonne and Demangeon had encouraged us to create the Geographical Union of the Humanistic Faculty, of which I was a co-founder; its purpose was to make a longer excursion again after the one which had been organized by the professor. Graduate students at the old Sorbonne found out that there was something special about geography: this was the first discipline where the professor's course was not everything. Thus some historians were more or less temporarily attracted to geography, Reinhart for example. Nowadays one has moved away from hiking excursions (about 40 kilometers on foot in Alsace and about 20 kilometers in the Pyrenees), and replaced them by automobile excursion. The car was a downfall since one 'saw' more when walking.

Monbeig: Excursions never focused exclusively on human geography even when Demangeon was the leader; during that era a human geographer could still participate in physical geography without making a total fool of himself. The excursions consisted mainly of visits to agricultural cultivation centers, analyses of rural landscapes, and visits to factories where the questions posed by Demangeon tried to work up enthusiasm for the art of surveying. Neither Demangeon nor de Martonne was involved in helping students to initiate research at the level of the *licence*. The inspiration came all by itself; the masters did not push, nor did they take one by the hand until the preparation of the D.E.S.

Bataillon: To become a geographer during the 1930s evidently took more of an original effort than it did to engage in classical university careers. For a discipline which considered itself synthetic and which, actually drew heavily on its neighbors, it must have been very important to maintain good intellectual and professional contacts with other fields. Half historians by profession, how did the young geographers experience the 'breakthrough' of those nonconformist historians who were on the staff of the journal *Annales* at the time, who were dealing first with economic and social history, later with *Economies, Sociétés, Civilisations?*

George: Every influence from Lucien Febvre's and Marc Bloch's school was very important for us. As a student I was far more fascinated by the extraordinary talents of Jérôme Carcopino than by any other comtemporary historian at the Sorbonne during that era, even though there were great specialists like A. Mathiez or P. Renouvin. Contemporary history, which was in effect the poor relative of university history studies at the Sorbonne of that era, was taken in hand by the *Annales* school. My conversations with Lucien Fèbvre were always extremely fruitful, but he was quite inaccessible except for the small cohort of historians who surrounded him. Dion, who was among the closest geographers to him, has always stayed outside of this order.

For us, common geographers, Marc Bloch was the main personality within the *Annales* school, and he died during the war. Even Braudel, who was much more in line with Lucien Febvre than with Marc Bloch, did not reestablish contact, not even with 'geo-history'. The teaching structures did not, of course, lend themselves to this type of contact; 'Sorbonnians' frowned on the practice of attending courses at Collège de France.[9]

Meynier: Within history the journal *Annales* has been important for us. Ever since its first publication in 1928, many geographers subscribed, thinking it could influence their research. Contact was established through Marc Bloch but also through Lucien Febvre, even if he made fun of the geographers. Ever since that time students have had an historical base for their research on agrarian structures, which is one thing I totally lacked at the time of my research. During that era Baulig in Strasbourg especially steered his students towards L. Febvre.

Monbeig: Juillard and Dion were closely connected to the *Annales* school, also Dion who almost became completely an historian. There were a few turncoats like P. Vilar and J. Bruhat, who eventually switched over to history. But there was also a certain number of human geography these which were very close to the *Annales*; I am especially thinking of Derruau and Papy.

Papy: Yes, I witnessed the birth and development of the *Annales*. That was the time when Lucien Febvre saw two distinct schools among geographers: the 'determinists' and the 'possibilists'. Naturally I was a possibilist! Lucien Febvre has helped to disseminate Vidal's thought which he has subtly analyzed. The *Annales* has had an undisputable influence on geographers through the new perspectives which it opened on history. I am grateful to the two founders of the *Annales* whom I have come to know through Robert Boutruche who encouraged me in moments of doubt.

Monbeig: Before the war the link existed. After 1945 when I got in touch again after my stay in Brazil, I felt that there was a very clear break, and geographers only rarely cited works on history. Students in my graduate courses in Paris scarcely knew about the existence of the *Annales*.

Bataillon: Nevertheless, history apart, how well were your absolutely necessary work relations established with respect to other disciplines, such as the natural or the human sciences?

Meynier: With archeology and toponymy, very well, yes. With botany, a little like with the natural sciences ... where one worked as autodidact like in geology. Sociology of that era was asleep following Durkheim's grand period at the beginning of the century. We had heard about Mauss. Economics was taught by the faculty of law and it was abstract and general. It was not until the 1940s that the economists became interested in regionalization and from that time on we have been in contact with them.

George: Economists essentially work on coded materials and with schemata which correspond to their own typical mental structures; in my opinion they dealt merely with abstract constructions. Certain currents within sociology, however, seemed to be much better at treating contemporary reality in its spatial context. I have always felt comfortable with Gurvitch's way of thinking in spite of certain difficulties of communication due to the great complexity of his thought. Certain demographers' techniques have also seemed to me very fascinating for a geographer. Naturally I don't think of those maniacs for demographic analysis, but of those like Alfred Sauvy who, thirty years ago, was turning demography into a social science of very ambitious scope. I have always admired his broad perspectives on the economic consequences or social implications of different types of population increase. A geographer should feel totally at home in this subject. It was a pleasure to work with people like Gurvitch, Duvigneau, Balandier, or with Sauvy and his co-workers. I have been struck by the number of population conferences which, in spite of their size, always have understood how to keep more of the spirit of synthesis about them than did geography conferences.

129

Birot: Relations with geologists have been rather mediocre in spite of geography's autonomy. Even after the official triumph of multi-disciplinarity it has been very difficult to establish relationships.

Papy: During the 1930s and 1940s one was very concerned about 'multidisciplinarity' and geographers at Bordeaux did collaborate closely with geologists, naturalists, and economists. I used to join in excursions led by the geologist Fernand Daguin and the geologists visited us. I had the opportunity to make a field trip to Morocco in 1946 with Daguin. Morocco – now there is a country where interdisciplinary work relations were especially close, to everybody's great advantage. I believe that men like Daguin and Dresch have contributed much to this success. It has often been easier to undertake teamwork outside of France than at home. Since 1948 we in Bordeaux have established research groups in Africa in which geographers, sociologists, ethnologists, and biologists were associated. I think this collaboration resulted in good work.

Dresch: I have done geology and history one after the other. I've also been intrigued by problems which involve geography, economics, and history – for example, the problem of capital investments which enables one to evaluate different types of imperialism. How can one understand problems of development unless one follows them both geographically and historically?

Bataillon: Your contact with ethnologists and sociologists – was that an important feature of your work overseas?

Dresch: French ethnologists have traditionally been fascinated by primitivism – which gives an indication of their grasp of politics – and it was a long time before they became interested in current problems, for example, how indigenous societies evolved after their contacts with imperialism, and later with colonialism. It came late with men like Balandier. Thus I have always felt a bit isolated with respect to my first standpoints.

Monbeig: I have felt the lack of an education in economics and sociology. But then I was very young, only twenty-one years old – the youngest *agrégé* in France in 1929. I had to learn by myself, with the help of friends, more than others have had to. In Brazil we were a small group of young people from different disciplines and what an experience that was. We would never have had such an experience if we had stayed at home in France in our *lycées*. We had idle periods in Sao Paulo. There were economists like Courtade, Froment, and Perroux, some sociologists and ethnographers.[10] When my thesis was published I was surprised to hear young colleagues opposing the fact that I had given so much emphasis to economic phenomena, for example, coffee prices in Brazil. And I was really annoyed that not a single member of the jury at my thesis defense made any remarks at all about my ideas on the psychology of migrants. I thought this was a novel and important idea, and it passed totally over their heads.

Bataillon: From 1941 and on, university geography became autonomous through the founding of a *licence* and then later an *agrégation* in geography.

How did you experience this transformation? How did E. de Martonne manage to carry it out?

Meynier: I had reservations at the beginning from a pedagogical point of view. My question to de Martonne was: were geography teachers at the undergraduate level going to have the students for only 1½ hours, i.e., probably have a negligible influence on them, just as was the case with natural sciences? The problem was partially solved when Carcopino increased the geography schedule to 2 hours in all classes. In fact one never really thought of separating the two subjects. Initially the curriculum for the new *licence* program was a nationally based one: it covered the regional geography of the entire world within four years. This curriculum gave teachers a knowledge of geography, but not sufficient to enable them to train other teachers. That was why I did not want to give morphology such an exclusive role. De Martonne, who was on friendly terms with the minister Abel Bonnard, played a major role in writing the curriculum. I was called to Paris to discuss it, but we talked more about the structure of examinations than we did about the curriculum.

Papy: I must admit that I was not too keen on this idea of separating history and geography. De Martonne's *agrégation* created some animosity – I remember that at the time of the meeting with the Minister of Education. Ultimately, however, I think it has benefited geography.

Birot: During my university studies, and later as assistant professor, I realized that three-quarters of the effort was applied to history, and there was not sufficient time left to assimilate the principles of geography. I am convinced that without the creation of an autonomous *licence* and *agrégation* in geography we would not have had this extraordinary development of our discipline between 1945 and 1970.

In fact, it was for this reform that I, for the first and only time in my life, have engaged in militant activity at the side of de Martonne and Cholley. Blanchard was wrapped up in a struggle between provincials and Parisians, a struggle which ended when he himself, after his retirement, became a Parisian. It is because of this unimportant fight and not for fundamental reasons that he found himself opposed to the reform initiated by de Martonne. Besides, one of his students, J. Blache, later was to preside over the *agrégation* in geography.

George: De Martonne showed the same obstinacy in organizing international research as he did in the organization of meetings. He also founded a discipline at the level of training teachers. His essential wish was to integrate the techniques of geographical research and especially the explication of maps into the education of secondary school teachers through the *agrégation*.

Dresch: I was working closely with de Martonne and have observed his evolution since 1941. Given the situation in French universities between the wars, and geography's subordinate status *vis-à-vis* history, it was necessary to create the material framework which could allow geographers to function normally. That is why de Martonne created all these institutions like the Association of French Geographers, the Geography Society, and he himself became a member of the Academy of Sciences – he tried everything to provide ways and means for geographers. He never really wanted a clear break from history, rather he wanted to give geography a *cursus propre*, followed by the *agrégation*. Naturally, Cholley agreed. To achieve this, special ties with the natural sciences were necessary. Physical geography was so successful – it scared off those human geographers and historians because of its apparently more accurate, genuinely scientific methods. Even during the pre-war period in 1939, when Demangeon and Sorre were developing relationships with social sciences other than history, there were no such dramatic tensions. It was only after geography had become autonomous that the problems became tricky. In retrospect now it seems that the methodological problems of the pre-war period were child's play, the growing precision and multiplicity of analytical methods in modern times have accentuated the divergences.

George: The new curriculum did not really change the structure of geographic education. What was required in the historical part was simply an introduction to historical methods and to the value of using historical documents to interpret a geographical situation. The unique character of the French School of Geography, which gave an important place for history, was toned down to fit the new specializations: a little less economics, a bit more sociology, or specialization in a branch of physical geography. It was believed that geography could deepen one's education. In reality one sees a dispersion of effort, and the loss of a guiding thread or the feeling of continuity.

Bataillon: Why did physical geography and especially geomorphology have such a privileged position in this new *agrégation*?

Meynier: The neophyte geographers at the beginning of the twentieth century were full of enthusiasm when they discovered that the geologists had not said it all. Geographers were then the sole masters of geomorphology. Besides, this field allowed you to develop the spirit of observation and education ...and also one had de Martonne's treatise of physical geography at one's disposal. Climatology did not really develop until the work of the Norwegian school became known in France and that was as late as 1927–28 even though its rise was tied to the First World War. Cholley had a more systematic spirit than de Martonne. Demangeon for his part always demanded that his students learn morphology, but quite early on he feared that the development of morphology would detach it from human geography. At Rennes I have put a brake on this tendency by balancing my courses. *A propos* my thesis on Ségala M. Bloch asked me: *Since your goal essentially was to explain the transformation of a poor agricultural region into a prosperous region, why this long discussion on the eroded surfaces?* I answered: *Because they interest me.* Everybody who got involved with morphology became fascinated after having had an education which put emphasis on the shapes of the relief.

Bataillon: **Is that one of the reasons why certain regions of France have been**

studied more than others following the lines of study by Thibault in the *Annales de Géographie?*

Meynier: The distribution of university centres may have been a factor along with the personal influence of their geography directors: Musset has only inspired few vocations and the west was not studied very much. The predominance of physical geography in Paris resulted in the geomorphologic map of the Paris Basin, whereas only a few studies of human geography treated the same region . . .

Bataillon: How can one explain why a good number of the geomorphologists who authored this famous map later specialized in urban studies, often applied geography: in particular J. Beaujeu and Ph. Pinchemel just like Barrère and Nonn more recently?

Meynier: Maybe geomorphology was a source of enthusiasm to begin with, but later it seemed like a dead end with only weak practical utility.

George: The formulation of laws dates much further back and appeared much more tangibly workable in the domain of natural sciences than in the social sciences. This could satisfy certain minds which were educated in research involving a simple logic, who found in physical geography something more comfortable than in human geography where it is necessary to improvise on facts that were not always so precisely definable. In physical geography it was thus easy to create autonomous methods whereas in human geography one kept running into positions which were already occupied by sociologists and economists; this is where the very interesting personality of Maximilien Sorre intervenes. He defines relationships between geography and sociology and searches for geographical originality in what he called the 'technical foundations', that is to show how modifications of the matrices of space through social collectivities proceed to successive acquisition of different techniques. This perspective protected the specificity of human geography better than Demangeon's views, which caused some confusion between economics and economic geography. Sorre responded incontestably to the specificity of de Martonne's physical geography by showing the connections between instrumental matrices and organization of space. He sought lessons which social geography could derive from the other social sciences without losing its own identity. Unfortunately geography developed in the direction of instrumental specialization, thus losing sight of the synthetic association between history and geography. A series of fractional studies no longer leads to a global spatial image, and one has turned an entire part of the analysis of situation and prospective over to the sociologists and the economists.

Bataillon: This creation of geography by de Martonne with its glorification of geomorphology through map explanation, was that the only possible solution?

George: The personal influence of de Martonne was extremely powerful throughout the country, since the main geography textbook for twenty years was his *Traité de géographie physique* (Treatise of physical geography). Blanchard's generally well-founded criticism of de Martonne's manual did not appear in the form of a manual of rebuttal. Besides, de Martonne who had introduced Davis's scientifically oriented geomorphology also later introduced climatic geomorphology, especially following his first voyage to Brazil; Cholley was to develop this aspect of de Martonne's work.

Birot: I quickly became fascinated by climatic morphology, when in 1935 I started reading German literature on the subject. In Germany where climatic morphology was initiated in 1900 no such rupture occurred. As a matter of fact, in both their undergraduate and graduate programs, there was balance between physical and social geography. However, for those students who lacked scientific education, physical geography was resented as a largely superior effort; human geography during this period did not include statistics. The practical exam on map topography was based on physical geography because, thanks to the geological map, a privileged document, it is considerably enriched. Other types of complementary cartographic documents (e.g., demographic, etc.) could not be regarded as containing *explanations* to phenomena.

Papy: In Bordeaux, even after the reform, we did not give more space to physical geography in our *licence* program than it had before. In the *agrégation*, of course, the role of geomorphology has been asserted. In retrospect, I see no reason to regret this. I think that it would have been disastrous to exclude physical environment from geography.

Monbeig: When I returned from Brazil in 1947 I had the impression that geomorphology had become 'sacred', but it was already a bit like that with de Martonne. Beginning with Cholley, however, it is no longer easy for a human geographer to understand the physical side, just as it is for an historian to develop a taste for geography such as it developed. I was quite surprised in Strasbourg by the animosity expressed by certain geographers with respect to historians. I systematically forced certain history students to take notes and that was no mistake on my part, since all three are university professors now. If I had not been there they would have failed their exams. In the interuniversity excursions in which I participated during that period I myself felt a bit outdistanced. When it comes to geomorphology I always have the feeling of understanding when one explains it to me but I feel totally incapable of discovery.

Bataillon: For a long time geography and history have been disciplines which were intimately tied to teaching. Now they have emerged as 'applicable' in the same manner as sociology and economy. How do you visualize this role for geography?

Birot: I am convinced that geomorphology is applicable, but only at a very general level, where the geographer could mobilize input from a variety of sciences, for instance, functioning as consultant to the regional prefect.[11] This belief is independent of any philosophical or political opinion.

George: I have always tried to avoid the role of public consultant. It is an illusion to believe that one can convince people at top levels of territorial management about measures which could be in the interest of the population at large. In fact, the motives of those who hold political power are totally alien to the logic and style of territorial organization and management which we might wish to defend. Having said this, one also has to consider the restricted educational outlets which exist today for our students; it is our duty to prepare them for this consultant's role, informing them, of course, that when one works for the machinery, one becomes an integral part of it. In any situation a government will choose as its politics that which seems realizable within the short term, and thus remove any possibility of giving importance to long-term projects.

Meynier: I have been interested in applied geography possibly sooner than many of my colleagues; I pushed Phliponneau, who saw the reticence in Paris on this point, in that direction. I have scarcely had the opportunity to work within this framework myself; before the proliferation of assistants, teaching was very time-consuming. The dramatic moment arrived around 1965–66 when the number of students increased rapidly without any growth in the foreseeable outlets in teaching, so they had to look for other employment opportunities.

Enjalbert: As far as I am concerned my membership in the C.O.D.E.R. of Aquitaine (1964–75) and participation in the National Commission of Regional *Aménagement* have allowed me to use my knowledge of concrete problems; this once more offers geographers an opportunity not to become too abstract, and I believe that my participation in regional and national life has rounded out my own general knowledge. This is what I endeavor to make my students understand; teaching must also be permeated with a spirit which makes the students participate.

Papy: I think that geography's perception of the world is useful for some purposes, especially now that technological initiatives are multiplying in ways which ignore humanity and places with the greatest damage to both. Geographers must certainly be aware of the limits of their knowledge and techniques. They should also avoid backing questionable enterprises by hiding intentions and interests which are difficult to perceive. In applied work, economists usually take the center of the stage: what they bring us is useful, but their view of things is often too abstract, don't you think? The geographer who wishes to grasp 'the spatial experience of life', doesn't he have a right to be heard? At international meetings where projects of aid to the Third World are discussed, a geographer may stand up to fight preposterous proposals which have been put forward by technocrats who know nothing about the countries and the people they talk about, and his intervention is often received with relief and favor by those involved.

Bataillon: Have the limitations of the university education environment given you the desire to have an activity besides teaching?

Monbeig: The world of university teaching always seemed a bit narrow to me. I happily accepted a position at the *Conservatoire des arts et métiers* (Academy of Engineering), when it was proposed to me. Now I could have colleagues who would no longer be 'faculty professors',–although at Strasbourg I had excellent colleagues like Julliard and Boutruche–but also it gave me different kinds of students. Later on, when I was invited to become a member of the administration of the CNRS it was an opportunity for another new experience.

Birot: I have never regretted my choice to become a geographer. Research in the field and in the laboratory were always combined with teaching. A great deal of my time was taken up in studying related sciences on my own through successive cycles (two or three years of atmospheric physics, two or three years of granulometry, two or three years of geochemistry, etc.). All that hardly leaves times for 'related' activities.

Dresch: I found satisfaction in my *métier* as geographer and university teacher. It is surprising how rare a situation that is. I love my contacts with students and have found, without exactly seeking them, opportunities to interact with diverse milieux: from Africa to Latin America, to the Near East. This constant discovery of the world, reminiscent of my first desire to go to Morocco, has convinced me that one must always check theoretical ideas through personal contact with people and things: for the geographer, critique can only be guided by contact with live reality and landscapes. I have spent so much time interacting that I have not had time to use all the notes I have accumulated.

Bataillon: Towards the end of the 1940s and the beginning of the 1950s the environment of university geographers, especially in Paris, was very markedly left-wing politically, a trend which was far from common in the old Sorbonne. To express it more precisely, the Communist Party attracted numerous geographers of which some are present here. What relations can we establish between this political coloring, the very content of geography, and the social environment from which geographers were recruited?

Meynier: It is important to note the influence in Paris of such men as Clozier, Dresch, and George. In Rennes there was a different atmosphere, and the political aspects were quite other.

George: One must distinguish between membership in the Communist Party in the very special political context of the Resistance Movement and the Liberation, and the integration of Marxist ideology in the *problématique* of the geography profession. Marxism did provide a certain number of answers to questions posed by geographers. De Martonne's school supplied a satisfactory logical construction for physical geography, but one was much less well equipped in the domain of social geography and that generation discovered that Marxism could contribute to the problematics of research. Some people have confused political movement on the one hand, and the ideological moulding and scientific responses introduced by Marxism on the other. The philosophical connection has been far more important than the political one for geography. Lessons have been learned from the new system of spatial organization in socialist countries. There was a dissatisfaction with the methods of pre-war geographers which had been a sublimation of the Anglo-Saxon world and capitalist economy, especially through Siegfried. Territorial management in the Soviet Union was extremely fascinating, something new to chew on. Later on, it is true, I discovered a certain number of hoaxes which have caused detachment on a political plane, without, however, renouncing those explanatory elements which Marxism could furnish.

At first, Marxist geographers practiced a common style of research, profession, and political involvement – a phenomenon which could not be developed in the same way with historians. However, this convergence of research *problématique* and political engagement only lasted during those few exciting years. Once this simple logic defaulted, the geographers' collective and the political collective both fell apart. If a Marxist felt some discomfort at being locked into the Communist Party's dogmatism, then he could either leave the party, or become a specialist in some sectoral interest in geography where political problems no longer interfere.

Monbeig: The reason a young person becomes an economic or human geographer is generally – I do know of exceptions – because he is attracted by certain social preoccupations, all of which can be reinforced once he gets into the field. The recruitment of geography students has always been more 'democratic' than the recruitment of other disciplines, for example, economics and even for the historians. This difference can be appreciated better if one looks at Saint-Cloud and the possibility of a *licence* in geography without Latin. I was a student during the era when Latin was required but the students who preferred geography to history were often of more modest origin and of more markedly left-wing political opinion than the others. By sentiment and family background they were much closer to manual labour, and geography, after all, is closer to manual labour than other fields. Among historians, too, there were possibly more young people who had a stronger religious education and had attended parochial school. Thanks to Demangeon, I was in touch with graduate students very early on while studying for the *agrégation*. It was clear that almost everybody who specialized in geography was of a more modest origin than other graduate students; everybody was already militant and belonged to leftist parties. And it wasn't Demangeon who encouraged them in this direction! Later on, as problems became more serious, tensions became more dramatic.

Birot: Up to 1939 Saint-Cloud educated college professors, whereas at that time very few graduate students from rue d'Ulm specialized in geography. When Saint-Cloud became a graduate school, it would recruit proportionately fewer *littéraires* from rue d'Ulm (with the Greek and Latin requirements) and more *modernes*. Until 1955–60 this seemed to go along with a socially more popular recruitment than in history for instance: school teachers were especially numerous and among them we found a seedbed of geographers.

George: Does one get involved in geography because one belongs to the lower middle class or because one is attracted by a certain type of action and participation? I tend to believe that geography engendered a sensitization to social reality and as a result, people moved toward that party which was the strongest and most socially dynamic during this era. Besides from 1945 the education without Latin can explain a certain feeling of disdain for historical studies on the part of geographers, since the attraction to history implied a certain cultural education which was not 'modern'. Geography with its special mechanistic aspects, especially during the era of physical geography, was more satisfying to those people who were educated in the sciences.

Birot: Hegelian dialectics satisfactorily describe the steps of almost all scientific progress. On the other hand, Marxist philosophy has inspired standpoints which in my opinion were arbitrary on concrete problems, such as the cycle of erosion, the critique of which was purely formal and did not propose a theory to replace the old one.

Dresch: As a young undergraduate student my scientific opinions were quite separate from my political stance for the good reason that on the scientific plane it was necessary to obtain a specialized education. I may have been into Marxism before I was into geography, since my political choice preceded the liaison which was established between my discipline and Marxism. Events pushed me in this direction far more than methodological reflection on, for example, Engels and the dialectics of nature. Even if I tackled Marxism in the framework of a critique of the environment in which I grew up, I considered that the clear-cut ideological position of economic and human geography should be contested. In 1945 the contestation led straight to the Communist Party. This experience of the Communist Party was an excellent education for some, not as much from a political point of view, in the measure to which they changed opinion, but from the point of view of scientific reflections.

Bataillon: The same preoccupations and the same reasons for taking a political standpoint are found in the profession of history. Yet we find no such massive movement towards the Communist Party in the history departments of France from 1945 on: How does one explain this?

Dresch: In France at least, the division of history into rigid time periods certainly limited the phenomenon since Marxism only directly concerns those who are working on modern and contemporary history. Pressures from the social environment, e.g., the fact that history is much more strongly imbedded in the academic tradition, may possibly exert a greater influence. Social pressure was much less strong for the geographers than for the historians. Geography, within the university, has always been seated on a folding chair in an uncomfortable position. Since Latin played an essential role in traditional French university education it is not surprising that there were so few geographers educated through the university system prior to the Second World War. On the contrary Saint-Cloud played a considerable role in the recruitment of geographers. This recruitment was not the same as for the *Ecole normale supérieure*[12]; on the whole it was more democratic. The students graduating from Saint-Cloud were more easily oriented toward geography and otherwise toward progressive positions.

Concluding remarks by Claude Bataillon

The institutional rupture in the organization of geography was not, in fact, reflected in a corresponding rupture in attitudes or contacts between history and geography. The *licence* in geography equipped one to teach in the *lyceum* where history and geography were considered one subject. These circumstances served in some respects as a selective filter for university professors, but they also set up geography as a discipline with didactic and speculative goals rather than practical ones. It is not yet an applied science.

This situation explains why physical geography, the foundation and *raison d'être* for the discipline's scientific value, does not become applicable: that would be like ecology becoming management, a trend which does not break through until during the 1960s. During the 1940s physical geography consolidates itself as an observational science with two objects: the landscape and the topographic map (which is essentially grounded in the geological map) thus a pedagogical exercise from the researcher's education gains autonomous value in geomorphology.

Human geography also grounds itself on the same twin foundations: landscape and the topographic map. The archeology of rural landscapes led researchers schooled in history to a particularly fruitful osmosis in the group associated with the *Annales*.

But how can *landscape* become the object of science? What could be more difficult to submit to laws of generalization? The geographer has no unique observational method at hand. He has an explicit theoretical but scarcely practiced object, viz., to explain all the steps in the epiderm of the earth. The way to apply this theory, so deeply anchored in professional mentality, was in large-scale studies of the landscape itself, or in survey, primarily the rural survey. Are these methods, these field-survey-in-the-landscape tricks about to be reinstated now?[13] Are they not the only ones to allow us to perceive the connections which are lost in offensive analysis?

One had to wait until the 1960s to see landscape, once identified with region, become an incontestable object of rational analysis (Juillard and Sautter). The region in the 1940s was at least the object of typologies. The *Students' Guide* by A. Cholley (1942) provided the foundations; they were to be applied in a collection of *lycée*-level texts edited by Baillère until about 1955–60. Regional classification and subdivision triumphed during the 1936–38 period, again directed by Cholley and co-authored by Clozier, Birot, Dresch, and George. This is a pedagogical tool following the scientific model of the *Géographie Universelle* tradition dating back to 1920–30. One could not identify 'laws' governing regions, but one could attempt to classify them in a typology which at least suggested fundamental regularities.

Cohorts of scholars thus embarked on a new course: the science of both nature and man. Geography belongs neither to the aristocracy of classical humanities (where the basic barriers of Greek and Latin stay until the end of the 1950s) nor to that of the polytechnical and mathematical push of the 1960s. Yet geography becomes the environment for an important social promotion for a stream of teachers, most of whom have graduated from l'Ecole normale at Saint-Cloud. A profession without strong precedent in the establishment grew larger in relation to history and those social sciences which had their origin in philosophy (sociology and ethnology) and law (economics).

Our survey has deliberately omitted personalities from that generation who differed from those trends established in the late 1940s. As dissidents or free-lancers they do deserve attention. The generation which became established after the reform would also have much to offer. They were rapidly promoted within the new system, a *pléiade* of geomorphologists, more often than not pupils of A. Cholley. Some were later to be attracted to problems of *aménagement* (e.g., Beaujeu-Garnier, Pinchemel, Tricart), for some, Marxism provided the pillar of theoretical support for their science. All have witnessed the virtual explosion of the geographical family since the end of the 1960s, and their reflections would be valuable indeed.

Notes

1. The French educational system, during the period under discussion was organized roughly as follows: After graduation from the *lycée* (secondary level education somewhat more advanced than American high-school) with a *Baccalaureat* one usually spent two or three years of university studies in preparation for a *licence*. Then followed a year during which one wrote a thesis for the *Diplôme d'Etudes Supérieures*, the credential required to enter the national competition for the *agrégation*. One could now teach in any of the state schools, and usually did for a period sometimes as long as ten years before undertaking research for the thesis of *Docteur d'Etat*, which qualified one to hold a professor's chair at the University. Paralleling the university system in Paris there was the Ecole Normale Supérieure de la rue d'Ulm, which recruited students by competition and prepared them also for *agrégation* and for research. Similarly, the Ecole Normale Supérieure de Saint-Cloud, but its clientele consisted mainly of young teachers. Once the requirements for Greek and Latin were relaxed, these schools became an important source of students for geography.
2. Throughout the text D.E.S. will refer to the *Diplôme d'Etudes Supérieures*.
3. The regions of Limousin and the Brive Basin are on the western part of the Massif Central.
4. de la rue d'Ulm.
5. 191, rue Saint Jacques has been the address of the Institut de Géographie at the Université de Paris since its foundation in 1926 by E. de Martonne.
6. Southwestern Massif Central.
7. Teacher's college.
8. Published between the end of the 1920s and early 1940s, under the initial direction of Vidal de la Blache and later of E. de Martonne and Gallois.
9. The Collège de France was an institution of higher education which did not grant degrees. L. Febvre, F. Braudel, and R. Dion have taught there.
10. Bastide, historians like Braudel, and especially C. Lévi-Strauss; see his *Tristes Tropiques*. Libraire Plon, Paris (1955), pp. 5–7, 37–8.
11. Administrative head of a province appointed by the Central Government (functionally equivalent to a State Governor in the USA.)
12. de la rue d'Ulm.
13. Reference to the 'new geography' of Anglo-Saxon inspiration.

Chapter 10

Fig. 10.1 Jacqueline Beaujeu-Garnier, France

Few experiences could possibly rival that of being introduced to France and to French colleagues by Jacqueline Beaujeu-Garnier. The scale and ingenuity of her brilliant work is matched by the charm, cordiality, and courage she shows to students and colleagues all over the world.

Initially attracted toward history, she 'discovered' geography in André Cholley's lectures on French regions. Challenged by the explicit assertion that 'morphology was no field for a woman' she briskly trod that very field, and with her *Morvan* study became the first female Docteur d'Etat in French Geography. In 1960 she became the first lady professor at the Sorbonne, and since then has been a pioneer in the fields of population, medical, and urban geography.

'A geographer does not have to be walking encyclopedia,' she remarks, 'but he/she can cultivate an encyclopedic curiosity and then know how to consult specialists in other fields'. Comparing the geographer's situation with that of the general practitioner of medicine, she demonstrates a fine-honed sensitivity to the human condition, energy to engage creatively in teamwork, blending inductive and deductive insight in the unraveling of practical problems. With a spirit as ecumenical as her mentor, Max. Sorre, she is as much at home with the humanistic leanings of the classical tradition as she is with the challenge of unraveling the mechanisms of spatial systems.

Autobiographical essay

Geography for me means more than just a profession: it means much more a way of understanding the world. A geographer looks at his surroundings in a special way: a vast mountainous region, for example, stirs not only appreciation of its esthetic beauty, it evokes curiosity about its underlying structure and the phases of its evolution. City slums evoke not only compassion and horror but also motivations to search for historical and social explanations for such miserable conditions. The geographer not only looks and observes, he or she automatically wants to understand. Direct field experience of land and people, an essential element of the geographer's *metier* ((profession) provides an openness to the lives of other people and a fuller awareness of one's own life. It is no accident, therefore, that every geographer has a wide network of

141

friends surrounding his or her place of work, and frequently assumes an active role in political matters.

Every individual can learn much from personal experience whether he or she cares to admit it or not. Experiences can modify preconceptions, open up new horizons, and stir one to fresh research adventure. The geographer's domain is the living world which yields not only information and enrichment, but also, I dare say, inspiration.

Career success

An accidental vocation

A calling to geography is a very personal matter. Having completed a marvellous study experience in high school I faced a choice of two possible careers: in science or humanities. I chose history, largely because of a brilliant and dynamic teacher. It was with that intention that I first came to Paris.

I detested geography. It was badly taught and seemed to call for simple memorization of insipid nomenclatures and endless details. In France, however, higher education in history and geography were inseparable. My first course in geography was truly a revelation. André Cholley spoke of French regions – suddenly everything became clear, demonstrable: here was a field which could be descriptive and concretely grounded, and at the same time vivacious and interpretative. Observations were supported by explanation: geological insight shed light on morphology, while history elucidated patterns of population, rural life, and urban development. At last the memory could take second place; and my intellect could be free to explore the concatenation of phenomena; logical reasoning replaced encyclopedic memorization. Enthusiastically I embraced geography and have never been able to leave it since then.

Six rather uneventful years – including one year spent at the Army Geographical Service after the outbreak of war – left me with some doubts about what subject to choose for a dissertation. Archeology was a tempting subject, but the decisive step came with a challenge from Emmanuel de Martonne who had the imprudence to declare, in my presence, that morphology was no field for a woman! Immediately I persuaded my adviser to steer me in this direction – hence the adventure with *Morvan et sa bordure*. About this time a second assistantship was opened up at the Sorbonne and it was offered to me. This provided the opportunity to work with one of the great masters of human geography: Max. Sorre. Ultimately, it is to him that I owe the essence of my own development in geography. A scholar of prodigious learning, he was at that time putting the finishing touches to his monumental *Fondements de la géographie humaine*. He would discuss his work with me, ask me to read pages from his final chapters, and include me in conversations with leading scholars in France and abroad. This whole period had a decisive influence: even if my thesis topic was pure morphology, my complementary thesis, a requirement at that time, showed a new learning – a regional study of the Brenner Valley.

University career

Five years as assistant at the Sorbonne, followed by one year at the Centre National de la Recherche Scientifique (CNRS), here was I, thirty years old, a Docteur d'Etat. On my mentor's advice I applied for a university teaching position. For one year I was received at Poitiers, a quiet little town with wonderful Roman churches and a university which was its cultural center. The following year I was appointed to Lille, a large industrial and commercial city where the university played only a minor role in urban life. Lille provided the crucial stepping stone to the eventual destination of the Sorbonne. The thirteen years I spent at Lille marked the burgeoning of that sensitivity to economic and social issues which has motivated much subsequent research. Max. Sorre had successfully planted the seminal ideas, Lille provided the fertile and highly variegated soil for them to flower: traditions and prejudices of localized familial 'patronat', the harsh lifestyles of miners and women in textile factories, a proud pioneering region of the nineteenth century confronted suddenly with the demise of coal and fiber-based economy, a ridiculously artificial political boundary severing European people destined for community coexistence, the convergence of immigrants from different periods, diverse races, and competition between Church and State on questions of education – these were my images of a world in transition, not sure of its own future. It seemed worth while to capture and document the physiognomy of this fascinating world in the *Atlas du Nord et du Pas-de-Calais*.

How can I define all that moved me to prepare that atlas? No doubt it was an enormous and exciting challenge to discern the 'anatomy' of a great European industrial region which was facing a crisis. Certainly the team spirit and friendly cooperation of colleagues and researchers at Lille helped enormously; the eager and efficient administrative support from Rector and Prefecture, and the engagement of politicians, whose opinions often varied, lent approval and strength to our project. It was in this context that I became convinced about a special vocation or ambition for geographers: to open up the university to life, to educate young people from experience as well as from books, and to share in dialogue with the widest possible array of people from administrative, political, and social sectors of society; to gather data without bias or blinkers.

Another extra-curricular activity assumed importance for me during this period: from 1949 to 1958 two successive directors of the general secretariat of the *Revue Politique et Parlementaire* encouraged me to familiarize myself with economic, fiscal, financial, and political problems at a national and international scale. The *Revue* dealt with real-life problems and sought close contact with different socio-economic and political milieux. I became aware of the difficulties facing not only academic people, but also those of businessmen and politicians who really wished to be *au courant* with up-to-date information. This awareness inspired a young colleague and me to launch *Images économiques du monde*, which now in 1980 celebrates its twenty-fifth year of continuous printing.

143

The Parisian years

After thirteen years of provincial life I was appointed to the Sorbonne – a much envied position at the time. Having been the first female 'Docteur d'Etat' in geography in 1947, I became in 1960 the first woman professor in that august department. Ever since then I have lived a few blocks away from the *Institut*, in a milieu full of memories of my life as a student and as a junior assistant. I assumed the chair in regional geography, succeeding my friend and colleague, George Chabot, a renowned scholar in urban research. He invited me to collaborate in his proposed *Traité global de géographie urbaine*, the first such work ever published and subsequently translated into eight other languages. This was an enormous education for me. About this time Pierre George introduced me to Jean Bastié, who was then doing a thesis on the growth of the Paris periphery, and this opened up a whole new perspective and a series of joint projects which still continue. Here it was: the great metropolitan region of Paris with its 10 million inhabitants, its global reach . . .

Joint research with Jean Bastié, graciously supported by Paul Delouvrier, first District Delegate of the Paris region, bore fruit in the monumental *Atlas de Paris et de la région parisienne*, published in 1967. Now my interests in the north of France re-emerged, but at quite a different scale. For many years my regional work had led me in the same direction: to establish a University Research Centre, which was later recognized by the CNRS. Thereupon followed numerous articles, several books, participation in diverse meetings, collaboration with regional administrative authorities first as consultant and then, since 1971, as a member of the *Comité Economique et Social de la region Ile-de-France*, one of our two Regional Assemblies. Within this committee, I preside over the Commission on Planning and Quality of Life.

These are some of the strands and textures of my professional work as a geographer. In my national and international activities I have had quite another life. At a national level my work has been mainly of a concrete and pragmatic nature, directly involved with political and administrative levels of planning. Through all these activities I have endeavored to develop an educational program for students in regional planning. This course is now officially recognized and the diploma is sought by students from outside geography, particularly from architecture and engineering and from foreign countries (especially Africa and the Middle East). The course involves collaboration with other disciplines as well as work with non-academic bodies, for instance, the *Délegation à l'Aménagement du territoire et à l'Action régionale* (DATAR). My urban interests continue: experiences in the north of France and later in Paris, combined with numerous travels through virtually all countries of the world have led me to a sense of urgency about the global dimensions of urban problems. My proposal to set up an IGU Working Group on World Metropolises at the Tokyo Congress in 1980 was a step in the direction of providing an international forum on which results of urban research may be shared.

And the Paris years, at least for the moment, continue . . .

A succession of interests

Confidentially

To read my bibliography may give the impression that I have never had a well-defined specialty: titles change from year to year and several fields have attracted my attention. To shed light on the inner consistency and continuity of my work, a few character traits must be mentioned.

When I am attracted by a subject, I plunge into it wholeheartedly. I love to find a new subject, or one which is just being discovered by geographers. I do not try to 'geographize' or 'annex' it, but try to find out what geography can contribute toward knowledge or understanding of this subject. I do not try to monopolize but to participate with other disciplines in the investigation. In population studies, for example, a geographer finds himself situated somewhere between the economist and the demographer, contributing to them – as well as to colleagues in medicine and sociology – an attempt at a global, non-sectorized perspective: to provide a framework for the analysis and presentation of results which gives each discipline an opportunity to share insights, emphasizing interrelationships rather than specialization.

I greatly appreciate interdisciplinary collaboration. Discussions with fellow geographers often lead to a sharing of the same preoccupations and information – whatever the subject, discussions nearly always demonstrate a common point of view. Let a geographer and a medical doctor discuss a problem, however, and this can lead to altogether new insight. It would be highly desirable to encourage such cross-disciplinary efforts.

How can the geographer realistically hope to master all the knowledge and analytical skill which are needed for his domain? That would be simply misplaced ambition. I have a story to relate on this point. In the context of a *sciences humaines* conference on recycling in Paris organized by mathematicians I gave them some articles by some of my Anglo-Saxon colleagues of brilliant quantitative fame. They read the articles and their verdict was 'what useless complications!' and 'this must be mathematics for humanists'. Geologists, demographers, economists, may have a similar comment on geography. A geographer cannot be a walking encyclopedia, but should be capable of undertaking an encyclopedic work, and this cannot be done without consulting other disciplines. Our situation is comparable to that of a general practitioner of medicine who needs to be in touch with different medical specialties. Which of us have not had the experience? What we recognize as being quite obvious in medicine, we hide for ourselves, but our situations may be very similar.

Now let me show how these personal preferences translate themselves into my scientific work.

Different and successive interests

Beginning as a geomorphologist, I did my first field work in the region which was suggested to me: the Morvan. Realizing quite soon my weakness in scientific training, I went to the *Faculté des sciences* and earned certificates in geology and mineralogy. I engaged in microscopic observations side-by-side with the most competent geologists in my area of interest. For quite some time I became preoccupied with the composition and disaggregation of crystalline rock and did much laboratory research on this. At the end of this period I brought two insights which were innovative at that time: (1) the persistent dynamism of ancient massifs and the importance of fracture, replayed up to relatively recent time; and (2) the differential disaggregation processes within crystalline rock following a certain number of observable characteristics in the crystalline forms studied in very thin plates. I added some insight on recent periglacial climatic factors – a subject which my friends Jean Tricart and André Cailleux had amply demonstrated in their study of the Paris Basin. Thus I helped French geomorphology to open up beyond its formerly 'structuralist' moorings.

For several years after my dissertation I did follow-up research on these themes in the Massif Central and the Limousin, with some other work in Brazil, Algeria, and USA. I have always had the 'morphologist's eye' and could scarcely look at a landscape without wondering about its underlying structure and hypothesizing about which factors had influenced its form. This habit endures. *Relief de la France*, published in 1972, summarized not only my own field work, but also reviewed other regional and local work done in France during that period. Geomorphology delights me for the logic of its reasoning, but also it brought me in touch with two of the greatest geomorphologists of our own era: Pierre Birot and Jean Tricart.

It was Max. Sorre's influence which led me to an interest in human problems. My first contribution here was in medical geography, a field which was virtually absent in French geography at the time with the one exception of Sorre's discussion in the *Fondements de géographie humaine*. Several medical doctors, however, were beginning to examine the statistical side (at the *Institut national de Statistiques et d'Etudes économiques*, INSEE), and also the question of genesis and distribution of disease (at the *Institut national d'Etudes demographiques* – INED, and the Faculty of Medicine at Montpellier and Paris). The real pioneer was a French doctor who had his career at the University of Hanoi, and became an American citizen after the war, Dr Jacques May. He was to become the first President of the Commission on Medical Geography which was initiated at the Washington Congress of the IGU in 1952, and I was the French representative.

For several years I pursued research at medical centers, at the INED, on projects related to the IGU Commission, and began to see how closely questions of health and disease were related to the environment – surely the geographer's perennial domain. This was what led me to my study of world population patterns in the major regions of the globe. To my knowledge, no such attempt has been made before or since, either in France or in any other country. The two-volume work, *Géographie de la population*, has now been reprinted and constitutes a basic record on world population patterns for the first half of the twentieth century. This project cost me years of work, reading, travel, from which I re-emerged with a definite commitment to human geography and an impressive stock of knowledge. Soon afterwards I tried to harvest all that experience and knowledge in *Géographie générale de la population*, with a subtitle, '*Démogéographie*'. In France it was published with the title *Trois milliards d'hommes*, translated into English as *Geography of Population*, in Dutch as *Des hommes par milliards*. A second edition, completely updated, appeared in English in 1978 and has just been translated into Portuguese. These successive revisions and reprintings have enabled me to emphasize the economic and demographic trends on this planet – trends which scarcely lead one to optimism.

Having completed the transition to human geography, I then pursued some joint work with Georges Chabot on his proposed *Traité de géographie urbaine*, and also inherited his teaching responsibilities at the *Institut d'Urbanisme* in Paris. I now had to educate myself on other aspects of urban life: politics, spatial problems, social segregation.... There was the attraction of novelty in this, but soon I realized how relatively incompetent I was in certain areas, and so I plunged myself into urbanism, urbanization, and everything one could do or say about cities. In this I see a link, however tenuous, with my previous demographic interests. It also allowed me to collaborate with new kinds of specialists: architects, engineers, planners, and politicians.

Urban problems, however, are multiform and complex, especially in a metropolitan center like Paris, within a country whose average standard of living was rapidly increasing and 'mentalities' changing even more quickly than technology; almost every day I could witness sudden changes in habitual behavior. Some were disturbing, some a delight. Public officials also worried about such changes, not knowing their potential consequences, and indeed, sometimes not even knowing about them at all. The Ministry of Commerce established a research program on changes in commercial structures, while various municipal and regional bodies were concerned with planning. Again I found myself at a strategic cross-roads in my own research, examining changes in the commercial infrastructure, consumer habits, the repercussions of competition within the urban region of Paris, and attempting to compare these with provincial patterns, or with overseas examples. This gave birth to the idea of a *Géographie du commerce*, written with Annie Delobez, which was based on numerous studies done by our research group.

Should one conclude that this list of successive interests indicates a lack of unity to my intellectual development? I do not think so. In reality my moving to a fresh theme was usually done within a broader array of choices, from one broad subject to a more particular one, but all my previous knowledge fed into the new research. An example is this new *Géographie urbaine* which appeared in 1980 and gives an overall synthetic picture of my own development, both conceptually and methodologically.

Parallel pursuits

Running in tandem with these major thematic interests within which I trust I have made some original contribution, is another stream of effort which has continued since 1948. These are the informational books in regional geography. First *Economie de l'Amérique latine*, followed a few years later by *Economie du Moyen Orient* and *Economie du Commonwealth*. The main merit of these volumes is that they are short and readable. Another work on *Les régions des Etats-Unis* has been reprinted several times. There I proposed a modified regionalization scheme for the USA on the basis of various criteria: climate, relief, history, and economic comparative advantage. This was my 'demographic' period, and then also my little book on *La population française* appeared, together with the human part of *Les Iles britanniques*, written together with André Guilcher, who authored the physical part. These were all 'informational' books.

Between 1956 and 1967, however, the major serious project was connected with the *Atlas regionaux*, a project I'd like to explain further. My sojourn at Lille coincided with a period of major economic change in the north of France. At the beginning of my stay, euphoria still reigned after the wartime restrictions, and later the recent planning efforts (First Plan 1946–51) gave priority to those sectors whose needs seemed overwhelming: it was fair wind for coal and textile interests. However, having studied closely the problems in Great Britain, I could foresee that when normal conditions were restored in the north of France, this region would not escape the fate of 'reconversion zones'. At the time this notion was heretical. There was no serious documentation to allow a detailed regional study: that was why I undertook the idea of a regional atlas in collaboration with the INSEE of Lille and other regional bodies of specialists. These annotated maps would provide a precise and living portrait of departments in *Nord-Pas-de-Calais*, combined within the economic region of the north. I must say that this initiative was enthusiastically supported: the University emerged from its Ivory Tower, and economic interests, for the first time, paid attention to these documents which were scholarly and yet accessible. Etienne Juillard, my colleague in Strasbourg, was doing a similar project in northeastern France. This was a turning point for French geography which had hitherto been confined completely to the University while only sociologists and economists counted in external activities.

The consequences of this move were far reaching and enduring: the *Atlas de Paris et de la région parisienne* was completed by thirty people in three years: 1967 saw the harvest of a collective effort. In both cases – in Paris and in the North – regional administrative bodies provided financial assistance while the INSEE personnel worked very efficiently with us. Even a bigger national consequence ensued – DATAR and CNRS accepted the responsibility for funding atlases of the same type for every economic region in France (there are twenty-two). This project which was launched with a vigor that varied depending on the region, now covers the whole country with the exception, unfortunately, of two of the poorest regions: Limousin and Auvergne. To bring this enterprise to completion and to procure the necessary funds, an association called UDARAR (*Union des Associations pour la Réalisation des Atlas régionaux*) was formed and I became its President. The French Atlas project became a kind of international model. The International Geographical Union set up a commission dealing with national atlases and I was a member.

As a kind of sequel to the Atlas of Paris, I edited two huge, well-illustrated volumes under the title *Atlas et géographie de la région Ile-de-France*. These volumes have been widely distributed and belong to a sixteen-volume collection, published under the direction of Professor Papy, which covers the entire country of France.

Thus was completed – provisionally at least – this series of publications on regional ensembles; they contain documentary information quite consistent with the French geographical tradition.

Development of geographical thought

Geography as 'information'

Early modern geographers were either explorers like Humboldt or library researchers like Réclus. I'm inclined to surmise that many people do geography without being aware of it. Excellent descriptions of travels, or fictional accounts in novels provide perfectly good descriptions of landscapes, people, towns, and rural life. To really gain perspective on a country, it may be more fruitful to read certain novels than to comb the statistical renderings of a scientific work.

As long as the world was not well known – which is still true of certain less accessible regions – simple descriptions constituted what could be called geographic information. Descriptive accounts, of course, shed light on problems, and one important aspect of classical geography was an attempt to explain also. The French School is particularly renowned for its famous regional theses which were virtually the rule of the day from the beginning of the twentieth century to the 1930s. Steeped in this tradition, I have endeavored to reflect that style quite faithfully in my own writings, especially in my work on regional geography.

A literary style of geography, which some regard as an archaic hobby, had the great merit of bringing the country and its people to life, and it required – one must emphasize – direct knowledge of the field and some special talent. Why should this style of geography vanish? Should it deserve the condescending disdain it gets from advocates of modern, so-called 'quantitative' methods?

Modernism in geography

Other winds have come from Sweden, USA, and later from Britain. A number of geographers, following the lead of econometricians, have pushed for a quantitative geography. Far be it from me to reject these currents. When they arrived tardily, in France, they generated both violent hostility and violent acclaim. The older generation and those who were not skilled in mathematics were trapped, while among the young, they were generally welcomed. Several French geographers have been reflecting on this theme, but

I'd like to mention just a few conclusions which are also based on my own recent experiences.

Modern methods can serve to make geographical thought more precise. They are also helpful for analyzing certain phenomena and for discovering certain types of relationships. Some precautions, however, are necessary in handling them. First, one has to get a good grasp of the problem (be it landscape, distribution, production, etc.). Then one has to write down – state verbally – the initial statement of the problem, then go out and get the maximum possible amount of data on that theme. After these two steps, one designs an operational strategy for accomplishing the analysis on that specific problem. Many contemporary geographers forget to do that preliminary (literary) suggestive description; the result is that they can get fantastic correlations and marvelous classifications, but these could be found anywhere in the world – Milan, Montreal, or wherever. Secondly, one can become a prisoner of one's data: sometimes the only way out is through a distorting kind of simplification; another is simply to substitute some equivalent measures whose relevance is not always demonstrated. However, these are at least two good reasons for taking these modern methods seriously: they certainly improve our potential for measuring and comparing phenomena, and they also force geographers to think carefully about their research.

What would help to improve the scientific quality of geography is not more quantification – after all, that has been done in a simpler way for a long time – but more *logical reasoning*. We need to place our highest priority on modes of reasoning and the development of a deductive approach in modern geography. Abandoning the ancient slogan about the 'unicity' of geography, a deductive approach takes the general logical view of a 'model' and then examines the degrees of deviation from the model, i.e., the individuality of each case. Certain precautions are also necessary here, especially in designing the model. This cannot be done without serious reflection on the already established graphical monographs done either by the researcher himself in its initial phase, or by his predecessors.

There is no real chasm between classical and modern geography, only progressive refinement. Just as new technologies more or less completely render others obsolete, it seems to be desirable to preserve whatever was of value in the first phase while enriching the second. Otherwise, geography will become a kind of subdiscipline of econometrics, and the geographer will disappear, to the profit of other disciplines who will not apologize for their concreteness, like sociology and ethnology.

Systematic reasoning

New methods have enriched geography but they should not change its nature. There is one method which I feel has particular value, i.e., the notion of systems. To examine the interrelationships among parts of a whole seems to me to be a fundamental aspect of geographic research. In fact, the formulae used by the most outstanding pioneers in our field articulated precisely this conception: recall simply Max. Sorre's (1952) comment on the city 'as an expression of relational life'. What is new is the idea of making 'relations' the core of research. This could mean that we abandon the idea of taking juxtaposing chapters on different sectors (as in some foreign texts, where each chapter is written by specialists from different fields), often in the same sequence: relief, geology, climatology, population. . . . The systems approach, by contrast, demands much preliminary reflection: one has to search for the fundamental elements of one's theme and especially the mechanisms and processes which sustain its links and equilibrium. Any use of the systems method therefore demands much careful preparation and impeccable reasoning. Research done on these deep-lying mechanisms allows one to move from the level of appearances to the level of structure, and it is to the unraveling of such underlying structures that geography should address its scientific efforts. In my recent *Géographie urbaine* (1980), I have presented a view of the city as a system composed of subsystems.

To adopt a systems approach one must go more deeply: one has to discover the kind of equilibrium which is characteristic of the system, and then study the reverberations of quantitative or qualitative changes throughout the system. Quantitative change does not threaten a disaggregation of the system but qualitative change can trigger that. Studies done with this approach can provide not only a much more solid base of geographic research, but also render it more amenable to practical applications. Once one has unraveled the system, one can attempt to reproduce it: regional planners can thus find it interesting.

I should have taken more time to elaborate on a theme which has always been a preoccupation of mine, i.e., relations between geography and regional planning. The geographer's role in planning has become quite significant, at different levels, in various lands – Brazil, Poland – and is only now developing in France. It is expressed sometimes via the biases of research groups and scientific publications, sometimes through direct engagement with different official bodies – municipal, economic or political assemblies. Present-day orientations in geography can only help to foster this trend, begun at least twenty years ago.

In conclusion of this autobiographical essay, let me recall something already mentioned on the first page. Young geographers should think of their discipline as an integration of research and teaching, of university life and participation in public action. With such a conception of their field, they can gain a sense of a full and burgeoning life. Perhaps the chain of events which has led me to this conclusion has been a matter of chance, but it was also a matter of using this chance with maximum efficiency.

Selected readings

| 1950 | *Le Morvan et sa bordure.* (Morphology of the Morvan Region.) Presses Universitaires de France, Paris. |
| 1955 (and annually since then) | (In collaboration with A. Gamblin and A. Delobez) *Images économiques du monde.* (World Economic Survey.) S.E.D.E.S., Paris. |

1956–58	*Géographie de la population.* 2 vols. M. Th. Genin, Paris.
1967	*Trois milliards d'hommes.* Hachette, Paris. Translated as *Geography of Population.* Longman, London.
1959	*L'Atlas du Nord.* Direction des travaux. (Atlas of Northern France.) Berger-Levrault, Paris.
1976	*Les régions des Etats-Unis*, 5th edn. (Regions of the United States.) A. Colin, Paris.
1971	*La géographie: méthodes et perspectives.* (Perspectives and Methods of Geography.) Masson, Paris.
1977	*L'économie du Moyen-Orient.* (Economic Geography of the Middle East.) Presses Universitaires de France, Coll. 'Que sais-je?' No. 473, Paris.
	Atlas et géographie de Paris et de l'Ile de France, 2 vols. (Atlas of Paris and the Ile-de-France) Flammarion, Paris.
1978	*L'économie de l'Amérique latine*, 6th edn. (Economic Geography of Latin America.) Presses Universitaires de France, Coll. 'Que sais-je?' No. 357, Paris.
1980	*Traité de géographie urbaine.* Reedition mise a jour. (Treatise on Urban Geography.) A. Colin, Paris.

Chapter 11

Fig. 11.1 William William-Olsson, Sweden

'How could people be so stupid as to start a world war?' A gripping puzzle this was for a twelve-year-old boy in whose cordial home foreign friends were welcome, and whose parents had lived and worked in England and had traded with many lands. One tentative answer for William William-Olsson was: 'They simply do not understand each others' history and geography.'

It was indeed a cordial home that I found on my first visit to Bromma with a host of curiosities about his own history and geography. Through subsequent meetings – for example, at the Sigtuna seminar – and especially through a lively correspondence, I have come to know William better, thoroughly enjoying his energy, occasionally abrasive humor, and never-failing wit. For most of his life he has worked in Stockholm, where first he was inspired by Sten De Geer and Hans W:son Ahlmann and where he began in the 1930s to develop a style of geography which was to be at once empirically grounded and theoretically innovative. Throughout his life and work rings the value of his childhood home milieu: trust in individual and group ingenuity; lightheartedness and risk-taking; scorn for materialism, intellectual compromise or greed; courage to follow one's own convictions; and encouragement to be daringly different.

My responsibility and my joy

To give a brief but nevertheless meaningful report on my activities as a geographer I have to go back to my family and to early stages of my life. It is also necessary to say something about my undergraduate studies and a period when I was a school teacher. As to my research I can only give a skeleton outline and refer to selected titles from my bibliography. I will conclude with some thoughts concerning creativity, geography, and research.

Family

My great grandfather was one of those clever farm-boys who through the centuries had been recruited to the Swedish clergy. He became minister to the

153

town of Filipstad in Värmland. Besides his ecclesiastic work he also helped in the foundation of a Savings Bank. After some years he was appointed rector of the parish of Kroppa south of the town and here he preached the Gospel, fought against inebriation and ignorance, founding the school system and the library of the parish.

My grandfather was his eldest son, born in 1825. He studied theology but became an agnostic and migrated to England where he married an English girl and had a big family. He was employed by D. C. im Thurn & Son and was in charge of its Scandinavian section until the crisis of 1875 when this firm went bankrupt. In this awkward situation, he started his own agency for the import of Scandinavian wood to England: Martin Olsson & Sons. Two of his boys, later three, joined him, and the proud father showed his appreciation by suggesting that the name of the firm be altered to Martin Olsson's Sons and Father. He was a jolly fellow with a great sense of humour. One of his London friends in his memoirs tells us that he met Martin Olsson in the Swedish Church on Christmas day. When asked how this tallied with his agnosticism, Olsson answered that he thought it his duty to wish God a happy Christmas.

My father was born in London in 1862, went to elementary school there until the age of thirteen when he and his brothers were sent over the North Sea to his old school in Karlstad in Värmland. At the age of seventeen he was back in London helping his father. It was a tough time, characterized by a servant girl: 'The hups and downs of this here family are something hextra hordinary.' He visited all parts of England selling wood and as quite a young man he went on business trips to Holland and traveled for the firm to Germany and Russia.

My mother was the daughter of Squire Bergman at Lundsberg (one of the small ironworks which was abandoned in the middle of the nineteenth century), and she brought elements of manorial culture to the marriage. She was an unusually well-balanced and strong human being, a devout Christian but not hidebound by dogma or by conventional ideas. My father claimed he was agnostic, but she considered him the most religious person she had ever met. I grew up with six older brothers and sisters and two younger, all of whom picked professions where human caring and cultural interest could be expressed. Our house was open for all our comrades and friends. Each of us played, or at least tried to play, a musical instrument. The spirit in our home was free, and lively discussions and music fill my memory.

During the first ten years of their marriage, my parents lived in London but moved to Sweden when my father left the family business. He became 'company promoter' with his work mainly in Sweden. In this role he became for a while a dominant power in the structural rationalization (to use the modern term) of Swedish economic life. An amazingly large part of Sweden's wood and iron industries, as well as mines, workshops, and transport industries passed through his hands.

He was not really interested in property and felt that children should not inherit. He was a promoter and designer of new things. He invested the first really big profit from his business in the Lundsberg school in 1896, a school for boys which was built on the unprofitable farm that he took over from his father-in-law. The idea behind this venture was not to copy English boarding schools but rather to give teenagers an alternative to the private board and lodging in little towns which he and his brothers had experienced. When the present schoolhouse was being built in 1906 he stood on its foundation and turned to the boys saying: 'What we are doing here is to be given to you chickens.' The school is still owned by ex-teachers and pupils. His reasoning was that if the school was good, then the old pupils would take care of it in the future, if not, it might just as well go down.

The economic crisis of 1907 dealt a serious blow to my father, leaving him heavily indebted. However, he continued his activities even if some people didn't raise their hats to him as high as before.

During the summers of 1910 and 1914 our family lived at the seaside in England. A journey through Germany and the sojourn on the south coast of England during the summer of 1914 brought the realization of what the war would mean. That journey home at the end of August between Newcastle and Bergen, on a blacked-out ship, overloaded with returning Russians, gave me an idea about the meaning of what was happening in 1914.

From the lower end of the dinner table I remember discussions among the grown-ups on electrosteel and the Svartö ironworks at Luleå, which my father had started, the problems of moving heavy manufacturing from Värmland to the west coast of Sweden, the floating of sawn wood over the sea, lignite in Devonshire, the competition between German and English shipping, and the problems of peace and war. It was fascinating and tragic to hear father telling about his contacts with Lloyd George and Sir Robert Cecil, Ballin and Helferich in his attempts to stop the war. He was banished from England.

In 1923 my father died, followed shortly afterwards by one of my sisters. At that time his children changed their family name from Olsson to William-Olsson to show everybody that we honored him in spite of his financial debacle.

I was born in 1902 – with club feet, a defect which nowadays is cured in the first month of a baby's life. Then it could have meant life as a cripple. At the age of three, however, I was operated on by one of the most skillful surgeons in Stockholm – on the table in our nursery. My mother served as anesthetist and 'her hand trembled'.

Some years of therapy by a skilled masseuse cured me and taught me not to be squeamish about pain. Although this defect was treated and healed from an external point of view it has played an important psychological role in my life. I could never prove myself in athletics, nor could I develop a sense of rhythm. The pronounced need to walk later in life, I explain as compensation for my childhood confinement. I even constructed and patented a shoe-brushing machine – it was probably also related to a fixation from that period. I could never be a mountain climber or a physical geographer because I have never been able to walk more than 20 kilometers.

My home gave me a basic safety and a deep sense of beauty, the impression that life was a gift, the habit of seeing the good within human beings and the

155

understanding that Western culture is rich and venerable. There was a great sense of humor in the atmosphere and that made one quite fearless in taking initiatives. The windows stood open toward Europe and the world.

School years 1913–21

At the age of ten I was sent to the Lundsberg school. The general feeling was that boys should not be spoiled in comfortable upper-class homes. Lundsberg was a tough milieu, and it placed great demands on the pupils – demands both from school work itself and from one's comrades. This period of my life is memorable mostly because of the experience of nature, the boys' collective, and the work discipline. The most positive experience of Lundsberg was the Värmlandian forest and sea landscape, its Nordic ski winters and the northern lights, and its genuine population of small farmers, charcoal burners, and iron-mill workers. The school collective was a hard grind, reflecting many of the peculiarities of that milieu. I took the initiative on promoting self-government among the pupils – my first experience with public performance of this kind. Discipline was strict: five to seven lessons each day became painful, although later one appreciates the amount of knowledge to be gained this way. However, it was the malaise I felt over school procedures together with my own difficulties with dyslexia which at the age of seventeen made me investigate literature on new ideas in pedagogy.

Undergraduate studies and teaching

I left school in 1921 with feelings of doubt, not being suited for natural science, nor for humanistic studies. Three choices of career presented themselves: priest, businessman, or teacher. The priesthood – a choice I would have made now – was eliminated because of a conflict I perceived between belief and knowledge. To be a businessman was tempting, but during the post-war years it was difficult, so I chose to be a teacher. This meant that I had to go to university in Uppsala and Stockholm which was financed by loan. Like most students I was poor and once I even pawned my cello, which felt almost like selling my soul.

My subjects were history and geography and I organized my program so as to shed light on the question which had puzzled me since 1914: how could people have been so stupid as to start the First World War? In history I focused on Bismarck, to get an understanding of the war, and the post-Napoleonic period after the Vienna Congress as a parallel to the Versailles Treaty and its consequences. In geography I focused on the Pacific from a geopolitical point of view – a study which later on made me appreciate, among other things, that the Japanese attack on Pearl Harbor was precisely what one should expect.

A year as tutor in Bergslagen, the iron country of Sweden, a Fall as substitute teacher at Lundsberg, and a voyage to England for language practice and visits to modern schools, interrupted my university studies.

Through my sister who was a Montessori teacher I contacted Carl Malmsten, a famous furniture designer, who had just started an experimental school for which he was making plans. I worked for some months on the program of this school and spent a summer in Sigtuna at his course for handicrafts, under highly competent leadership practicing my favorite hobby, carpentry. Here I also met my wife. Our marriage a year later became the cornerstone of my life.

In the fall I began teaching. Malmsten was the 'leader' of the school and after a year I was its headmaster. I had definite ideas about the program and my time was spent in intense and happy efforts to find ways in which children could be made responsible for their own activities – a trend which was later incorporated in the Swedish school reforms after the war. However, I did not carry sufficient weight to force my ideas through and opposition eventually made me leave the school in 1933.

Research

My first contact with scientific work was in Uppsala in 1922. The time of school and textbooks was over and we were thrown headlong into research through a system of pro-seminars and seminars patterned after doctoral dissertations. As a student of history I was appointed opponent and the subject was the first parliament under the first Bernadotte, Karl XIV Johan. After ten days of work at the university library I listed ten objections to the respondent's presentation. They were all discussed in the seminar except one which I had excluded because of inadequate proof. The leader of the seminar, Gottfrid Carlsson, concluded the session by saying that he had never before had the experience of finding all his own objections anticipated by a student. He just added one point and that was the one I had excluded. In spite of this encouraging result I still did not feel ripe for serious academic work. Ten years later I was.

At this time my health was poor – an ulcer, diabetes, and symptoms of Bright's disease at the same time. On top of it I limped because of tendonitis. Since then my health has been excellent. But I was declared an army 'wash-out' at military inspection.

Already as an undergraduate I had listened to Sten De Geer, known to American geographers for his studies on the Manufacturing Belt.[1] To the horror of conservative professors he sent his students out into the streets to study the distribution of milk shops. He mapped shop intensity by measuring and comparing the display windows along the shopping streets in Stockholm and also accomplished other kinds of fundamental work on cities and population. He was much before his time but many of his colleagues, experts in their fields, judged him as more or less crazy. He has always been an inspiration for me. He moved to a chair in geography in Gothenburg where he died prematurely.

Hans Ahlmann was appointed in Stockholm. He was a morphologist and glaciologist who was also responsible for human geography. Hesitatingly he entered the field of urban geography but switched to high gear when he came across the nine-volume work *The Regional Survey of New York and its Environs* 1924–28 by Haig and others.[2] Town planners and architects were interested and the best place to execute a similar investigation was at the department of geography, where earlier De Geer had been active. I joined a group of three students who together with Ahlmann published the first volume of the Stockholm investigation, *Stockholms inre differentiering*.[3] Primary material was data on every house, household, shop and all gainfully employed population in various occupations.

There were two ways of treating this wealth of information: either to concentrate on the houses or on the people. The latter way was chosen since the buildings are really only instrumental. The leading principle was empirical – to go down to location and site on the one hand, and to individuals and firms on the other, mapping them to analyze the conditions of every branch. Generalizations were avoided until the final stages of summaries.

During this work we collaborated with town planners and through them with foreign movements – for instance the German *Bauhaus Bewegung*, from which German experts came as refugees when Hitler took over. For me this collaboration with architects and town planners was easy since my eldest brother, Tage, was one of the leaders. He was a teacher in town planning at the institute of technology and later head of the town planning office of Gothenburg, and from him I gained the necessary respect for the work of town planners.

Already at this stage I made a lifelong plan of research: (1) the shopping streets of Stockholm; (2) its town area and suburbs; Greater Stockholm and its development; (3) its relationship to the rest of Sweden; (4) different parts of Sweden; (5) towns of Sweden; (6) European towns and countrysides; (7) total picture of Europe; (8) population of the world; (9) description of the world (geography).

In 1937 my doctoral thesis, *The geographical development of Stockholm, 1850–1930* was published.[4] In this book I formulated a theory on the inner differentiation and changes in the city (points 1 and 2 above). Those responsible for the long-term planning of Stockholm asked me to broaden my outlook toward the future and offered me a post at the statistical office of the town with a competent statistician as collaborator for demographic calculations. My formulation of the program was to investigate the interplay between Stockholm and the rest of Sweden; their wish was to get a population forecast for Greater Stockholm. The result was published in 1941,[5] *Stockholms framtida utveckling* (point 3). In this treatise I formulated a theory on the interplay between town and countryside. It was based on the concepts of basic and non-basic activities. These concepts are rightly rejected as short-cuts to population forecasts but here we are not aiming at projections but at a careful study of the function of well-defined places and regions.

From my previous work I knew that even with such a vast quantity of material it is possible and necessary to go into details since generalizations at an early stage in social research blur the discussion and distort results. By necessity official data on towns and regions are such early generalizations of patterns which the geographer has to investigate. Even at the risk of being tiresome I list here some well-known censuses since in their regions and classes a wealth of information is hidden which most regional geographers miss. It is too hard to delve beneath the official numbers.

Some of the main statistical series used here are: the development of population in all parishes (2,400) since 1805; size and occupational structure of every agglomeration with more than 200 inhabitants (1,210); the countryside divided up in sixty regions and the development and structure of each one of them; numbers of employed in every manufacturing branch separately accounted for in every place with more than ten employees. Missing generally in Sweden as well as in other countries are data on people in the civil service and employees in a vast and growing horde of social and semi-official occupations which have been playing a key role in the development of big cities. The chief way of getting data on them is from the pay rolls.

Using these sources it was possible to characterize and classify all rural parts of the country and their development as well as every urban place. The latter were classified according to their dominating economic and demographic features: mining and metal industry towns; wood or forestry industry towns; wood and metal industry towns; textile industry towns; other one-sided manufacturing towns; diversified industrial towns; service towns.

For the planners in Stockholm who wanted a population forecast a basic product was delivered, the main categories of which were major sectors of civil service, financial activities, leading cultural functions, wholesale trade, exclusive retail trading, printing, textiles and dressmaking, fashion industries, and machine industries carrying into effect new inventions. The nonbasic activities were treated as functions not only of the basic ones but also of general sociological trends. Demographers had all data necessary on the families and other dependants ready at hand.

In 1940 the economic problems of *Swedish Norrland* became alarming. I submitted a proposal to the government for an investigation, and it was approved. This was an excellent opportunity to test my theories on two-thirds of the surface of Sweden with different landscapes, countrysides, agglomerations, and towns.[6]

Following the publication of this work one town after the other asked for my help, and with the aid of students I analyzed their economic structure and development and made population forecasts which thirty-five years later were checked against reality and verified (point 5).[7] These forecasts were not made through extrapolation of population graphs. They were calculated with alternative and realistic conditions and based on the theory of basic and nonbasic activities.

At this stage it was evident to me that human as well as physical geography has to be based on empirical research which necessitates a rejuvenation of cartography and an overhaul of international terminology. Since we cannot accept places and regions delimited on political and physical grounds we have to fall back on our own maps and since we have to cope with a multitude of languages we had better agree on an international terminology.

Resistance at home showed that the time was now ripe to go abroad. If one's results do not happen to agree with ongoing political choices, then a fresh consultant team can be called in to document the government's viewpoint. This is what happened in my case, I was not even consulted, and my results were ignored because they pointed to a reality which was in potential conflict with politically stated goals.

The Swedish Airlines asked for my assistance in planning their service networks in Europe and gave me and my assistant free flights to all parts of Europe. This situation did not last long but it allowed me to start the work on my *Economic Map of Europe* which was published in 1953 (point 6).[8] The work on this map took several years. Colleagues in foreign countries collaborated generously but I was also made aware of political resistance. Just as at home I had learned that unbiased social research was restrained if its results did not agree with predominant political interests, I now met criminal methods to suppress unwelcome knowledge.

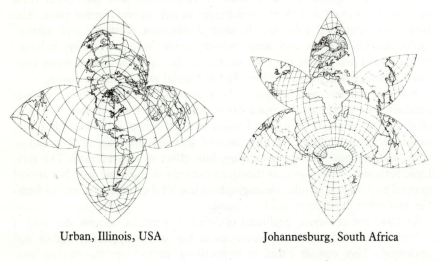

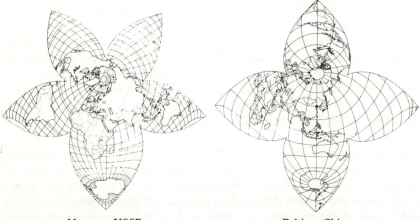

Urban, Illinois, USA   Johannesburg, South Africa

Moscow, USSR   Peking, China

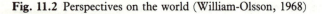

Fig. 11.2 Perspectives on the world (William-Olsson, 1968)

Some time after the IGU Conference in Washington in 1952, where I had shown my map of Europe in proof without its explanatory text, my office and that of my secretary were burgled. Nothing was stolen but every paper had been checked, evidently to find out how I got hold of certain data shown on the map. In one way it was flattering but it was disheartening too. My calculations had evidently been dangerously correct but I had to exclude the Soviet Union from my concept of Europe. Contacts with colleagues from some other countries behind the Iron Curtain were also obstructed since a letter from me was a serious risk for them. One of them disappeared for some years and another beseeched me not to write to him.

For eight years I was the chairman of an IGU Commission on a world population map in the scale of 1:1,000,000, originally an idea of Sten De Geer's. From my point of view it was a pleasant way of approaching point 8 in my program.[9] Another was the bringing together of social and economic statistics by province from all European countries into a manageable form. A first edition was published but the project was too large to follow up, in particular since the statistical methods were revolutionized by new data techniques. A way of approaching point 9 was the construction of azimuthal equal-area projections adapted to the fact that every country is the centre of its own world.[10] Even if people are content with maps showing their world with fantastic distortions of surfaces, distances, and directions, we have to give them world maps showing correct surfaces and at least approximately correct distances and directions from their point of view.

My own research centered on international interests from 1950 on but I returned to Stockholm in order to prepare for the International Congress which was held there in 1960. Between 1965 and 1969 I also published several papers on the planning of Stockholm and Sweden. During the last ten years I have tried to solve the problem of regional geography by writing a book on Europe (point 7).[11]

Social context and work environment

The above sketch on my research activities needs to be supplemented by at least some information on people, comrades, and fellow workers as well as antagonists.

In Uppsala students from different faculties used to meet regularly at mealtimes, and this was an excellent organization for widening one's outlook on life. The freedom and personal responsibility one felt in such settings was a relief after the regular lessons at school. At the University of Stockholm the pattern was similar. We read and discussed face to face and in regular seminars, criticizing and helping each other. Of particular importance for me was Ivar Ekstedt, a brilliant person who gave up his studies to be cured from alcoholism and became a smallholder. During and after these years Hans W:son Ahlmann was a reliable adviser and helped me especially in establishing contacts abroad.

In 1937 I was appointed 'docent'. A docent was a person successful in

161

research who had the competence required to lecture at the university. He was paid a small salary and kept for the purpose of breeding new professors. I became secretary of the Docents' Union. We organized a lunch table to which intellectuals from all fields were welcome and I persuaded my colleagues to participate in a survey of all Swedish docents on the need of expansion in their different fields. The aim was to show the necessity of enlarging the Swedish university system instead of reducing it as suggested by the authorities.

In 1946 a professorship was established for me at the Stockholm School of Economics. Now I had opportunities to put my planned program into practice. The department of geography, in beautiful premises, expanded, attracting teachers and students. Besides myself, there were one or two equivalent professors. I was never 'head' of the department. It was organized on an equal footing. As far as possible students took part in our research. An important part was played by refugees from the Baltic states who were paid by the government for their work at the department. They made it easier to handle the vast amount of material which we had to manage. One of them was Janis Rutkis, a refugee from the University of Riga, who measured the local relief by covering Europe with circles of 100 square kilometers (r 5.67 kilometers) and from large-scale topographic maps noting the highest and lowest points within each one of them, the difference between the two values being the relief (*lokalrelief* or *relief-energi*). The result was mapped at the scale of 1:4,000,000.[12]

At the IGU Conference in Stockholm 1960 I thought it possible to enlarge the department to cover regional knowledge of large parts of the world. This came to nought and when I retired in 1969 the department was drastically reduced and so was the study of geography. There were several reasons for this change, the principal being the general trend in all social sciences toward positivism and quantification. Geographers were also influenced by this tendency to the point of annihilating the subject. During two critical decades it was almost impossible to forward arguments based on verbal reasoning and thereby the whole subject was suppressed. *Ymer*, a main Swedish periodical for geographic debate was changed into a yearbook, which caused me to leave the board of our Geographical Society.

I also diverted my attention from my main line of research, compelled again to enter the discussions on planning in Sweden in general and Stockholm in particular. The planning was warped by fantastic ideas on future growth and affluence, supported by calculations of the type mentioned above. Empirical research was declared irrelevant and social scientists stood by, blindfolded by their theories.

Creativity

Psychologists dealing with school children roughly distinguish between receptive and creative talents. When having to solve a problem with many solutions the receptive child will quickly find one of them and finish it. The creative child will be slow but will find many solutions. At school and later in life the receptive child will be successful. The creative child is likely to be unhappy at school and as an adult will show more interest in ideas than in money. This can easily be exemplified with inventors unable to exploit their own inventions. Applying this distinction I definitely belong to the creative category.

To understand the individual we must take into consideration inherent gifts as well as environment. This is the excuse for writing so much on my family and home. Evidently they were exceptionally favorable for a person with a creative mind. The stress was on ideas, to see through the phrases. Money was important but that which could not be bought with money was more important. This attitude is essential for creativity; so is belief and self-confidence.

My family history is certainly dramatic, with a tradition of seeking new ways, new creations, and risk-taking. I have obviously taken over those interests. It seems to me that my basic sense of security comes from my parents' ability to take both adversity and success with the deepest calmness. My mother told me that my father slept like a peaceful child even on those nights when his financial situation broke down in 1907.

This freedom from success as a measure of value, from profit-hunger, is, I think, the key to creativity. It is rooted in reality and fantasy – imagination. The hunger for money is an enemy to new creation, attachment to authorities another. One who was brought up in a milieu where the family father or teacher or the superiors were regarded as more or less infallible directs his endeavor to imitate these authorities. This structure might be helpful to the beginner, but is generally an obstacle to the creativity of the individual. The present arrangement within Swedish university studies, whereby new students are allocated to inexperienced teachers and directed to read 'basic knowledge' in more or less casual compendia with mechanical tests (exams) is an illustration. A youth, eager to learn, who has been processed through the school-mill should, of course, be allocated to the most clever teachers who by experience know what research is, and he ought to get heavy scientific literature in his hands.

The Lundsberg time was filled with work, but at the same time I knew that the school was an experiment that naturally could and should be improved. At the universities in Stockholm and Uppsala individual teachers were inspiring, but above all, studies were still free, organized so that the students themselves had to search their own way – with necessary assistance but never 'mastering'. The completely informal discussions with fellow students were probably the best inspiration we could have had. It was natural for a young man who knew that teaching was a field in need for remodeling to become a teacher. Nor was it surprising that fresh from the university I took part in writing a program for a radically new school; it seemed to be an obvious consequence of years of discussions in my youth. It was a blessing in disguise when I had to leave that school, the reason being that I was more interested in ideas than in individuals, a disposition which later on in life was a disadvantage for my pupils, and for myself a cause of self-reproach. Could it be that such a fundamental character trait explains why creative professors sometimes have few pupils?

My interests as a geographer were inspired by the activities of my father. He gave me an understanding of the fantastic drama which remodeled our society from self-sustenance to exchange economy. He also inspired in me a sense of responsibility in this drama. When I met it in scientific literature and discussions at the university I sketched out the plan previously described.

Geography and research

The word 'geography', description of the world, is adequate. In this capacity it has its special place among the sciences. Its fundamental purpose is to reduce this huge object of concern to human proportions. To do this we divide the task into sectors and branches, and thus circumscribe neat and manageable problems. The difficulty is to join them together again in order to get the overall picture of countries, continents, and the world. We need such a picture, preferably before we destroy it through ignorance or in fury blow ourselves and our planet into pieces.

The trouble with geography is that so many of its practitioners become specialists. This is not surprising, it is inherent in science, but if we forget that the fundamental function of geography is to give competent surveys of the world, we have to pass the term 'geography' over to those who are geographers in the literal sense.

In all geography we encounter the problem of reduction and generalization. These problems must be solved in accordance with the scale chosen. If the scale is altered we have to alter the generalization too. This is evident in map-making, but it is just as important in the text. Regional geography has neglected this, and that is probably why it is so often dull reading.

Geography must also be considered in terms of its relationship to time, space, and language, more especially the mathematical language. In my previously mentioned work on Stockholm, an effort was made to gain a time perspective on the processes of change by constructing a series of pictures of the city at fifteen year intervals. It seemed feasible to assemble these pictures in a film that should show the differentiation of the city both in space and time – how the 'building areas' spread, how the city region pushed away the central settlement, how the old wooden houses in the periphery were wiped away by apartment blocks and factories. . . . With the aid of *Svensk Filmindustri* we produced such a film, one picture for each year between 1880 and 1930.

I have seldom been so disappointed. One could see *nothing* of the changes. It was too much to comprehend at one instant. Immanual Kant was right in his assertion that the human psyche is incapable of grasping both spatial and temporal differentiation at the same time. One should not try to make moving models of big contexts. We must distinguish between history and geography, which naturally does not mean that one shall prevent oneself from letting the one 'conceive' the other.

Everyday experiences and concepts in one country may be rare in other countries. This makes it necessary to define our concepts and make up our vocabulary with internationally valid terms. We have to go back to the old definition of geography as the interaction between nature and man and forget everything about regions.

Our first concern with nature will be the landscape, just as a newcomer sees it. This will call our attention to light, landforms, water, and vegetation, not as scientific problems but as impressions perceived by unsophisticated onlookers. When we have visualized the landscapes it then becomes meaningful to give our scientific explanations and open the door to the fascinating game behind what we see. This will lead into neighboring sciences.

Among human geography's primary concerns are the different ways in which mankind supports itself. The fundamental difference is between those who gain their living on fields, forests, and water, spreading out over the surfaces (areal production), and those who work in shops, factories, and offices and settle around them (point or *stigmal* production). The two categories are not the same as rural and urban activities but properly scrutinized they answer the riddle of urbanization, depopulation, and growth and decline of cities.

It is vain to organize the material in regions. Economy, literally speaking, refers to men utilizing their part of the world. Man is active and it is an appalling simplification to talk of regions as if they did this or that, as if they were successful or suffering. Such simplifications are common in political talk and often used by demagogues, but in social research they have to be condemned. It is always men – human beings – who are active.

To form a basis for generalizations we must go to the very root of our economy, which is human activity. Once we discard the notions of region, we can apply the concepts of basic and non-basic activities and thereby explain localization patterns. So doing we find the types of settlements which are characterized and analyzed in my early Swedish work and accounted for in my maps of 1953 and 1974 mentioned above. They all can be derived from the two digit level data of modern census reports and show basically the categories of places which the visitor encounters: villages, mining towns, manufacturing towns of various kinds, small and big service towns, and capitals. To describe the history and explain the localization of these agglomerations can thus become meaningful, just as it was meaningful to explain the different types of landscapes in which they are situated and through which the visitor has traveled. Here again one touches on the domains of neighboring sciences, particularly that of economics. Economists have done much work on the location of manufacturing and other industries. Gratefully accepting their fundamental theories we must share what geography can offer, i.e., information needed for the fitting of their abstract surfaces into the world of realities.

Notes

1. De Geer, Sten (1927) The American Manufacturing Belt, *Geografiska Annaler*, **9**, 233–358; also his *Befolkningens fördelning i Sverige. Beskrivning till karta i skalan 1:500 000*. Wahlstöm & Widstrand, Stockholm (1922) (The Distribution of Population in Sweden. Description of a Map at the Scale of 1:500 000) and Storstaden Stockholm ur geografisk synpunkt, Greater Stockholm in Geographical Perspective *Svenska Turistföreningens Arsskrift*, 1922.
2. Haig, R. M. (1927) *Regional Survey of New York and Its Environs. Vol. 1, Major Economic Factors in Metropolitan Growth and Arrangement*. McCrea.
3. Ahlmann, H. W:son *et al.* (1934) *Stockholms inre differentiering*, (Internal Differentiation of Stockholm) Stadskollegiets utlåtanden och memorial, Bihang 51, Stockholm, pp. 67–98, 99–120.
4. See Selected readings, 1937.
5. See Selected readings, 1941.
6. Utredning angående Norrlands näringsliv. Förberedande undersökning verkställd av 1940 års Norrlandsutredning, (Investigation on the Economic Life of Norrland. Preparatory study by the Norrland Commission of 1940) *Statens offentliga utredningar*, **39**, Stockholm.
7. Nya och gamla befolkningsprognoser, (New and Old Population Forecasts) *Kommunaltidskrift*, **3**, 1970.
8. See Selected readings, 1953 and 1974.
9. See Selected readings, 1963.
10. A new equal-area projection of the world, *Acta Geografica*, **20** (1968), 389–93.
11. See Selected readings, 1981.
12. See Selected readings, 1974.

Selected readings

1937   *Huvuddragen av Stockholms geografiska utveckling 1850–1930*. Akademisk avhandling, Stockholm. (The Geographical Development of Stockholm 1850 –1930.) Summary essay (1940) Stockholm: its structure and development, *Geographical Review*, **30**, 420–33.

1938   Utvecklingen av tätorter och landsbygd i Sverige 1880–1935, *Ymer*, **58**, 243– 80 (The Development of Urban and Rural Settlements) in Sweden, 1880– 1935.)

1941   *Stockholms framtida utveckling*. Stockholms kommunalförvaltning, Stockholm. (The Future Development of Stockholm.)

1946   *Ekonomisk-Geografisk karta över Sverige*. Nordisk rotogravyr, Stockholm. (An Economic Map of Sweden.)

1953   *Economic Map of Europe* (In English, French, German, and Swedish editions) *at the scale of 1:3,250,000*. Nordisk rotogravyr, Stockholm.

1960   *Stockholm: Structure and Development*. (IGU Congress, Uppsala), 96 pp.

1963   The commission on a world population map: history, activities, and recommendations. *Geografiska Annaler*, **45**, 243–50.

1974   *An Economic Map of Europe West of the Soviet Union* and *A Map of the Relative Relief of the Western Half of Europe. Scale 1:4,000,000*. Nordisk rotogravyr, Stockholm.

1975   A prelude to regional geography. *Geografiska Annaler*, Ser. B., **57**, 1–19.

1975   Vår världsbild, *Ymer*, **95**, 174–88. (Azimuthal equal-area projections.)

1981   *Europe: Nature, Population, Function. A Regional Geography* (in preparation).

Chapter 12

Fig. 12.1 Hans Bobek, Austria

From the city of Wittgenstein, Freud, Schoenberg, and other luminaries comes Hans Bobek who, perhaps more than any other twentieth-century geographer deserves the title of Renaissance Man. Born into a family of highly diverse background his boyhood trips throughout the Austro-Hungarian Empire awakened curiosity about landscapes and people at an early age. To scale the heights and decipher the structures of Alps and Elburz, to unravel the dynamics of cities and hinterlands, to 'read' the landscapes of Occident and Near East in terms of social structures and cultural values – for such remarkable achievements has the name of Bobek become renowned throughout the world.

A privilege indeed it was when in 1965 he graciously welcomed me to Vienna and listened to my queries about social geography. We discussed French and German traditions, rent capitalism, and he opened my eyes to the geographic significance of Judeo–Christian value biases. Little did I ever imagine how colorful the social geography of his own life experience had been. Fifteen years later he reveals the story and my only regret is that it cannot be presented here in German, as he wrote it, together with the full impressive list of his published works.

Some comments toward a better understanding of my scholarly life-path

Introduction

At first glance, a comprehensive record of my written work could give the impression of a certain lack of purpose, for I have dealt with many and very different themes. I would like, therefore, to touch and enlarge upon two aspects of my academic life-path.

First, there were significant external impulses which have caused radical shifts and changes in my scholarly work. Nevertheless, there has been an internal continuity in two completely different lines, which can be understood only in the context of geography's role in those days, quite unlike the specialization common today.

167

Early stimuli at home and at school

I was born on 17 May 1903 in Klagenfurt, in the Kärnten region of Austria. My father, Karl Bobek, came from Brandeis-am-Adler in eastern Bohemia. The son of a carpenter, he attended the Realschule and was a book-keeper, later one of the chief inspectors of the South Railway Co., which controlled an important part of the old monarchy's railway network. My mother Agnes (maiden name Petz) was the daughter of a farmer in Brunndorf, near Marburg an der Drau (southern Steiermark). My grandfather also owned an inn, and became the mayor of this aspiring place which was later incorporated into the City of Marburg.

I was the youngest of four children whose birthdays were only five and a half years apart. I had but one sister, two years older than myself. The next closest in age was my brother Josef who shared with me a common love for drawing and painting; sketching people or landscapes was our favorite preoccupation in school and especially during the holidays. I was serious enough about it to consider entering the Academy of Art. Of course, such a 'hopeless' intention would have caused difficulties with my father, who already bore the considerable financial burdens of supporting academic studies for three other children and of recently building a house. So I forgot about it. I still do not doubt that this talent which I later had to abandon under scholarly stress was relevant and contributed to my pleasure and ability to observe. Throughout my entire life I maintained a strong inclination toward the visual and a rapid comprehension of what is essential. But perhaps much deeper and all-embracing is the mental stimulus and enrichment which grew out from two very different circumstances.

First of all, since my father was a railroad employee, he was entitled to a certain number of free tickets for himself and reductions for his family. As a result, our family undertook a fair number of journeys throughout the years, even after our move to the distant Innsbruck (Tyrol). These travels included the annual summer vacation with my maternal relatives in Brunndorf, from whence we embarked on other visits to more distant relatives in the area of Marburg. Less frequent were visits to paternal relatives in Brandeis-am-Adler, or other places in the Bohemian area, or even further in Vienna. We also traveled to other areas in the empire, specifically to the Adriatic coast. These early experiences with very different landscapes and people were important since both my parents came from areas of mixed languages – we had Czech relatives on my father's side in eastern Bohemia, and those of my mother in the German–Slovenian population of southern Steiermark. Understandably, this had led to several mixed marriages. As a young boy, I may have been shocked to be greeted by a distant relative in an unintelligible language, but was nevertheless interested later to watch the behavior of people communicating with one another though they spoke different languages. I would observe their professional positions and their behavior concerning the nationalistic conflicts which were characteristic for the period when the Austrian monarchy was waning. After our move to Innsbruck in 1909, all these experiences, as well as my observations of people in Tyrol who had a completely different mentality, triggered many incentives to think and all in all brought me great enrichment.

University and first scholarly activity in Innsbruck

I attended the Volkschule and the humanistic gymnasium in Innsbruck, passing the final exams in 1921 with high honors. At the university I chose my major fields in history, geography, and social science. There were five historians (O. Stolz, H. Steinacker, H. Wopfner, R. Heuberger, and F. Lehmann-Haupt), but only one geographer (J. Sölch) who, after being active in the middle school for some time, had just been promoted. He was one of the last students of A. Penck, and taught primarily geomorphology and regional geography.

Although I took my studies very seriously, I was also a member of a student fraternity which had a liberal, nationalistic[1] morale, and cultivated music through choral singing. At diverse meetings of the fraternity I met a fair number of people from Innsbruck society. I also participated in their various summer trips and winter ski tours, and so became acquainted with certain sections of the Alps.

Originally I wanted to write my dissertation in history. I turned to Professor Steinacker for advice, since I held him in high regard. He gave me a theme entitled 'The development of the Habsburg domains in the Alsace region before the French occupancy', but I realized as I was doing my preliminary research that this topic had already been studied for some time by a French Abbé. I turned to Professor Sölch when Professor Steinacker, for some reason, would not propose another topic. Professor Sölch's suggestion of 'Natural frontiers' was indeed extremely topical in the post-war era. However, I decided that the problem of political boundaries could be successfully approached only in the context of political factors and power relationships, not really in terms of spatial characteristics. Thus I offered to do another topic which dealt with Innsbruck and, to my delight, Sölch agreed.

Innsbruck at that time was a city of about 50,000, the capital of its surrounding region, the Tyrol. It was a fascinating object to study in every respect. Urban geography was just beginning to attain some stature. The discipline was still influenced chiefly by O. Schlüter's *Kulturlandschaftslehre*, so the city could also be explored as part of the cultural landscape – or, in fact, as a landscape of its own – because of the uniqueness of its phenomena and the wealth of its patterns.

For me this view was inadequate, since from my early youth I had been trying to study people and their activities. By reading French works, such as R. Blanchard's book on Grenoble (1912), and above all, through stimuli from the discipline of economics (M. Weber), I found that I was not alone in this view.

I was primarily interested in those activities which lead to the rise of cities. My research on this problem suggested that in any given society with a particular division of labor, certain activities would develop which were primarily oriented toward supplying other people with goods and services, and would therefore be forced to concentrate on specific and easily accessible locations. With these and some further considerations about cities and hinter-lands – which were originally meant to form the theoretical introduction to my thesis but had to be published separately because of its length – I antici-pated the nucleus of Christaller's *Central Place Theory*.[2] Christaller admitted this himself in a footnote in his book. My thesis on Innsbruck was accepted and published in 1928; it received lasting acclaim as well as some criticism from the parties concerned.[3]

Had I continued to work on this theoretical foundation life would have been much simpler, although I would have needed some further encourage-ment to do so. However, things developed in quite another direction. Soon after I had passed the examinations in history and geography I took over a position at the secondary Realschule. After one year in this position, Pro-fessor Sölch, before his departure to Heidelberg to succeed Alfred Hettner in 1928, unexpectedly arranged for me to be appointed as assistant at the Geo-graphical Institute at the University of Innsbruck to replace Dr Kinzl, who followed him to Heidelberg. He also cautioned me: 'If you expect to succeed in academic geography you must work in geomorphology.' This warning was not exactly the encouragement I needed but it was quite in line with the predominant view among German geographers at the time, viz., that one should have a thorough background in geomorphology.

This earnest advice was not really a hard blow for me then because I had meanwhile done quite well with publications. Very probably I would have needed new discussions and strong stimuli to continue in the same vein. However, neither the new chairman, Friedrich Metz, nor other people at Innsbruck, got active in any way in this direction. In fact, I still felt motiv-ated because of many mountain trips and mostly because of a 'Glacier' course led by the famous Professor Sebastian Finsterwalder on the Berliner Hütte in the Zillertaler Alps in 1925. The following summer, after my pro-motion, I became temporary assistant to Richard Finsterwalder – Sebastian's son – when he began his photogrammatical measurement of the Zillertaler Alps for a new map for the German and Austrian Alpine Society. Through this work I learned much about particular parts of this mountain range, and I also gained some training in the method of terrestrial photogrammetry, which proved very useful later on. I thus followed Sölch's advice and began to study Alpine geomorphology as well as the geology of the Zillertaler Alps. Already I felt that this mountain range would be my next research project. With this major shift in my major field of study, I could supplement the work of Pro-fessor Metz, who had little interest in the physical branches of geography.

In 1929 I married Helene Procopovici, a young but successful teacher in English. Helene's Romanian heritage came through her father, who orig-inated from Czernowicz and had worked in the Austrian District Adminis-tration of Rovereto in Tyrol before his early death. In the summer of 1930 we visited the extended family, some of whom lived in Czernowicz, and some in the area around the city. I thus had the chance to observe the extensive social stratification in Bukowina: the intellectual upper class and landowners were Romanian and German, tradesmen and middle class were Jewish, and the peasants were Ruthenian (Ukranian). I had been able to make equally interesting observations about settlements and lifestyles of various nationalities during an earlier trip through large parts of Siebenbürgen.

Berlin (1931–39/40)

In the winter of 1930–31 I received an unexpected visit from Professor Norbert Krebs, who abruptly asked if I would be prepared to take on a job as assistant in the geographical institute of Berlin (of which he was director). One out of his six positions was vacant due to the departure of W. Panzer to China.

My wife and I deliberated at length, weighing the undoubtedly better academic possibilities in Berlin against the favorable living conditions in Innsbruck (we lived in a beautiful area of the city in my parents' villa). I nevertheless opted for Berlin, although it meant a lengthy separation from my wife and my new-born son.

My commencement in Berlin started out on a melancholy note, due to a blunder I made in my first lecture. Professor Krebs had completely neglected to make clear to me that I was invited to Berlin as an urban geographer, whereas the majority of his assistants worked in morphology. I did not, however, want to discuss my past research published three years previously, but rather my recent work in the Zillertaler Alps, which involved trying to clarify the role of fluvial and glacial erosion. I was interested in intensifying research by means of the systematic construction and evaluation of numerous exact valley profiles on the basis of the excellent relief sheets of the already mentioned photogrammetric map. To my great shock the lecture received sharp criticism, because people in Berlin (where A. Penck still lived – he was also present at the lecture) still abided by the concepts of his famous work 'The Alps in the Glacial Age'.

Although both my early works – on Innsbruck and on the Zillertaler and Tuxer Alps – had been printed in the distinguished scholarly series, *Forschungen zur Deutschen Landes- und Volkskunde*, and were both well accepted by the public, they did not gain me the official recognition as an academic lecturer that I had hoped for.[4] I had meanwhile either published or prepared several other geomorphological works in Tyrol, most significant of which was a study which delineated a totally new representation of the decay of the Würmglaciation in the Inn Valley on the basis of a major system of glacier-rim patterns.[5] I proved that the 'Final Glaciation Period', as postulated by the much respected geologist O. Ampferer, was unnecessary. My interpretation has been universally accepted since then. This work, which also had been approved by A. Penck, was the positive basis for my 'Habilitation' at the Faculty of Natural Sciences of the University of Berlin by the end of 1935. This, however, did not mean that I was entitled to hold lectures: new regulations issued by the new National-Socialist Government required also a positive judgement about one's political trustworthiness which, in my case, obviously did not exist.

Already before this unpleasant event another, most important turning point in my scientific path had taken place: the enlargement of my field of research into the Middle East. The opportunity to take research trips had been a major incentive for my move to Berlin, since in Germany at that time, a great importance was placed on successful field trips abroad. My choice of the Orient was to some extent an intuitive one, stemming from the desire to learn about one of the oldest civilizations and the various influences it had on the countries and populations of the Austrian–Hungarian monarchy. Iran I chose particularly because few geographers had worked there. Luckily, I also found an excellent professor of Persian who taught me the language in such a short period of time that I was soon able to travel and converse on necessary questions.

By 1934 I was already able to carry out a seven month research trip to Iran, with the support of the Prussian Ministry of Science. There was to be a second extensive journey to Iran in 1936, but due to a nocturnal flood in the mountains which carried away all my equipment, it was shortened to two months. An expedition with some students, supported by the German and Austrian Alpine Society, was conducted into the mountains of the southeasternmost part of Turkey in 1937. These journeys were primarily devoted to mountain ranges which were virtually unknown at that time, and especially to the research of their present and former glaciation. At the same time, whenever feasible, cultural and other geographical studies were pursued. It was still possible to have the conclusions of these studies published before the war or at its very beginning.[6] All these studies and publications led to more journeys after the war, and to critical insights which had not been achieved previously.

During this entire period, my interest in urban geography had continued, although to a lesser extent. While still in Innsbruck I produced a critical literary report on 'The North American small towns and their development'.[7] To my knowledge, I was the first to recognize the importance of Christaller's 'central place theory' although I already pointed out some specific weaknesses in his work.[8] In my lecture at the International Geographical Congress at Amsterdam[9] I tried to prove that Christaller's 'central places' could not be regarded as ubiquitous phenomena, but were primarily limited to the West European–North American civilization; whereas in other, older cultures, different city models are prevalent whose functional character and relationship to the surrounding countryside is largely dependent on the respective social structures. During that time I published other studies which certified my persistent interest in cultural and social problems.[10]

Through the intercession of one of my wife's old friends, who had some links with the National Socialist Party, the obstruction to my tenure was lifted at the end of 1937, although I myself did not join the party. In 1939 my appointment to the position of 'Diäten Dozent' (reader) ended my duties as administrative assistant at the library of the Institute, and various other educational training courses so that I could have more time for my own scholarly works.

The war years (1940–45)

In the middle of 1940 I was drafted into the army, and after some short basic training I was transferred to the military-geographical department of the High Command of the army (OKH-Mil-geo). Here I was entrusted with the directorship of a work group which was charged with the responsibility of reconnaissance and military survey mapping in Middle Eastern and Near Eastern countries. Over the next three years, my co-workers and I compiled an extensive dossier of both cartographic and literary information on more than a dozen countries; booklets were printed in great numbers for possible use by troop commanders. This period gave me the much-desired opportunity to study almost the entire collection of Oriental literature available in Berlin, not only from a military standpoint, but also (in my free time) according to my own scholarly views. I was able to test – in a more specific and exact context for the entire Orient – those trains of thought which I had expressed at the Amsterdam Congress. In this manner I discovered for the first time the extraordinary and fundamental importance of the various social structures for the shape and functioning of towns and settlements. I produced an extensive manuscript entitled 'Social landscapes of the Orient.' The research I conducted for this study, which regretfully has never been published in its detailed form, was in fact the precursor of those concepts which I later called 'social geography.'

In the middle of 1943, I voluntarily returned to the troops. The special tasks to which my work group had been assigned were finished, and had in the meantime become irrelevant in the light of unforeseen changes of the war situation. It might be interesting to learn, though, that in the summer of 1942, when the German troops had reached the edge of the Caucasus, an order of highest urgency requested my group hurriedly to prepare a study on 'The landpassages to India'.

Now, however, Allied forces had landed in Italy, and it appeared that while the appropriate 'Mil-geo' studies had been produced for North Africa, there were none for Italy. So I proposed to send out several 'motorized' Milgeo groups who could, as quickly as possible, collect the necessary information. The proposal was accepted but, contrary to my expectations and my good knowledge of Italian, I was not chosen to be among the group. I was deemed dispensable, so I reacted by requesting a return to the troops which had to be granted.

As long as I was a military official with Mil-geo I had a captain's position but now I found myself working again as lance corporal of the survey troop. We were located at the front in northern Russia, which at the time ran slightly east of the Lithuanian boundary line. Our job was to measure specific points in the large forests, immediately behind our front, which could serve as orientation points for our artillery in case of further retreat. Although – or rather because – I had to call myself with good reason an idiot, it was a time for much beneficial self-reflection.

In May 1944 I was claimed by a small special unit, founded by Major Schultz-Kampfhenkel to provide quickly needed information to troops about difficult tracts of land. This unit consisted of several competent scientists – geographers, vegetation experts, and soil scientists – all familiar with the interpretation of air reconnaissance. My specific task was to handle the cartographic representations of our results. Our first assignment was concerned with the so-called 'Wedge of Kowel'. Soviet troops had pressed back the German Front in the area of the Pripet swamps in the form of a deep wedge, and this was targeted for elimination. Within one week our maps had marked several ridges and other areas suitable for tanks. Other assignments of this small detachment took me for special duty to Yugoslavia and northern Italy, and finally to Prague in the spring of 1945. The war ended for me when I was taken prisoner by the Americans, and I stayed in various camps along the middle Rhine until my release on 12 December 1945.

This experience of almost one year in that research detachment had much scientific value for me particularly because of the collaboration with ecologically oriented botanists such as Dr Heinz Ellenberg and the geographer, Dr Josef Schmithüsen. These discussions contributed greatly to my awareness of how important geo-ecological research is in the geographical frame of reference.

The immediate post-war period (Freiburg i.B., 1946–48)

I could not return to Berlin for several reasons. Our home was destroyed, my position at the university had been terminated, and the confused conditions and heavy damage in the city robbed it of any attraction. My wife had already left Berlin with our son in 1943, since she had luckily been offered an assistantship at a branch of the famous school in Salem in the Bodensee area. Thus her little makeshift apartment, located in a hamlet near the school, had to accommodate me as well for several months. In May 1946 I was offered the geography chair at the University of Freiburg in Breisgau, since the French Zone Authority banned Prof. Metz, who had long since moved from Innsbruck, from teaching. The Geographic Institute was, however, totally destroyed and I had to carry out the entire operation on a provisional basis almost without books. Still, I greatly enjoyed this time because of the students. It was sheer pleasure to teach these early matured young men who had returned from the murderous war, or from the prison camps, finally to prepare for their respective professions. I enjoyed taking excursions with them to regions unfamiliar also to me.

Two factors influenced my own scholarly preoccupations during this time; first of all, I was practically cut off from my own books and working materials so that I could hardly continue working on my previous research. Secondly, I was still strongly under the influence of the predominantly geo-ecological atmosphere in which I had been working for the last years. As a result, I produced several ecological studies about the extended Freiburg region.[11] Over the years, too, my ideas about the basic importance of various social structures within the geographical framework had ripened to the point that I published my first article about the 'role and significance of social geography'.[12]

On the first German Geographentag (after the war), which took place in 1948 in Munich, I summarized my unpublished work on the 'Social land-scapes of the Orient', which I had originally written in Berlin under the title 'The formation of social areas in the near Orient'.[13] Finally, a theoretical article, 'The landscape in terms of the logical system of geography' (1949) was written together with J. Schmithüsen during this period in Freiburg.[14]

Professor (1949–71) and emeritus in Vienna

In the summer of 1948 I ended my activities at the University of Freiburg i.B. because of a conflict with the reappointed Ministry for Sciences in Baden; they wanted to force me to work again with my assistant, who had been of very little help to me before, after he had returned from a six month stay at a psychiatric clinic. Shortly thereafter I gladly accepted an offer for a full professorship in economic geography at the School of Economics in Vienna, a vacancy caused by Bruno Dietrich's death. I held this position which I thoroughly enjoyed, for a relatively short time since in 1951 I was appointed as successor to the distinguished human geographer at the University of Vienna, Hugo Hassinger, who had just received emeritus status. Thus I became the younger colleague of my former professor, Johann Sölch, who occupied the physical geography chair at the same institute. Unfortunately, however, he died from a heart attack in September of the same year. His successor was Hans Spreitzer, whom I had already met at the end of the war when he was professor at the University of Prague.

In Vienna I was confronted with new tasks. Prof. Hassinger, who unfortunately died in a traffic accident at the beginning of 1952, had become increasingly concerned in his last years with questions of space and spatial patterns. He contributed much to the promotion of planning (*Raumforschung und Raumordnung*), a concept which had met with a certain amount of distaste immediately following the war.

Thus, I was soon faced with the dilemma of whether I should take part in these efforts and the ensuing research questions. On the one hand I realized that I would not be able to continue work on my previous areas of study to the full extent – and I had still some tempting scholarly goals in mind, which no doubt would be affected. On the other hand, I knew that only practical work in this new direction, at least in preparatory research for land-use planning, would open up interesting professional opportunites for future graduates of geography, and ensure its appropriate participation in this crucial social concern. In addition, it appeared that no other geographer was available who could relieve me of this task. Aware of these conditions, I took part in the foundation of the Austrian Society for the Promotion of *Landesforschung und Landesplanung* (1954) – the title of which was changed in 1965 to *Raumforschung und Raumplanung*. I was one of the science representatives on the Board of Directors. When I was later elected a full member of the Austrian Academy of Sciences, it was this same motivation which led me to assume the directorship of the Academy's *Kommission für Raumforschung*, which Hassinger already had established a few years later.

Given my teaching responsibilities and my own research activities I did not want to enter the actual planning field as Hassinger had done at times. It seemed to me, however, that well-understood social geography of the kind I was trying to develop, was especially well suited for spatial studies. Empirical research would, of course, have to be done primarily in Austria, especially in the region surrounding Vienna, which was largely unfamiliar to me. Frau Elisabeth Lichtenberger, an assistant of superior scientific qualifications, was immensely helpful in this work. She had studied geomorphology previously but quickly trained herself in social geography during our countless excursions. Soon I could let her do the courses for the beginners which dealt with specific problems in the landscape and with the systematic gathering of data on villages and small towns.

My lectures were devoted primarily to the basic theoretical questions and their relationships, which I endeavored to clarify with case studies. Thus the maturation of social geography, which had started before the war and continued in Freiburg i.B. and Vienna, proved useful. This process was manifested in several publications.[15]

Meanwhile my Persian research continued as I worked up and published older research materials and undertook a new seven month trip with limnozoologist Heinz Löffler through the southeastern and southern parts of Iran in 1956. Finally I was able to make valuable supplementary observations in a cultural and social-geographical context during a year which I spent as guest lecturer at the University of Iran (1958–59). The rest of my publications in the 1950s were concerned mainly with the climatic and ecological conditions in Iran.[16] In 1955 I presented an article which disproved the prevalent theory that the periods of augmented glaciation were also 'pluvial' periods.[17] I showed that, apart from the decline in temperature, ·the climatical conditions had been even more arid than in the present day. I was pleased to see this conclusion confirmed later by pollen profiles taken by a group of scientists under the leadership of W. Zeist from lake deposits on the western face of the Zagros mountains. They indicated the complete lack of forest during the last glacial period because of climatic dryness and the much later penetration of present-day forests (about 6,000 years ago). Thus, in a later publication, I was able to explain the present widespread phenomenon of forest decline and soil deterioration or erosion in Iran, which had frequently been falsely associated with the alleged post-pluvial aridity, with the effects of human encroachment and misuse during thousands of years.[18]

In 1956 and again in 1958/59, an extraordinary opportunity to further my knowledge of Iran was presented to me by Mr B. Mostofi, director of the Iranian Oil Company, as he generously gave me permission to study the complete collection of photographs of the US stereo-photogrammetric aerial survey of Persia, to be used at my discretion. In addition, I was invited to join two expeditions undertaken by the geological department of the Iranian Oil Company, which led me into various parts of the Great Kawir and of the Masileh (Qom) Kawir. From field observations made during these expeditions, as well as from information acquired from these aerial photographs I produced a work on the Great Kawir and Masileh.[19] Another study of the southern Lut was based almost exclusively on the scrutiny of the full series of aerial photographs.[20] It was quite evident to me that the fill of various basins in the Great Kawir and Masileh had never before reached its present level; it was also clear that in the southern Lut Basin, after an extremely long dry period of powerful wind erosion, an actually more humid phase with increased water supply from the surrounding mountains had produced several smaller salt pans in the lowest parts of the basin. Furthermore, there are indications of the existence of some other such humid phases, with similar consequences, within the long dry period of predominant wind erosion.

My studies in Tehran on the aerial photographs were following also further research lines. Thus I also designed a working map for the Lake region of the Hilmand in eastern Persia, similar to the one I made on the southern Lut. The map was also to be accompanied by a text which, however, was never written, since I did not find the time to make another journey into that area, which I deemed necessary. It seemed to me that an outline of climatomorphological regions in Iran would be worth while and also possible with the use of the aerial photographic material. A detailed map of the hydrographical areas, which I considered to be a significant contribution to such studies, was prepared. During many excursions with my students, we made numerous surveys of agricultural land use. I was also especially interested in the different types of field patterns and their connection with different forms of land tenure. It took me years, however, to find the necessary time to work out this material.

The many cases of painful delays or complete resignation were of course a natural result of the necessity to take over more and more work for *Raumforschung* within Austria. By the end of the 1950s, further research excursions to Iran were practically eliminated. Thus, the number of my publications on Iran declined rapidly in the 1960s and they underwent also a change in character.[21] Only in 1976 – when I had already emeritus status – was I able to publish my postponed work on field patterns in Iran and other physico-geographical studies.[22]

My activities in 'regional research' were closely connected with two main institutions. The *Kommission für Raumforschung* at the Austrian Academy of Sciences had begun to compile an atlas of Lower Austria, while Hassinger was still in charge. E. Arnberger continued his work as editor, and it was completed in 1958. Even before this time, I had decided to devote the commission's work to the project of producing a major thematic atlas of the entire Republic of Austria.[23] The major reason for this resolution was that almost all the provinces (*Bundesländer*) of Austria already had, or soon would have, similar thematic atlases at their disposal. There were, however, extraordinary discrepancies in their thematic outlines and technical qualities. Although planning in Austria is foremost a matter of the *Bundesländer*, it seemed highly worth while to create an atlas of Austria, which would allow true comparisons between the different *Bundesländer* or parts thereof. The objective was to examine and present in maps all relevant regional structures of Austria, natural as well as human, which are important for the society's life, presently as well as in the near future, so that the 'unique' and the 'particular' could be related to the whole. The essential questions about design and subject matter were immediately discussed by a panel of commission members and other experts; a small committee should do the editorial work. However, when the first delivery of the work was presented in 1961, it had already become evident that there had been serious difficulties in the allocation of editorial tasks. So I was forced to undertake all these tasks myself. The support of E. Arnberger as a specialist in the technical side of cartography was indeed a great help. He was finally relieved of his duties by Marianne Fesl for about the last quarter of the maps.

A main part of my work for the atlas was devoted to the recruitment of qualified authors for the intended 120 map sheets. Of these, 25 were concerned with natural features, 34 with the settlement and population, 41 with the economy and transportation – including tourism; 20 were devoted to administration, education, health facilities, mass media, urban centres and their specific regions. Even on a relatively small budget, it was possible to win over scientists of excellent stature. They were to work on specific maps or groups of maps, with the help of many young academics. Among them were a few full-time staff members of the *Kommission für Raumforschung*, and to that number came many assistant professors and doctoral candidates from various institutes at the universities or other state institutions.

I attended personally to the design of almost fifty maps, usually the more complex ones. I was able to sustain the burden of editing the atlas largely because I took such enjoyment in visual representation and design. The atlas had been planned to be finished in about twelve years but as a consequence of some unforeseen delays near the end the whole period of production was stretched out to twenty years. Instead of six planned deliveries their actual number had risen to eleven.

Naturally, the turn toward regional research and the work on the atlas affected my publications during the 1960s and also after my retirement on 30 September 1971. It was no wonder that my oldest scientific interest in cities, their functions as well as their structures, returned to the foreground. Aside from that, especially within the frame of the atlas, regional structures of various kinds also played an important role. So already in 1959/60, supported by H. Helczmanovszki, I started work on the central places in Austria which, however, as I soon discovered was based upon too small a number of representative services.[24] Upon closer examination of the individual centers and their areas, deficiencies became apparent. Accordingly, I organized a new survey together with Dr Marianne Fesl in the early 1970s. It proceeded along the same principles, but much more sophisticated materials were used for the ranking of central places and for determining their corresponding areas of influence. These principles had been presented in 1967 in a lecture at the German Geographers' Convention in Bad Godesberg and other shorter essays followed.[25] The results of the new survey were used not only for new atlas maps but also appeared in a stately volume in 1978. In addition to an extensive quantitative and qualitative analysis of the system, this survey contains statements about changes since the first survey (1959/60) – the results of which were reconstructed according to the new method – as well as some prognostics for future developments.

Another equally time-consuming project I had tackled earlier with the help of Dr E. Lichtenberger was an analysis of the structure of Vienna. This survey was to grasp and to analyze the physical as well as the functional and social information of the whole city. It should thereby offer to the planning division of the city council a hitherto largely neglected basic resource. The mapping had, after a common commitment to all vital elements, been carried out in the years 1955–60 by the appropriately informed participants of the culture-geographical seminars under the direction of E. Lichtenberger. All types of residential structures, public and other types of construction, and the remaining land uses were covered. The date of the buildings was deciphered partly from the façades or from the general type of construction, partly from an official survey of houses done in 1951, which also gave data for the size of the interior living quarters. Time dimensions were separated into five periods: prior to 1840; the early, high, and late *Gründerzeit* (which is the name of the main period of industrialization); between the wars and the post-war years.

The mapping results were handed over to the city planning division in maps of a scale of 1:10,000 and also, more generalized, at 1:50,000, to be used while establishing the intended main concept of city regulations and the new land-use plan. Also a sensible subdivision of the twenty-three districts into approximately 220 smaller units could be made for a better future representation of housing and population data.

The volume of Vienna[26] was designed primarily to offer guidance and primary data clarifying the emergence of the random mosaic of buildings and land-use types. There was need for much further material – political, socio-economic and demographic – which was not at all complete for the five periods nor easily accessible. It was Mrs Lichtenberger who gathered most of the required data, which allowed us to identify not only the different contributions made by each period to the complete mosaic, but also to understand them as products of the various and sometimes very markedly changed political and social conditions during these five periods. The influence of respective spatial conditions on the expansion of the city could also be evaluated. In their closing comments, the authors offered – from their extended knowledge – a critical view of the 'Master Plan of Vienna'. Their work was widely accepted and a second, unchanged edition appeared in 1978.

A number of publications were conceived and carried out in close connection with specific maps of the atlas. In 1975 I prepared a guide to the entire atlas (with the exception of the few maps of the very last delivery) in the form of an essay which contains, after a short introduction on the concept and

179

organizational structure of the atlas, an explanation of the most important maps or groups of maps, what they show and how they illustrate single or aggregated regional structures of the country.[27]

In closing, I want to refer to a series of works which are particularly dear to me, which, despite lack of time, were written as a continuation of some older ideas. The starting point must be 'The main stages of social and economic development'[28] since it gained much attention and was soon translated into English. These theses and interpretations, especially on the subject of 'rent capitalism' and its effects in the stagnation of ancient highly developed societies were extended in several other essays.[29] Finally, reference should be made to my 1976 article in which substantial connections between field patterns and social structure (i.e., conditions of land ownership) are disclosed.[30]

In my private life, I suffered a great loss when my wife Heléne died in December 1976, after 47 years of marriage; she gave and meant much to me through good as well as difficult times. A year and a half later I married my long-time and highly valued co-worker, Dr Marianne Fesl.

Concluding remarks

In conclusion I should like to reiterate the point made at the beginning of this essay: that continuous lines of thought are discernible in my research activities in spite of several radical changes in topics.

The first and most important line certainly is that issuing from my doctoral thesis in urban geography. To characterize this I may, in part, resort to Peter Schöller's words in his *laudatio* on the occasion of my being awarded an honorary doctorate at the University of Bochum (West Germany) on 1 June 1978: 'In exemplary geographical integration two research perspectives have been developed, which, meanwhile, have steered not only urban geography, but cultural geography as a whole into new directions, namely the functional perspective and the socio-geographical approach.' The essential function of cities for their respective surroundings, already explicitly discussed – though the term 'centrality' was not used – in the theoretical introduction published separately in 1927, has been 'proved statistically in many and complex ways and fixed cartographically' for the case of Innsbruck. The socio-geographical approach had elucidated the building fabric of this city, as 'types of building complexes and housing densities were paralleled with problems of immigration, age structure, and socio-economic groups of the population'. In addition an attempt was also made 'to establish the relationship between social segregation and election results'.

The next essential step in this line was prepared by experiences in the Near East and intensive perusal of literature. At the International Geographical Congress in Amsterdam in 1938 I discussed a few functional city types whose interrelations with their surroundings could by no means be labeled 'central' in the sense comprehensively defined by W. Christaller in 1933. It was mainly the search for the various backgrounds of these different functional city types that induced me to study intensively the social and economic conditions of Near East countries during the last years before, and the early years of the Second World War. This resulted in an extensive manuscript on 'Social landscapes of the Orient'. Unfortunately, it remained unpublished, however, due to the war. Only a much shortened version, the text of a lecture, was published.[13] In this context the concept of a 'social geography' – not as a novel discipline, but as a new approach – was developed, which, in P. Schöller's words, *was and still is sound and extremely fecund today*. Two other papers (1950, 1959) crowned this line of development.[31] Of course my approach was a socio-geographical one, actually setting forth the important worldwide problems of stages in the cultural and economic development on 'a new scientific level' (Schöller) and, thus, providing new insights. This study – together with the complementary paper on 'The specific position and achievement of the Occident within the framework of the high cultures of the world' (1960) – offered an original answer to the question, why the last, truly world-transforming upsurge of industrial development with all its concomitant and after-effects did not originate in the Near East, which had been leading the way for such a long time, but in Atlantic Europe. In the former area, where the older, traditional type of cities had developed and prevailed to the present time, the corresponding economic system of 'rent capitalism', as I called it, for several reasons, largely refrained from investments apt to bring about an increase in productivity, whereas in the latter 'productive capitalism' could develop in connection with a growing political importance of the citizens. It was definitely within the substructure of this hermeneutical theory that at the same time I have become interested in the problems of underdeveloped countries, both in general and especially in the case of Iran, and published a textbook-level booklet on that country.[32]

This approach of urban and social geography was also continued within the framework of applied research for regional planning. The book on Vienna[26] and the intensive research into the central places of Austria and their problems, as well as a number of other works, cartographic and textual, connected with the *Atlas of the Republic of Austria*, fits more or less into this line.

A second line of development, followed for a great part of my scientific career, was also started at a very early date: the well-meant suggestion by Johann Sölch in 1928 actuated geomorphological studies in the Tyrol and resulted in a number of publications. It was continued by research work in Iran and southeastern Turkey carried out during the time I stayed in Berlin and also later. The situation of recent and Pleistocene glaciation were then practically unknown so the traces of the latter were studied. Moreover I had planned from the very beginning to arrive at a registration of the ecological structure of Iran in as much detail as possible. Field observations during a number of exploratory journeys were to form the basis for it, together with an interpretation of the existing scientific literature as well as the data of public survey services for the first time available then in some variety and acceptable quality. Whereas it was possible to publish the findings as to the glaciation of the last phase of the Pleistocene and that of recent times, either just before or at the beginning of the Second World War, the studies regarding ecology could only be completed and published during the 1950s and

181

even later. The short period of work in the field of ecology which I had had in Freiburg i.B. (Western Germany) definitely was beneficial for this.

Further stimulus came with my discovery that there had not been a pluvial but a dry period in the Iranian highlands during the last ice age, as proved by a comparison with the recent glaciation, a result that was to be confirmed by other scientists by means of pollen profiles later on. In my paper 'Climate and landscape of Iran in prehistorical and early historical periods' (1955), I represented the respective inferences as to vegetation development as well as to the conditions for settlement.[17] Another paper discussed the topic of 'Devastation of the vegetation and soil exhaustion in Persia and their connection with the decline of older civilizations'.[18] Both studies were highly appreciated by experts in prehistory.

In my subsequent studies on the interior basins of the Great Kawir (1959, 1961) and the southern Lut (1969) I was able to provide further evidence for one major and several minor climatic changes. These publications marked the end of the second line of development, however, as I have not been able to find time for longer journeys in Iran since then, although there was no lack of interesting problems and even of valuable materials and information about them.

My basic conception of geography, characteristic of the times, was the same 'classical' view espoused by A. Hettner, O. Schlüter, and many other important geographers of the first half of the twentieth century. I have therefore considered the geosphere, countries and landscapes, as the 'object' of geographical research. From the outset, however, I have felt a strong urge to reveal formal or causal relationships within this complex object or at least in those sections studied more closely, and in most cases this yearning was stronger than the wish to describe such sections as completely as possible, namely in the way of regional studies ('Länderkunde'). Fact is that I produced very little in that field during my life-path.

This is why I may say that problems as such were the main foci of almost all of my research work. Theoretical concepts, at least tentative ones, constituted its essential results or provided a basis for an elaboration later on. Thus the anticipation of the idea of centrality, namely the concept of a relationship between a city and its surroundings in my paper 'Basic problems of urban geography' (1927), formed the base for the description of highly diversified connections between cities and their surroundings according to the economic structures of the respective societies (1938) and, subsequently, for the creation of a 'social geography' (1948) and the formulation of the theory of 'rent capitalism' as opposed to 'productive capitalism' (1959).[33] Founded on an underlying set of problems there is a similar concatenation to be found between my different studies and their respective results in my publications in the field of physical geography and ecological studies.

As far as research methods and the techniques of information treatment are concerned, the bulk of my publications date back to periods before the reception of quantitative analytical methods in the German-speaking countries. Not having sufficient mathematics myself prevented my own active participation, but of course I am fully aware of the very considerable improvements made possible in the important field of hypothesis-testing and theory-formulation.

Another fact I have been aware of since the early 1950s was that an engagement in the newly emerging tasks of regional planning was absolutely indispensable, if geography wanted to retain its position, at least by contributing appropriate regional research. In this connection improvements in hypothesis-testing were imperative, as the tasks to be undertaken inevitably would involve prognoses.

My attitude toward this necessary change of paradigms in geography can be summed up shortly. I should like to quote those words here that concluded my farewell lecture at Vienna University late in 1971: *So far geography has been the domain of people who arrived at holistic syntheses – and in many cases very impressive ones – by way of more or less systematic observation combined with a great deal of intuition, imagination, and sensibility. If logical reasoning was one of their strong points, very often their theoretical concepts had a high degree of objective, intersubjectively verifiable validity. Let us hope that those systematists who think primarily in analytical terms and are taking geographical research by storm now will have enough knowledge of real life and sufficient sensibility to make the simplifications inevitable in models and theoretical formulae in such a way that the results and, even more so, the interpretations and inferences, are neither ivory-towered nor awkward.*

Both approaches, an urge to perceive and interpret 'Gestalt' intuitively, essentially related to the sphere of creativeness and art, on the one hand, and a yearning to ascertain scientific laws governing everything and, thus, to control the forces of nature on the other hand, are deeply rooted in mankind and have equal rights. A wise combination of them might, however, constitute an ideal solution for the existing discrepancies.

Notes

1. 'Nationalistic' here means that – as a German by birth and language – after the defeat and fall of the Austrian monarchy, one sought union with Germany – this was also officially resolved by the First Republican Austrian Parliament, yet prohibited by the Allied Powers.
2. See Selected readings, 1927.
3. See Selected readings, 1928.
4. Schlusseiszeit oder Rückzugsstadien? (Diskussion mit O. Ampferer), *Pet. Mitt. Gotha* (1930), 227–35; also Die Formenentwicklung der Zillertaler und Tuxer Alpen, *Forsch. z. Dt. Landes- u. Volkunde*, **30** (1933)
5. Die jüngere Geschichte der Inntalterrasse und der Rückzug der letzten Vergletscherung im Inntal, *Jb. d. Geol. Bundesantalt Wien* (1935), 135–89.
6. Die Rolle der Eiszeit in Nordwestiran, *Zschr. f. Gletscherkunde Berlin* (1937), 130–83.
7. Die nordamerikanischen Kleinstädte und ihre Entwicklung, *Mitt. Geogr. Ges. Wien* (1930).
8. Eine neue Arbeit zur Stadtgeographie, *Zschr. Ger. Erdkunde* (1935).
9. See Selected readings, 1938.
10. Um die deutsche Volksgrenze in den Alpen, *Dt. Archiv. f. Landes- u. Volksforschung* (1937), 734–48; Zur Judenfrage in Rumänien und Ungarn, *Dt. Archiv. f. Landes- u. Volksforschung* (1940), 137–40.

11. Südwestdeutsche Studien, *Forschungen z. Dt. Landeskunde*, **62** (1952).

12. See Selected readings, 1948; see also: Kann die Sozialgeographie in der Wirtschaftsgeographie aufgehen?, *Erdkunde*, **16** (1962), 119–26.

13. Soziale Raumbildungen am Beispiel des Vorderen Orients, *Tagungsbericht Dt. Geographentag München* (1951) Landshut (Amt F. Landeskunde).

14. Die Landschaft im logischen System der Geographie, *Erdkunde*, **3** (1949), 112–20.

15. Aufriss einer vergleichenden Sozialgeographie, *Mitt. Georg. Ges. Wien*, **91** (1950), 34–45; Die räumliche Ordnung der Wirtschaft als Gegenstand geographischer Forschung, *Der Österreichische Betriebswirt*, **1** (1951), 25–39; Gedanken über das logische System der Geographie, *Festschrift* Hans Spreitzer, *Mitt. Geogr. Ges. Wien*, **99** (1957), 122–45; see also Selected readings, 1959.

16. Die Verbreitung des Regenfeldbaues in Iran, *Wien: Geogr. Studien, Sölch Festschrift* (1951), 9–30; *Die natürlichen Wälder und Gehölzfluren Irans Bonner Geogr. Abh.*, **8** (1951), 62 pp.; Beiträge zur Klimaökologischen Gliederung Irans, *Erdkunde*, **6** (1952), 65–84.

17. Klima und Landschaft Irans in vor- und frühgeschichtlicher Zeit, *Geogr. Jb. aus Österreich*, **25** (1955), 1–42.

18. Vegetationsverwüstung und Bodenerschöpfung in Persien und ihr Zusammenhang mit dem Niedergang älterer Zivilisationen, International Union for Conservation of Nature and Natural Resources. 7th Technical Meeting (Athens), *Proceedings*, **1** (1959), 72–80.

19. *Features and Formation of the Great Kawir and Masileh*. Arid Zone Research Centre, University of Teheran, **2** (1959).

20. Zur Kenntnis der Südlichen Lut-Ergebnisse einer Luftbildanalyse, *Mitt. Österr. Geogr. Ges.*, **111** (1969), 155–92.

21. Tehran, *Festschrift Hans Kinzl*, Schlernschriften, **190** (1958), 5–24; Die Problematik eines unterentwickelten Landes alter Kultur: Iran, *Zeitschrift des Nah- und Mittelöst Verein Hamburg*, **2** (1961), 64–8, 115–24; see Selected readings, 1962.

22. See Selected readings, 1976; see also Vegetation, in W. B. Fischer (ed.) (1968). *History of Iran*, vol. 1, *The Land of Iran*. Cambridge U. P. 280–93.

23. Bemerkungen zur Ermittlung von Gemeindetypen in Österreich – Beiträge zur Ermittlung von Gemeindetypen, *Schriftenreihe der Österr. Ges. zur Förderung von Landesforschung und Landesplanung*, **1** (1955), 15–38.

24. *Die zentralen Orte und ihre Versorgungsbereiche*. Raumordnungsgutachten für die Österr. Bundesregierung, A, B, and C (1967–68).

25. Die Theorie der zentralen Orte im Industriezeitalter, *Tagungsbericht und Wissenschaft.* Abh. Dt. Geogr. Tag (1967), 199–213; Die zentralen Orte und ihre Versorgungsbereiche, in *Strukturanalyse des Österr. Bundesbebietes* (hg. v. R. Wurzer) Rahmen der Schriftenreihe d. Österr. Ges. f. Raumforschung und Raumplanung, **2** (1970), 475–504.

26. See Selected readings, 1966.

27. See Selected readings, 1975, also with J. Steinbach and E. Ehrendorfer, *Die Regionalstruktur der Industrie Österreichs. Beiträge zur Regionalforschung*, Kommission für Raumforschung d. Österr. Akademie d. Wissenschaft, **1** (1975) 80 pp., attempts to define the different factors of location according to their importance. A similar study, together with A. Hofmayer, (1981) Gliederung Österreichs in wirtschaftliche Strukturgebiete, *Beiträge zur Regionalforschung*, **3** (1981). Others are still in progress.

28. See Selected readings, 1959.

29. Die spezifische Stellung und Leistung des Abendlandes, *Wissenschaft und Weltbild* (1960), 169–78; Über den Einbau der sozialgeographischen Betrachtungsweise in die Kulturgeographie, *Deutscher Geographentag Köln 1961*, Wiesbaden (1962), 148–65; Zur Problematik der unterentwickelten Länder, *Mitt. d. Österr. Geogr. Ges.*, **104** (1962), 1–24; Erwerbstätigenstruktur und Dienstequote als Mittel zur quantitativen Erfassung regionaler Unterschiede der sozialwirtschaftlichen

und kulturellen Entwicklung, *Münchner Studien zur Sozial- u. Wirtschaftsgeographie*, **4** (1968), 119–31; Zum Konzept des Rentenkapitalismus, *Tidjschrift voor Economische en Sociale Geografie*, **15** (1974), 73–8; Rentenkapitalismus und Entwicklung in Iran: Interdisziplinäre Iran-Forschung, Beiträge aus Kulturgeographie, Ethnologie und Neuerer Geschichte, in G. Schweizer (ed.) (1979). *Beihefte zum Tübinger Atlas des Vorderen Orients*, B, 40 Wiesbaden, 113–23.
30. See Selected readings, 1976.
31. Aufriss einer vergleichenden Sozialgeographie, *Mitt. Geogr. Ges. Wien*, **91** (1950); see also Selected readings, 1959.
32. See Selected readings, 1962.
33. See Selected readings, 1927, 1938, 1948 and 1959.

Selected readings

1927   Grundfragen der Stadtgeographie, *Geogr. Anzieger*, 213–24.
1928   *Innsbruck, eine Gebirgstadt, ihr Lebensraum und ihre Erscheinung.* Forschungen zur Deutschen Landes- und Volkskunde.
1938   Über einige funktionelle Stadttypen und ihre Beziehungen zum Lande, *Comptes Rendus Congr. Intern. Géogr.*, Amsterdam, **32**, 88–102.
1948   Stellung und Bedeutung der Sozialgeographie, *Erdkunde*, **2**, 118–25.
1959   Die Haupstufen der Gesellschafts- und Wirtschaftsentfaltung in geographischer Sicht, *Die Erde*, **90**, 259–98. Translated as The Main Stages in Socio-Economic Evolution from a Geographical Point of View, pp. 218–47 in P. L. Wagner and M. W. Mikesell (eds.) (1962). *Readings in Cultural Geography*. University of Chicago Press.
1962   *Iran. Probleme eines unterentwickelten Landes alter Kultur.* Moritz Diesterweg, Berlin.
1966   With E. Lichtenberger, *Wien. Bauliche Gestalt und Entwicklung seit Mitte des 19. Jh.* 394 pp. 10 maps. Kommission für Raumforschung der Österr. Akademie d. Wissensch., Wien.
1972   Die Entwicklung der Geographie – Kontinuität oder Umbruch? *Mitt. Österr. Geogr. Ges.*, **114**, 3–18.
1975   Österreichs Regionalstruktur im Spiegel des Atlas der Republik Österreich, *Mitt. Österr. Geogr. Ges.*, **117**, 117–64.
1976   Entstehung und Verbreitung der Hauptflursysteme Irans – Gründzüge einer sozialgeographischen Theorie, *Mitt. Österr. Geogr. Ges.*, **118**, 274–304.
1978   With M. Fesl, *Das System der Zentralen Orte Österreichs. Eine empirische Untersuchung.* 310 pp. 10 maps. Kommission für Raumforschung der Österr. Akademie d. Wissensch., Wien-Köln.

Interlude Three

If there is one event common to the life contexts described in this book it is unquestionably the Second World War. Most of our essayists and discussants were well launched into their professional careers at the time, and yet many speak of 'the war' in parenthetical and oblique fashion. No doubt the significance of this event varied depending on the nature of one's involvement in war-related service, on the national school with which one was professionally affiliated, and especially on which side of the Atlantic or Maginot line one experienced it. Someone whose impressions of the Second World War have been mediated via film, novel, and textbooks, as mine have been, can only speculate on what it may have meant for geographers in the great text-producing nations, for example, to see their maps reshuffled, to gaze on those landscapes over which *excursions* trod and theses were pondered, now shattered and torn. In deference to our essayists, it seems better to withhold comment on the event itself and instead seek to discern certain changes of meaning and metaphor which may have been associated with the changed milieu of the post-war world (see however, Gottmann 1946; Troll 1947).

Although the Second World War resonated through a wider spectrum of technological, social, economic, and ideological interests than any other event of the twentieth century, war *per se* and military concerns are no strangers to geography. A recent interview with Foucault suggests, that an 'archeology' of of geographic language might reveal the extent to which our commonly used concepts and terms such as 'front', 'region', 'field', 'displacement' and others are rooted in strategic concerns of war, peace, and the 'balance of power' (Hérodote 1975). One could scarcely envision armies marching without a map, navies sailing without a chart, or political dominion without an organization of space. But for airborne assault or defense, did one not also need the more flexible techniques of operations research and systems control (Morse and Kimball 1951)? In terms of metaphor, the Second World War seems to mark the beginnings of a profound transformation of thought style from 'map' to 'mechanism', paving the way for an eventual integration in the new 'paradigm' of the 1960s. That prose and poetry of 'natural regions', of 'national identity', of 'organic bonds between man and land', all of which had become associated with an unsavory kind of nationalism, now appear embarrassing. The chorological alternative and the field-based mapping of regional pattern and form also seem frail and impotent in the face of technologically transformed landscapes and the increased centralization of power and decision making. The world from the air, seen by many an enlisted geographer, opens up new vistas not only for photogrammetry and cartography, but for traditional world images as well. Over the mosaic of areally differentiated regions one could now discern systems of spatial interlinkages and a topological surface of forces.

The map, of course, would remain the key metaphor within which debates over new conceptual and methodological ideas would be framed. In one generation it would become the language for exploring abstract issues of geometric form and topological surface as well as for the elucidation of practical problems. But this was a world of dynamism and change; and the map itself would change. Aeronautical charts, polar and azimuthal projections reflecting the strategic concerns of an air age would compete in the classroom with Mercator and other representations of 'true' area, shape, and contour as well as the great circle trading routes (Henrikson 1975). The world could now be projected as manipulable, open to conquest of time- or cost-distance, and evoking a consciousness of perceptual space and relative locations. On the ground, too, if applied geography were to be scientifically based as well as practically relevant, it should address these two apparently irreconcilable imperatives: the need to accommodate diverse human and physical reality within bounded regions on the one hand (mapping), and the need to exploit and monitor trans-regional and trans-national spatial processes (mechanisms) on the other.

Through the rearview mirror, too, an explorer into disciplinary history observes some profound transformations in personal meanings and value positions expressed among practitioners of geography during the post-war period. One notices a bolstering of faith in scientific objectivity, positivist conceptions of truth, a more intense search for explanatory rather than descriptive language, and an unapologetic thrust toward thematic specialization and problem orientation. A growing ideological commitment to applied research is also evident, either in promoting post-war economic development and regional reconstruction, or in redistributing wealth and access to resources and services according to the canons of the Welfare State. Did science and rationality not promise objectivity, a vanguard against the subjectivities which had led to the horrors of war? And were these not the lessons to be learned from the more 'successful' social sciences at mid-twentieth century?

Attempting to identify international trends during the immediate post-war period, however, feels somewhat gauche. What seems to have happened in European schools at least was a concentration on one's own particular situation, to re-create and renew the theory and practice of the profession at home. Schools vary in their emphases on physical or human aspects of the field, depending in part at least on how national funding was allocated; they vary in their orientations to 'pure' versus 'applied' work. By the 1950s a remarkable recovery is evident everywhere, and a second wave, a plethora of new journals, institutes and associations appears, reaching out across national boundaries and maintaining a lively interaction (Taylor 1951; *L'Information Géographique* 1957; Harris and Fellman 1960, 1971, 1980).

For America the post-war drama took quite a different course. For those involved, by their own admission, war wrought benefits in myriad ways (Ackerman 1945; Deasy 1947, 1948; Rose 1951; Stone 1979). Veterans returned as heroes, cheer-led on for the sequel to a great western movie very soon to be replayed in the Far East. Marshall Aid and NATO could now take care of European interests; there was much still to be done by the 'Army of the Potomac', not only in the State Department, but in large-scale environmental projects and the management of natural resources (Wilson 1948; Hart 1979). And there was more than ample challenge in the rapidly growing halls of Academe. Veterans and their prolific offspring would swell the schools and co-ed campuses across the land with generous support from GI Bill and student loans. Geography would expand, diversify, and become keenly involved in the dramatic evolution of the American landscape itself. The supply of professional geographers trained during the pre-war era, however, was quite short in meeting the teaching demands of these exponentially-increasing student enrollments on the one hand, or those of the growing job market in public service and planning offices on the other hand. Unlike university geography in European schools, then, this was not the scene where an 'old guard' returned to occupy former positions, but one where improvisation and experimentation were more a necessity than a luxury. Students entering American graduate schools in the 1950s were also older, their undergraduate education may have been in fields other than geography, and their non-academic work experiences may have made them less impressionable to the authority claims of *Maîtres Penseurs* than their counterparts of the pre-war period might have been. A second generation would seek to get on with their careers, seek cognitive certainty or at least methodological efficacy, and more practical relevance for geography in society.

It is certainly not easy for someone who was a high school student in Ireland during the 1950s to grasp fully the *esprit* of that decade in America. Hollywood images and the beginnings of 'anti-imperialist' rhetoric scarcely matched those which came from direct contacts with 4-H Club visitors from the American Midwest, or the record of Marshall and Kellogg largesse in helping put Irish agriculture on its feet. It was from this background I came to America and to American geography, during the early 1960s in a Boeing city bustling with excitement about its First World's Fair. My perceptions are thus based on limited experience of one particular tradition and on readings from others.

My first impression, gained mostly from urban work, was that this 'new geography' was a marvelous illustration of ethnoscience—a bold attempt by natives to explore those very processes which were transforming American life and landscapes since the 1950s. Edward Ullman's courses on transportation and urban geography opened my eyes to the 'realities' of pattern and process on the American landscape, and to the intellectual challenge of combining formal and functional approaches to geography. With the Railroad Map of the USA he would recount the story of break-in-bulk points, twin cities, entrepot and port (Ullman, 1949a, b). Cigar in hand and from his sometimes ill-prepared lecture notes he would spell out the drama of corner grocery store yielding to supermarket, of local highway and canal yielding up their commodity and passenger flows to interstate freeway and airline, all pointing toward that growing concentration of life and energy around key nodes in the circulation system. Within these nodes, too, he would show urban areas sprawling into metropolitan complexes, commuter traffic and market-capturing goods filling the interstices in a daily dance energized by cheap petroleum and steel consumerism, all inspired by the myth that what was good for General Motors was good for the whole economy (Ullman 1940 −41, 1953; Breese 1949; Taaffe 1962; Higbee 1960). How could one resist the challenge of re-writing one's geography, re-drafting one's maps, and eventually seeking explanation for that dynamism underlying this vibrant and ever changing milieu?

A sense of fresh challenge, new opportunity, and the momentum of a vastly increased professional body, rings through the discussion on 'American geography in the 1950s' which comprises Chapter 13. Chaired by George Kish (Michigan), this discussion offers insight into settings where post-war influences seem indirect, but where research and teaching underwent significant change. Fred Lukermann (Minnesota), Duane Knos (Iowa), Bill Pattison (Chicago), and Dick Morrill (Seattle) share experiences and opinions on the origins and significance of these changes within the departments with which they were affiliated during that time. The so-called 'quantitative revolution' is regarded by many as a phenomenon of the 1960s, and a common image is that it had primarily a methodological meaning. What emerges from this discussion is that its origins can be traced to the 1950s at least, that its roots were at once conceptual, political, and demographic, its energy stemming from esthetic as well as social concern, and that mostly, its forms were varied.

Chapter 13 does not, of course, attempt to give a fully representative picture of the 1950s. Was this not the decade when foreign area studies flourished, cultural and historical geography stepped to the 'Foreword' challenge (Sauer 1941; see also Brown 1943; Malin 1947; Wright 1947) and Berkeley basked in its halcyon days (Spencer 1979)? *Man's Role in Changing the Face of the Earth* (Thomas, ed. 1956) was surely the most widely acclaimed monument of the decade. Was there not evidence too of a 'cognitive renaissance' (Whittlesey 1945; Wright 1947; Kirk 1951; Boulding 1956a; Bowden 1959, and 1980; Lynch, 1960)? And, as George Kish remarks in this discussion, much of the responsibility for promoting developments could be ascribed to the reconstituted Association of American Geographers which proceeded to democratize, or at least decentralize, power among its membership (Kohn 1979). Presidential Addresses would broach issues of thematic and practical concern (Trewartha 1953; Whitaker 1954; Sauer 1956), the Association would encourage an *Inventory and Prospect* of the entire discipline in America (James and Jones, eds. 1954); Ackerman would challenge the research potential and scientific status of geography in terms of policy relevance (Ackerman 1953), and Hartshorne would reiterate a chorological perspective on the nature of the field (Hartshorne 1959). A fascinating question indeed, raised by J. K. Wright in 1956, was 'How American is American Geography?' (Wright 1956)

Consensus and conformity, one reads, characterized American society at the time. A neo-conservative mood among historians was construing the history of the United States as an unbroken journey toward liberal politics, problem-solving pragmatism, and 'middle-class consensus' (Viereck 1953; Kirk 1954; Grob and Billias 1967). American, born free of feudal legacy, was the land where Lockean politics could flourish. From colonial experiences, had Americans not learned to be self-reliant, 'doers' rather than thinkers, generalists rather than specialists, and to thrive on adaptation to changing circumstances (Boorstin 1953, 1958)? Historians, it seems, showed little desire to find class conflict. Down-playing ideological extremism of the New Deal era in the 'homogenizing of history', they would also celebrate business competitiveness, and the security of a lifestyle which America enjoyed before it became a super power and champion of the free world (Hartz 1955; Higham 1959).

My fellow graduate students in Seattle in the mid-1960s showed little interest in books like Nisbet's *Quest for Community* (Nisbet 1953) and might have considered the radical back-to-the-land idealism of the 1930s and 1940s to be quite reactionary (Borsodi 1933; Baker *et al.* 1939; Bowers 1943). Trivial and irrelevant for many were such speculations about national character (Mead 1942; Kluckhohn 1944; Festinger 1957) or worries over poverty, class, or prejudice (Myrdal 1944; Riesman 1950; Frazier 1957); who had time to squander on soul searching about prejudice or the personality of the American soldier (Adorno *et al.* 1950; Bettelheim and Janowitz 1950)? There was so much to *do*, to solve practical problems and forge new methods. The geography profession in the 1950s began to demonstrate the values of *The Organization Man* (Whyte 1956)–that creativity was best promoted by group projects and that professional 'belongingness' ranked high. One might have had time to discuss the *Power Elite* and the military industrial complex (Mills 1956; Gouldner 1954), McCarthyism might cause some bitter intrafamilial annoyance, but on Cold War politics and national security and welfare, there would be little dissent. Compared with compatriots in history and sociology, American geographers in the 1950s were neither neo-conservative nor conformist, but compared with their colleagues in European Schools, they showed a remarkable degree of ideological consensus around those values deemed 'liberal' by historians at the time even if a few had difficulties assenting to Republican normalcy.

It would take more temerity than I can muster, however, to speculate on distinctions between what Wright would consider exotic and Americanistic for this period (Wright 1956). Berkeley had always kept its doors open to Germanic thought at least, and had encouraged its students to do field work abroad (Spencer 1979). Chicago, Minnesota, Clark, and other schools had welcomed the occasional foreign visitor, and not a few American geographers took Fulbright years abroad. Even for this so-called quantitative movement, so often dubbed the most characteristically (Midwestern) American innovation, European roots are more than vaguely discernible. But J. K. Wright does open window on meaning and metaphor for that period when he cites Kouwenhoven, 'America is process' (Kouwenhoven 1961, pp. 37–73). It does seem notable that those same geographers who had practiced and preached a geography of pattern and form in the 1930s begän in the 1950s to emphasize process and function, whether in regional (Platt 1935, 1957; James 1926, 1952), cultural (Sauer 1925, 1956), urban (Mayer 1942, 1955), or political (Hartshorne 1935, 1954; Whittlesey 1935, 1954; Jones 1943, 1954) geography.

In fact, in most of the thematic specialties inventoried in 1952–54, is the prospect not usually couched in terms of process orientation (James and Jones 1954)? Like other social science disciplines during the 1940s and 1950s geography was on its way from formal to functional approaches to its agenda and was beginning to see value in the use of quantitative measures (Moreno 1951; Merton *et al.*, 1959; see also Berelson and Steiner 1964). Appelbaum developed a method for analyzing consumer behavior and retail organization (Applebaum 1940; Appelbaum and Spears 1951; Appelbaum *et al.* 1961); in political geography one sees a movement away from concern over boundaries and territories to curiosity about fields and forces (Jones 1954; Jackson 1958, 1964; Hartshorne 1950; Sprout and Sprout 1965). In cultural geography a tension would be discerned between so-called 'genetic' versus 'functionalist' interpretations of process (Platt 1952; Sauer 1956; Wagner and Mikesell 1962) and European precedents would still be cited to defend formal, historical, as well as functional perspectives (Sestini 1947; Kant 1953; Hägerstrand 1953; Sorre 1953; Bobek 1959).

A goodly number of Americans participated in the IGU Congress at Stockholm in 1960 and were well represented at that ground-breaking symposium on urban geography (Norborg 1962) as well as on the newly fledged Commission of Applied Geography. The 1960s could open with convincing statements about space and process (Blaut 1961, 1962) and unequivocal pointers toward a 'systems' approach (Ackerman 1963).

It seems to me, then, that this 'process' orientation, quite as much as quantitative methods, was the touchstone for the innovative *esprit* of the late 1950s and early 1960s in America. It has been rightly described as a conceptual revolution (Aay 1972; Davies 1972), for indeed the full working out of its methodological implications had to wait, and to some extent, is still waiting. Awareness of 'process' and 'function' among European geographers may have come through their closer associations with climatology, geomorphology, demography, and history, but in America, it would come primarily through association with economics and transport studies, and only much later with sociology and psychology. Physical geography, by and large, would seek the company of earth sciences (Pattison 1964). 'Process' would also connote a somewhat different meaning to scholars of the spatial tradition than it would to those of man–milieu orientation, a distinction which would survive throughout that subsequent exciting romance with logical positivism and general systems theory. For the former would look to short-term dynamism of synchronic process, whereas the latter would look to longer-term dynamism of diachronic trends in relationships between cultures and natural resources. It seems worth while to consider the milieux in which this more basic curiosity about process was nurtured, then, for it may shed light on those queries raised by Wright in the 1950s.

Such insight could perhaps best be gleaned from a consideration of milieu, experience, and expertise of those geographers whose careers spanned the wartime period (see *AAG Annals* 1979). Ullman, for instance, like his teacher and later colleague, Whittlesey, studied at both Chicago and Harvard. In lectures he would often refer to the 'Harvard economists', Galbraith and Milton Freedman. There he also met Lösch and Isard, with whom he no doubt discussed Alfred Weber's ideas on locational factors in economic organization as well as his brother Max's theories of 'action-oriented' and *verstehen soziologie* (Weber, A. 1909, 1929; Weber, M. 1947; see Gregory 1978). Names such as Kurt Lewin, Stouffer, and Zipf, may have been familiar (Lewin 1952; Stouffer 1940; Zipf 1949). By the time Ullman boarded the train back to Chicago to do his doctor's degree, he must have already been well aware of what colleagues abroad were doing and also imbided a 'functionalist' way of thinking. In the host milieu there was already no doubt some receptivity to this: were Platt and Philbrick not already recognizing the salience of 'areal functional organization' in their field studies (Platt 1942, 1957; Philbrick 1957)? Had Colby not already published that ground-breaking article on 'centripetal and centrifugal forces' in urban life (Colby 1933) and Mayer become deeply involved in the Chicago Plan Commission (Chicago Plan Commission 1943; Mayer 1955, 1954)? With Ullman's conceptual imagination, Ackerman's discipline, and Harris's impeccable scholarship, in a city more studied than any during the interwar years, was this not the place and time for the convincing statement on a functional approach to the 'nature' of cities (Harris and Ullman 1945; Harris 1943, 1979; Mayer and Kohn 1959)? Wartime service in Washington, involvement with national transportation interests, teaching urban and regional planning at Harvard and elsewhere, all provided ample challenge and opportunity for rehearsing a practice of geography as science of spatial interaction, before he arrived in Seattle in 1951.

Chicago, as Pattison describes it in the discussion, would witness another major conceptual innovation in the late 1950s. But first one needs to visit two other milieux: Iowa and Seattle, whose common interests can be ascribed, among other factors, to the enthusiasm over logical positivism. Wartime events bring Bergmann and Schaeffer, articulate spokesmen for the Vienna Circle version of positivism, to Iowa where they meet young geographers during the early 1950s. Their approach to knowledge, shared in coffee breaks, seminars and long soirées of interdisciplinary discussion, laid emphasis on method, on the verifiability of statements, on the accuracy and statistical representativeness of data (Ayer 1959; Bergmann 1957; Brodbeck 1968). An irresistible appeal, no doubt, was offered for a 'value free' way of doing research in a milieu where intergenerational battles about the nature of geography were noisily aired. They would present their innovation still in a 'mapping' metaphor, albeit with aggregate statistical data, and would meet arguments about their lack of field grounding, their abstractness, or even the literary quality of their presentations. Schaeffer's bravado on the logical implications of making geography into a science of space would generate a kind of non-communication which is reminiscent of the Bernard–Abelard conflict of the twelfth century (Schaeffer 1953; Hartshorne 1955; Bunge 1979; Sack 1980). Orthodoxy would win the battle of the press, but scientism would win the war for a substantial number of the succeeding generation. Logical positivists of the Iowa School would initially stage and win their campaign *wholly within* the metaphorical stance of 'mapping'. McCarty would present his arguments for a systematic economic geography with a short logical and substantive cause and then proceed to illustrate his position in the context of United States regions (McCarty 1940; McCarty *et al.* 1956). Studies of wheat and corn yields, school locations, population forecasts, and market potentials for recreational services, all would follow a basically formistic approach. 'Value-free' and methodologically challenging, these Midwest studies voiced a delight in technical efficiency, and a perceived emancipation from the ideographic and merely descriptive (Hartmann and Hook 1956; Thomas 1960; King 1961, 1979).

Myopic indeed it would be to comment only on the methodological aspects of this movement. There was definitely a concern to render geography more respectable as science, but there was also a concern to render the discipline useful and applicable to the resolution of societal problems (Garrison 1979; Morrill personal communication). This was evidently quite as true in Iowa as at Seattle which welcomed both Ullman and Garrison in 1951. Conceptual imagination, well grounded in neo-classical interpretations of United States economic history, joins with the methodological skill and social concern of eager students. Enthusiasm mounts for theories of location, interaction, and regional organization: a full-blown spatial tradition could burgeon. On featureless plains one could postulate theoretical schemata to sharpen one's analytical focus and avail of more powerful models borrowed from engineering and mathematics. Time-honored methodological conventions and ideographic goals of chorology could now be transcended and new horizons of topological surface and geometric form could be sought (Garrison 1959; Berry and Garrison 1958a, b; Tobler 1961; Taaffe 1962). Mimeographed papers would flow between Seattle, Evanston, and Iowa, while on the East Coast a 'social physics' approach was being proclaimed (Isard 1956; Stewart 1947, Warntz 1966). The time was ripe for creating a Regional Science Association, with *Papers and Proceedings* to sustain the morale of its members and growing clientele. The decade was crowned by a visit from Torsten Hägerstrand who was not only a leading promoter of change in Swedish geography but also a crucial personal link between North American and European developments. If the Salisbury seminar and Chicago summer school were Mecca for mapping enthusiasts in the 1920s, then Seattle, Northwestern, and Lund became Mecca for enthusiasts for spatial analysis, geo-coding, and model building in the 1960s (Garrison 1960; Berry and Pred 1961; Taaffe 1979).

So far, this burgeoning curiosity about process and function and a more dynamic approach to industrial location, land use, transport, and urban networks, rested on assumptions about 'rationality' of a peculiarly entrepreneurial and individualistic type. Cost minimization, profit maximization, 'least effort' on the actor's part could be taken-for-granted, but there was also an equally peculiar desire to find physical laws modeled on physics and engineering to describe action on the landscape. But there were other ways for construing process. At Chicago, as Pattison describes, a fresh surge of energy followed the arrival of Wagner (1955) and of White (1956). Wagner, and later Mikesell, could bring their Berkeley training to meet the more functionalist orientations of their Midwest colleagues, eventually to prepare an outstanding selection of *Readings* which would place cultural geography in a wide international context (Wagner and Mikesell 1962). A similar meeting of traditions would bring verve and excitement to the Minnesota department. In their own key, or several keys, cultural geographers would reflect on cross-cultural pattern and process, form and function, opening the way for disciples to explore the methodological implications.

But Chicago has kept another tradition alive. Barrows's homily that geography should concern itself with man–milieu relationships, with earth stewardship, was not only preached but practiced as well throughout the New Deal and wartime years (Barrows 1923, Whitaker and Ackerman 1951; see also Marsh 1864; Brunhes 1902; Van Hise 1910). For White and his disciples, this was the core, and they were less dubious about environmental determinism than they were about the cost-minimizing, profit-maximizing interpretations of economic man. With colleagues in sociology, ecology, and psychology, they would look to the diverse manifestations of 'bounded rationality' which the behavior of societies and individuals internationally seemed to express (Simon 1957). A methodology would soon be at hand– systems analysis (Boulding 1956b; Von Bertalanffy 1951, 1968). From careful scrutiny of planning projects of the interwar years, too, they would conclude that a fresh approach to the study of environmental management would be an obvious corollary (White *et al.* 1958; Kates 1962). A twofold innovation is thus on its way: one is the analytical horizon of cognition, perception, and behaviour *vis-à-vis* natural hazards, and the other is a critique and eventual revision of Federal policy on flood plain management (US House of Representatives 1966). At the horizon of their hopes, too, lay a 'systems' methodology; processes of perception, behavior, and eventually control of man– milieu relationships could be framed.

While colleagues of spatial orientation would push forward on modeling the geometry or topology of spatial form and unraveling discrete mechanisms underlying trade areas, transport networks, agglomeration and scale economies in urban concentration/decentralization, however, the ecologically oriented Chicago researchers seemed to be reaching toward a more organic conception of world reality, to link theory and practice within an explicitly articulated social philosophy (Boulding 1956b; White, *ed.* 1974). Intellectually, they would look increasingly toward models used in probability theory and risk-taking decisions, psychology, biology, and more ecologically oriented sociologists (Duncan 1959; Simon 1956, 1960; Schiff 1970). Pragmatically, they would count on institutional and legal systems to implement their recommendations.

In the wake of quantification and process orientation, thus, the old distinction between man–milieu and spatial traditions now seems to reappear in preferences for organic versus mechanical approaches to systems applications. The former would point toward an integration based on dialectical tensions resolving themselves into progressively more cohesive and complex systems and the implicit world view is one of an organic whole. The latter would elucidate discrete mechanisms underlying particular processes, but what would beckon on the horizon—*Gödel, Escher, Bach* (Hofstadter 1979)?

Many and varied are the opinions and experiences of the 1950s in America. The early post-war generation made little attempt to disguise their faith in economic rationality, in business enterprise, and showed no qualms about Federal support for research on highways, industrial location, and the rationalization of public services, but they were by no means univocal in political stances. Kollmorgen would pray 'Deliver us from Big Dams' (Kollmorgen 1954) and mock US Corps of Engineers hubris on straight cost—benefit terms; White, from a different value position, would work toward reforming Federal policy from within, also arguing on cost—benefit terms (White *et al.* 1958, White 1966). The first generation of urbanists, too, showed a direct concern about planning problems, but the second generation seemed to be, initially at least, less concerned about ideology and politics (see, however, Bunge 1962, 1971). *Summum bonum* by the end of the 1950s for the second generation seemed to be theory and value-free technology; diversification and expansion of theoretically sound thematic specialties. For some, too, the interests of downtown and planning became theoretical problems. Was it not this neglect of the practical which became the Achilles heel that later generations would pierce with arrows from diverse stances?

Like many other 'revolutions' in thought, perhaps it was this very element of experiential groundedness that made the stockpile of techniques and models so inappropriate for indiscriminate diffusion overseas. What was good for General Motors may have seemed good for the whole USA during the 1950s, but the principle was soon to lose its lustre even at home. And perhaps it is this connection between thought and context (between *genre de vie* and *genre de pensée*) which geography may not, with impunity, cede to the imperializing claims of scientism or ideology. Does *pays* geography work best where there really are some *pays*, ecological zonation/succession models where urban land uses obey the laws of neo-classical economics, and 'quantitative', process-oriented models of spatial interaction most relevant where development strategies are transforming spatial patterns into topological surfaces of forces?

Textbooks on the 'revolution' begun in the 1950s would be written later, some across the Atlantic (Chorley and Haggett 1967; Bartels 1968; Harvey 1969; Beaujeu-Garnier 1971; Hard 1973; Wirth 1979), pushing aside older metaphors just as the interstate and supermarket would render highway and corner grocery store obsolete. Did one not need a 'new paradigm' and a *grand espoir* to face the challenge of social and environmental reconstruction during the second half of the twentieth century? Our last three essayists speak of this challenge and also share their reflections on what has been learned in the practice of geography during the post-war period.

Chapter 13

American geography in the 1950s[1]

Fig. 13.1 (Left to right) Professors Bill Pattison (Chicago), Duane Knos (Clark), George Kish (Michigan, Chair), Fred Lukermann (Minnesota), Richard L. Morrill (Seattle) (Netche Studios, Lincoln, Nebraska)

George Kish: Our topic is American geography in universities in the 1950s. It was a time of change, a time of growth. All of us around this table were involved with it: at Chicago, Minnesota, Iowa, Washington, and Michigan. Our main purpose here is to discuss some aspects of that change as it affected the directions in which geography was moving, its programs and personnel. Let me suggest that we look upon each of the departments with which we were associated at that time, and look back some twenty-odd years later upon those times in which we participated and which in many ways were the formative years of what American geography is at the present time.

I'd like to start with Dr Pattison who was at the University of Chicago then, as he is now.

Bill Pattison: Well, George, in reflecting on the Chicago department earlier today I found myself adopting what might be called a conception of 'faculty determinism', that is, a belief that who was on the faculty at a given time was indeed of decisive significance. When I joined the department as a student in 1950, the faculty on hand created a rather special atmosphere – one that was to be dispelled only a few years later. Robert Platt had just become chairman;

Henry Leppard and Edith Parker were shortly to be leaving; Chauncy Harris was becoming securely established; and the following were all new: Wesley Calef, Norton Ginsburg, Harold Mayer, Allen Philbrick. There was a pervasive feeling that a new geography had come to Chicago as of 1950. In fact, a distinct phase in the life of the department had opened, one that is well worth remembering. But it was a brief phase, of an essentially transitional character. I believe during that time Robert Platt was trying to create a Chicago school of thought. He was working under two prevailing conditions, the removal of which was to bring the phase to an end after a few years. The first condition was one of continuity. Norton Ginsburg was a graduate of the department, as were Harold Mayer, Allen Philbrick, Wesley Calef, and Chauncy Harris. So too were the soon-to-depart Henry Leppard and Edith Parker, and Platt himself. The second condition was one of intellectual isolation, which applied perhaps most of all to Platt. There were great dangers in this. I have learned since then that the dissolution of our department was seriously considered at that time by the university. I wonder now whether this wasn't because of an awareness of isolation from other fields of knowledge. In any event, the department survived – in an atmosphere, by 1955, of what might be called baffled hopefulness. Platt's leadership had allowed old Chicago traditions to be pointed in promising directions, and a talented younger faculty to become established, but the enterprise was not realizing its potential. A great break in departmental history, which brought with it a release of energies and a rush of forward movement, began with the arrival of Philip Wagner, in 1955. To the fresh stream of ideas supplied to Wagner were added, in 1956, those of Gilbert White (a Chicago graduate to be sure, but one who had incorporated, during his many years of absence, points of view external not only to the Chicago department but also to the entirety of American geography). Not long afterward, and largely at the instance of White, Brian Berry and Marvin Mikesell arrived. White had become chairman; the intellectual isolation of our department had been breached and there was widespread intellectual re-sorting and reconsideration. Harold Mayer, in his recent reminiscence on Chicago in the Annals, has written that when Gilbert White was asked what he envisioned for the department at that time, he said: 'The program of the department should be the sum of the programs of the individual members of the department.' He brought to an end the resolve to compose a single school of thought. Under his aegis a gentleman's agreement was struck, an acceptance of many departments within one, which continues to the present day. And there was an encouragement of intellectual exchange and exposure, which also had a lasting effect. To my way of thinking, then, geography at Chicago showed two distinct phases within the decade of the 1950s. Of the two, the earlier one was real to me as a student, the second was real to me by inheritance: it is the one to which I returned later as a faculty member.

Kish: It has often been claimed that the Midwest is the heartland of American geography. It's also been said that American geography is within 500 miles of Chicago – apologies to Seattle. We will move on now and I turn to our colleague, Dr Knos, who at that time was at the University of Iowa.

Duane Knos: I started to work at the University of Iowa in summer school in 1949. I was a school teacher, disillusioned with painting houses in the summertime. It was about three years after the department of geography had been established in the College of Liberal Arts. That it was a new department is interesting, I think, because of what subsequently became an experience of much innovation. In a very real sense that department socialized me into this geography community. The department centered around Harold McCarty, the first chairman. He had received his degree at Iowa when geography, sociology, and economics were all together in the College of Commerce, as I understand it. He and a number of his compadres – a couple of economists, Woody Thomson and Clark Bloom, a sociologist named Harold Sanders, and a social psychologist, Mannfred Kuhn, were sort of the young Turks in developing the College of Commerce. And when they split the sociology and geography departments out of the College of Commerce and into the College of Liberal Arts and made them separate departments, that association was maintained in a probably more intimate association than you normally see across disciplinary lines. When I came as a graduate student I found myself suddenly not just with geographers, but with economists and sociologists as well. We were all friends and talked to each other, we had the same metaphors and the same language. That's one aspect which I think was important in the kinds of work people did at Iowa in the middle 1950s. Another had to do with the town of West Branch about 20 miles east of Iowa city. It was a Quaker community in which the Friends brought displaced scholars from Europe essentially as a staging center. A number of those scholars ended up at the University of Iowa during the war period, among them Kurt Schaeffer and Gustav Bergmann. Schaeffer and Bergmann were good friends. Bergmann was a member of the original Vienna Circle and probably the premier philosopher of logical positivism in the US. Schaeffer was incorporated in the geography department, Bergmann in the psychology department, but he had a tendency to integrate himself with the geographers, sociologists, and economists, with an interest in philosophical problems of the social sciences. The third element, I think, is McCarty himself and the rest of the faculty. I mentioned Kurt Schaeffer, but also George Hartmann and Walt Wood who were rather remarkable people, particularly because of the intellectual community that they formed. They had offices on the third floor of the library at the west end and went to coffee together, along with whatever number of graduate students wanted to go, every morning at 10:00 a.m.

Kish: Time-honored geographical pastime.

Knos: Yes. And also on the third floor of the library were the philosophers who used to go to coffee at the same time. I look back with great fondness at that regular half hour to an hour of general discussion where I probably learned most of what I've learned. In this context particularly the experiments with statistics took place. One of McCarty's mentors was a business statistician named George Davies, and another friend was C. Frank Smith.

That whole group was familiar with statistical analysis. And while they didn't do much of it, they had the resources to pursue it. But the effect of Bergmann on that community was important. We did statistical analysis, regression models and that sort of a thing. But I would suggest that we were really dealing with positivism, and a view of science in which statistics was just a very handy way to play that type of philosophy out. And in that rather incredible ferment you had not only geography but also psychology, philosophy, sociology, and economics with a philosophical commonality. From that environment most of the innovation at Iowa originated and was carried on.

Kish: One of the features in our earlier discussion was the expansion of geography during the 1950s. One of the departments with that expansion very much in evidence was Minnesota. So I now turn to Fred Lukermann.

Fred Lukermann: Expansion, yes, but more important still, it was a period of transition, and we all recognized it as such. I suppose I was as conscious of that as anybody since I walked into Minnesota as a freshman in September of 1940, left it for the war, came back in 1946, and the department was completely new. Davis was about to retire, Brown had only two years left of his life, Hartshorne had gone to Madison. Weaver came in 1946 just fresh out of the AGS,[2] and the Arctic Atlas. Broek came in 1948, Borchert the year afterwards. So it was a complete transition between the war period and the beginning of the 1950s. In the 1950s, I think, probably the greatest change was the enormous efflorescence of geography, both in faculty and in numbers of students. If I remember correctly the enrollment in introductory courses started climbing almost 20 per cent a year. As graduate students, fewer in number, we suddenly had to service this group. We became faculty and we weren't told we had choices, we taught what they wanted us to teach. So my memory goes back to a number of transitions. The old department with Hartshorne, Brown, and Davis had a strong tradition of research. It was said that Davis locked the door at noon in the department, kept all students away and told the rest of them to get down to their research. I remember that when Davis retired he was writing his second edition of *Earth and Man*, and he was in the back room behind the only lecture room we really owned, a cartography lab. He used to counsel us as graduate students, 'Don't get married until you get your Ph.D.' He was an old Yankee from Michigan, as you know. Weaver also brought a change. Although he was young and originally had been in Madison, he actually came out of Berkeley. Broek, who had been at Berkeley all during the 1930s had gone back to Utrecht for a year after the war, and came in as the new chairman. It was a two-man department at that time. Then came Borchert, the all-American boy from Wisconsin . . . I remember that very distinctly because his style was something quite different again from the old tradition and the Berkeley tradition that had been imported by Weaver and Broek. I remember the first time I heard him review his thesis on the prairie triangle – it was a new physical geography. I think that was one of the first things I ever heard John say – how happy he was to get away from Wisconsin and not to do 'Finch and Trewartha' any more: he went into a resource geography. Other changes came with consciousness of what was happening in the discipline nationally. It was a period of enormous interest in the outside world; the area studies programs were booming, we had gone into police action in Korea – that's what they called it – there was a vast influx of GIs and people who had delayed their education – enormous intellectual activity. And then you had all those technological changes around the country. The Twin Cities area, for example, had just broken out of its political boundaries and started its urban sprawl – the kind of thing Borchert was interested in – and we were essentially into applied geography: studies of public roads, demography, changes in technology in the American Midwest, the Corn Belt, and so on. And then, of course, what affected us most was another wave of thought which came out of Washington, Northwestern, and Iowa, as Duane described it – that hit us in the late 1950s. We were very critical about that. We had grown up with the Hartshorne and Brown economic, political, historical geography of the 1920s and 1930s. Then we had on top of that the Berkeley tradition, the cultural, historical tradition which we welded together with Brown, and we had this quantitative theoretical model-building school. It was enormously exciting: I mean, you knew that things were changing, and you knew you were part of it, and we could do what we wanted to. I think that's the best memory I have, that we were in charge.

Kish: We've just said that geography is a Midwestern field, yet I think it would only be fair to say that things were happening out on the West Coast as well. We're very fortunate to have a witness of what was happening in the Northwest: Dr Morrill.

Richard Morrill: It would be true to say that the University of Washington had no reputation nationally before the 1950s. Then rather suddenly it exerted considerable influence in the late 1950s. The influence came via Chicago and Northwestern from that Midwestern heartland, and not from our rival to the south in California which even today seems many thousands of miles further away than Chicago and Northwestern. It's interesting that none of the early group of quantitative students had any such intention when they went to Washington. I think all of us had a very traditional background. Mine was in physical geography at Dartmouth. So our conversion came after we arrived from a combination of people and circumstances.

Kish: On the road to Seattle?

Morrill: No, after we got to Seattle. I think the most important influence or at least the first, was the chairman Donald Hudson who came from Northwestern. Among other things, he challenged us not to accept the Midwest traditions and encouraged us to question even Hartshorne. He encouraged us to worry about the status of geography and its relatively low esteem – that never occurred to us until he told us. He also told us that geography could be practical and influential which gave us encouragement. But I suppose the most important thing he did was to bring both Garrison and Ullman in 1950 and 1951 to Washington. Ed Ullman was the second big influence on us. He never understood the slightest thing about statistics, of course, and didn't want to, and yet it was Ullman who was the fountain of ideas which we then used statistics to test. And it was really Ullman who introduced the concept of geography as a spatial science, who introduced Schaeffer to us, who introduced models and theories to us and led us into this questioning of Midwestern traditions. So his influence was much greater perhaps, than we realized at the time. But the greatest influence, at least on me, and probably on most of us, was Bill Garrison. He was a charismatic, prophetic leader, and I use these words because I have to confess that we almost fell into a discipleship relationship with him. The message was almost a Messianic one: that we needed to assault the citadels of Midwestern tradition, to go out to convert the heathen.

Kish: In a way it's interesting to reflect now at a twenty-five plus years' distance that the greatest single impact since that of William Morris Davis, one that gained influence for American geography worldwide, was precisely that message which came out of the University of Washington. Now in my case, to put it very briefly, Michigan probably changed less in that particular ten year period than any of the four institutions we've been talking about, partly because the natural process of attrition moved rather slowly, while our enrollment increased. We had, as Fred Lukermann said, this great influx of people enthusiastic for more knowledge, enthusiastic to get a degree, thinking of the years they had lost, and it had a very direct impact on the quality of our graduate group. Yet the staff didn't really begin to change until the first apostle from Seattle, John Nystuen, arrived in 1959. One of the points that comes out here is the interconnection in cross-country influences. Dick Morrill's remark that in Seattle, Chicago and Northwestern were much closer to them than Berkeley or the City of the Angels makes me wonder how you today look back upon the role of such influences – the printed media other than journals, the movement of these human interests in men, the controversies. I'm thinking for example of all the controversies around Schaeffer and their influences on the general atmosphere of American geography in the 1950s.

Morrill: We perceived ourselves out there as fairly isolated: it was difficult to get to meetings, we were unlikely to be published fairly readily. So I think one of the interconnections which had importance for us was the beginning of the Discussion Paper series in January 1958. This was quite explosive and became an underground circulating series too, especially to Iowa, but also spread to the dissidents at Wisconsin, Minnesota, and many places.

Lukermann: I think it was the visitors, George. We were moving so fast that we kept one main position open for visitors. Sauer came in 1950 to dedicate the new building we were in. Waibel was there, Pfeiffer, William-Olsson, and American geographers as well, and if you add to that the influx of new graduate students, the situation was fluid. Perhaps we think too much of the historical myth of Midwestern isolationism. We really did get out into the field, and the West Lake conferences, I think, were instrumental. We'd fill a bus and go . . . and the interchange with other graduate students was just as important as reading the journals, and the literature.

Kish: And going to the annual meetings, the formal, and in retrospect far more important, the informal postludes of sessions at the AAG.

Lukermann: The AAG just expanded in the 1940s.

Kish: When we get right down to it, the credit for one vital part of this whole changing atmosphere may be laid squarely on the doorstep of our professional association which after a long period of forty years suddenly underwent a complete change in 1948.[3] Look back on that last Wisconsin meeting in 1948. After that, nothing was ever quite the same. It is not just a matter of numbers, it's a matter of the atmosphere, the greater freedom, and that no one, I think, after that date was excluded the way, I dare say, one used to be. Sometimes it was like international conferences where the heads of delegations sat at their tables, rows upon rows, and in the background were those who really knew the answer, but they could only speak when called on. The senior citizens of the profession held forth, the rest being more or less excluded. All that, I think, changed after 1948, or shall we say beginning with the 1950 meeting, and that in itself to me was a very important ingredient of the new atmosphere.

Lukermann: There were only two or three journals also. You could keep up very easily, and it got you into the field. There weren't the enormous monographs which came out of my own field of history. God, you couldn't comprehend what was going on in that field, but in geography you could grab it, and you could do it yourself.

Knos: The network was also highly personal and highly oral, and there weren't that many people at meetings. You heard about them, read about them, but you also heard personal anecdotes, and there was a great deal more intimacy in that. I look back on those West Lakes and divisional meetings with an incredible fondness, because much of that paranoia we felt about statistics and so on was played out in a graduate student's living-room at Wisconsin.

Pattison: This may be the right time to speak in more general terms about relations between the university departments, on the one hand, and national meetings of the AAG and regional meetings on the other. New ideas were issuing from the departments for interchange under the auspices of such organization as West Lakes and the AAG itself. It seems to me that the AAG was coming forward more assertively than before in exercising its functions in spokesmanship for all of us, in setting standards – especially through the *Annals* – in the general fostering of communication, and in coordinating the work of individuals and groups. But the responsibility for production, for intellectual creativity, reposed in these changing departments.

Kish: Let's go back for one moment to something that Fred Lukermann just brought up earlier – the changing perspectives of American geography in the 1940s and the 1950s as a result of our history. There was a great upsurge of interest in the rest of the world, a tremendous amount of energy suddenly was channeled into a new set of directions – Japanese studies, Russian studies, South-East Asian studies. You were speaking earlier of intermixing and meeting on a fairly regular basis with colleagues in other fields. At Michigan, we did not have the good fortune of Iowa. Our interconnections within the campus came largely as a result of studies – urban studies, the Detroit project, the Flint project – where sociologists, psychologists, language people, literature people, political scientists, and historians, discovered that there was such a thing as geography. And staff, graduate students, all became increasingly involved to the extent that doctoral theses may have been written largely outside the department, actually under the leadership and supervision of a geography staff member but with a very much larger input than ever before from other disciplines.

Pattison: George, I wonder if we could ask Dick Morrill about the significance of area studies as an occasion for interdisciplinary encounter. Ed Ullman wrote in the early 1950s that he found representatives of other disciplines, working on area studies problems, who defined to his own satisfaction what he represented as a geographer.

Morrill: We had very close connections with several other departments, especially in programs on China, Japan, and the Soviet Union. I suppose in fact that over the past thirty years, maybe half of our graduate students chose those specialties. Another very important connection, for Garrison's graduate students especially, was the relationship with civil engineering, because this was our first introduction to large-scale research projects; most of us became involved in highway impact studies which continued to influence us for a long time to come.

Lukermann: I think that's very important. Weaver's project on crop combinations in the Midwest was an enormous resource for getting graduate students to do things together. You didn't get off and do your thesis, you were actually working with other people on bits and pieces.

Morrill: It probably was a change from individual research to group participation in projects.

Lukermann: And jobs were the result. I think at one time, probably toward the end of the 1950s, half of the planners in the Twin Cities were geographers.

Kish: If I look back on it, one of the truly striking differences between the pre- and post-war periods was, as Dick Morrill just stated, that a doctoral thesis used to be a lonely thing. You went out, you did your field research, you met with your thesis chairman, and occasionally you met people from other disciplines and you were chronically short of funds. The big difference is first the very large increase in support, but more than that, is discovering and being discovered by outside funds for research support, which certainly was not the case prior to the 1950s. The impact of that on our attitudes and values, and the speed with which we were working, I look back upon as a double-edged weapon. It's sometimes not easy to slow down your own thesis work, yet it gave you possibilities you used to dream of.

Morrill: We especially owe a lot to the Office of Naval Research, not only for sponsoring many of these projects, but later on for sponsoring symposia in which people from around the country got together for the first time.

Lukermann: What was the Navy doing in the Midwest?

Morrill: A good question. Supporting us.

Kish: As a matter of fact, another cover word we could use in the 1950s is mobility in our profession, which was mobility worldwide. Geographers suddenly found their ways through programs like the office of Naval Research to do doctoral theses in remote corners of the world. But I also think of mobility as demonstrated by the number of people attending divisional and national meetings of our association. We had the interest, we were excited, things were happening, and while I agree that our publications were still limited, we were anxious to see what the other fellows were doing, and we could afford to move – which was not the case before.

Lukermann: I want to bring up one more thing. We tend to talk about names and individuals and this sort of thing. But I think we as graduate students fed each other as much as many of the distinguished faculty did. That was possibly because of our increase in numbers. I think there was another point about context: our society in the Midwest was enormously mobile and the national meetings brought us together also. That more than anything else made you feel you were part of a society with your own age, coming out of the same background as yourself, and you were not some sort of disciple.

Pattison: Let me take up a special aspect of mobility: in the decade under examination, Chicago experienced a great renewal of movement. Having been long accustomed to being a sender, that is, a point of origin for faculty members around the US, we became a significant receiver. I think again most particularly of Marvin Mikesell, Brian Berry, and Phil Wagner. A certain disturbance was set up in the department by the encounter of unlike frames of reference. Phil Wagner says introducing his book, *The Human Use of the Earth*: 'A great inspiration and much of my training are derived from Carl Sauer and his colleagues of Berkeley, under whom I did my graduate work. Later, at Chicago, I was confronted with a new and unexpected kind of geography and was stimulated to an attempt to reconcile it with the tradition I follow.' That is, he held to his convictions, and yet he never was the same once this encounter had occurred – and the consequence was a heightened creativity.

Kish: There's another aspect of mobility; a European scholar may look upon American universities as far more in flux than the European ones, where appointments made are usually lifetime ones. I think that in our own discipline, the majority of the staff came out of two institutions in terms of their academic training (Chicago and Michigan). Yet even from the 1960s on and very much more now, faculty, staff, and graduate students came from a very widely varied background. In part I presume, this is because of the means available, in part because of the expansion in university enrollments. It was not as spectacular at our place as in others, where to teach introductory psychology courses, they hired the biggest hall in town, an auditorium of 4,500 seats! But there was a great expansion through the 1950s, expansion of budgets, of employment possibilities . . . just about everybody was involved, so that we believed there was no end to it. We know better than that now in retrospect, but then it certainly seemed we would just keep going.

Morrill: Not only could we add faculty, but as Fred said, we enjoyed a series of visitors. In Washington we may have seemed from afar to have a narrow quantitative focus, but in fact we had tremendous internal debate on the subject, and we had a series of visitors such as Richard Hartshorne, Preston James, and Jan Broek even during the most extreme years. And I suppose the most influential person, at least in my career, has been Torsten Häger-strand who visited from Sweden in 1959. His visit, of course, influenced many of us for years.

Lukermann: It was also an open society, George. I think my first publication was with Weaver. I remember Knos and McCarty, I remember Morrill and Garrison. That kind of collaboration, although it may be frowned upon as using graduate students, was tremendously helpful to us. And there was no lack of opportunity to publish those papers – there was enough demand for this sort of thing in the Occasional Papers, the Discussion Papers and so forth. Thank God for the xerox.

Kish: But there was also the staff seminar where everybody got his shot on an equal basis – the intellectual interchanges in those seminars were hot and incredibly heady experiences. Those were some of the most exciting times in my life.

Lukermann: Anne Buttimer calls this 'social construction', I think.

Kish: Yes, I think it will go under that heading. We started out by saying that geography expanded in staff, it expanded in students, there was a changing emphasis in the thrust of the field as against, let us say, the 1930s and the introduction of entirely new aspects, techniques, and points of view.

Pattison: It seems to me that it was new for the already resident faculty in the Chicago department of the early 1950s; they were, in effect, waiting for techniques by which a follow-through on current designs and intentions could occur. For example, Allen Philbrick was talking about decision-making with optimism and even excitement, and yet it wasn't really resolving into something operational. Similarly, Robert Platt was enunciating basic insights on the differentiation of regions and organization of settlement systems, but I believe he had reached the limit of his ability to put them to work for re-search purposes. It wasn't until the arrival of Gilbert White that decision theory became available, making it possible for the department to act productively on Philbrick's general orientation. As to Robert Platt's general orientation, not until Brian Berry's arrival from the Garrison group was there a local capability for doing something theoretically interesting about it.

Kish: We've spoken of various meetings, the chance to get people together in the field. We were all aware of the famous Chicago Midwestern field conferences of the 1930s, which was one of the creative influences at that time. And for a short while, in the 1950s, there was an attempt made for people within a subsection of the discipline to get together. We had it for a while in the late 1950s and into the 1960s, people interested in quantitative techniques at Michigan State, Wayne State, and the University of Michigan. We also had from the late 1940s into the later 1950s, a successful attempt to get people together who were interested in maps. Some were cartographers, some were interested in graphics – it was there that I first heard of computer graphics – some, like myself, were interested in the history of map-making. We usually got together in Chicago, once or twice in Ann Arbor, and it was a microcosm of a particular interest group. Now we have it, of course, on a national scale, but the fragmentation is almost too great for me to identify with. Previously, we had the get-together within the campuses of our region, within driving distance of each other. It resulted in some publications, some papers, and some very good informal discussions, which in the long run is the one thing we will all remember.

Pattison: This is what I was calling on another program, earlier in the day, an 'invisible college'. But I wonder: mightn't there have been *several* invisible colleges developing at about that time with exceptional vigor, and on a regional basis?

Lukermann: We had the NDEA institutes,[4] we had field work in the Ozarks, and out west, Mather, Salisbury, and myself. I think there were different forms in which we did these things, and they might have been smaller. It wasn't a question of getting all the profession out in the field as this survey group was in the 1930s. But I think we 'had it' in terms of various types of mechanisms to get the job done.

Kish: Did you have the impression in the 1950s that geography was gaining ground, getting to be more important, more popular, getting to have a better and a stronger place in the field of general education in this country?

Lukermann: Yes, I think it was called general education at Minnesota more than liberal education, but that's what it was. We expanded in undergraduate enrollments, there were significantly high proportions of graduate students, but that still was a very small number. I think we spread to interdisciplinary, multidisciplinary, and cross-disciplinary directions out of area studies after the war and then into planning and engineering. What we were doing with other disciplines is strikingly different, it seems to me, from what I hear about the 1930s the 1920s and the 1910s. I think geography has influenced its sister disciplines. That's as important as anything that's been done to geography itself.

Kish: There were also certain changes in educational circles. Bill, you can correct me on this, but it seems as though much of the undergraduate enrollment which came during that time was exactly in those departments where colleges of liberal arts were putting in core distribution requirements, and geography was put in to satisfy either the physical or social sciences – or maybe both requirements. Perhaps a lot of students would rather take geography as a science than physics for example; we then gained a clientele that we could do something with and develop.

Lukermann: I think the other disciplines wondered where in the hell we were coming from.

Kish: I do too. Now gentlemen, the title of our conversation was the changing context of geography within the wider frame of universities. One interesting thing which we have not really touched upon is that many practitioners were somehow involved in the defense effort, and not just for the period of the wars. After the Second World War and after Korea, perhaps there was a trend to recognize geography in the halls of planning and government. This is something I'm not entirely sure about myself, but it is certainly part of the context. Do you think that these developments brought a greater number of employment possibilities? I hesitate to say greater influence.

Lukermann: Yes, I think there is and there was.

Kish: City planning, for example, had an incredible explosion in demand for people, and relatively few people to take the jobs. It seemed like that was one of the really rich areas where a number of people who were interested in urban studies went.

Lukermann: I don't know what the field work was like in the 1920s and 1930s, but the development of the tools in the 1950s, I think underlies what we are doing now in environment, pollution, perception. I think that's when we came of age.

Kish: In a sense what we see today is really the projection of the 1950s, if we could look at it for just one moment. The arsenal of research tools, such as remote sensing, is a direct outgrowth both of the quantitative approach and of a bigger renaissance, which I think we haven't touched upon, of mapmaking and new ways of presentation in American geography. If you look upon the maps of our predecessors and those of the 1950s and the 1960s, I think there is a very substantial qualitative change. And in a way, that word underlies this whole decade. Having concentrated our attention entirely on one decade, it seems to me justified to say: Well, how do we look now upon the two following, since we are almost at the end of the third one? How do we look upon the events of the 1960s? Have we continued the same trends and the same changes, have we introduced a new thrust, or are we still more or less following what happened in what seems to be the one big time of change, the time of the 1950s? Do you think we are still moving along those particular channels and lines?

Lukermann: Yes, but on many fronts. I don't believe it was a quantitative revolution, but what happened was quantitative and we didn't have any other names. Now we are doing it in five different thrusts. It seems to me it's much more complex, but it has the same nature.

Pattison: Perhaps it is time, then, to remind ourselves of the title for this session given when invitations were sent to us: 'Contexts for Creativity'. It appears that all of our departments, whether in the 1950s or in the decades since then, have demonstrated that certain conditions must exist if creativity is to be brought forth. One is the availability to students of strong guidance combined with a permission to depart from current conventions. Another is access to more than one conceptual frame of reference, for comparison and confrontation. And a third is openness to the best contemporary research methodology.

Lukermann: I'm glad I reached the 1970s, George.

Kish: We'll end on that happy note. I thank you gentlemen.

Notes

1. Edited text of a videotaped discussion held at Lincoln, Nebraska, April 1978. Courtesy of NETCHE Studios.
2. American Geographical Society.
3. Association of American Geographers.
4. National Defense Education Art.

Chapter 14

Fig. 14.1 Gerrit Jan van den Berg, Holland

'I've finally figured it out,' Gerrit Jan van den Berg said at the Sigtuna meeting, 'you're forcing us all to have a spiritual sauna!' Such a prospect might have caused another less enterprising person to flee from the scene, but not so this practically minded adventurous soul. Van den Berg is acclaimed throughout Europe for his innovative efforts, already in the 1950s, to get people to assume an active, responsible, role in planning their own environments. Even in this society, so positively disposed toward planning, he has had to suffer from being so early with such radical ideas. Confrontations, conflicts, the critical self-awareness which can turn failures and setbacks into learning experiences – these are the stuff, in his own view, of which creative careers are forged. Perhaps this explains his subtly ironic response to my nit-picking editorial suggestions on his first draft of this essay: 'I think I understand my own ideas better now.'

His life story is a lucid statement of faith in the power of functionalist and systems approaches to geography in practice. Be it art or science, geography is for him a medium for practical problem-solving; he is understandably skeptical over conceptual and philosophical nuances. It is from experience as well as logic that he derives those convictions which guide his teaching and his actions.

Some personal reflections on geography

Introduction

My interest in geography awoke when I was a boy of only six or seven. Then I was greatly immobilized by a physical handicap that kept me from most physical activities and all sport. I started to read and I became very fond of it. At that time, geography was to me a way to world orientation, extending from the very local to the global measure. From the very beginning, then, geography has always had an instrumental meaning for me.

My interest in geography was closely interwoven with my interest in history. World orientation is never purely spatial nor purely historical. History concerns people in their time, geography concerns people in their space and place. From the very beginning, I perceived that geography could be really instrumental only as human or social geography. That feeling has not left me since.

Stamp collecting fed my interest in geography and history. Postage stamps tell tales, they carry symbolic images. I trained myself in deciphering those images, and they also became a source of information on world orientation. I still collect postage stamps. . . .

My secondary education was rather broad. In addition to history and geography, I studied Greek, Latin, French, German, English, and Dutch as well as mathematics, physics, chemistry, and biology. After matriculation I wanted to study geography and history as main subjects, but that proved impossible at that time. I decided to take (human) geography as main subject and history as minor. But I kept my university study as broad as feasible, and engaged myself intensely in students' clubs, conferences, and practical youth work.

This essay deals with the outcome of such broad-ranging interests. The outcome has been molded by my personal predilection – inherited from my father – for a professional job in the civil service in close contact with university research. When I left home at the age of eighteen, my father counseled me to be helpful to everybody who would ask or invite me to help them. As a civil servant I always tried to preserve a great personal intellectual independence, but in assisting other people my time became filled with numerous very diverse activities covering a wide field of concrete practical problems. Almost all of those problems have been brought before me by employers or by other people, very few of them were selected by preference or out of personal scientific interest. That has been the weak point in my whole career: failure to develop and implement a private research program for myself. I never wrote a Ph.D. thesis, but I enjoyed the privilege of a wide range of personal contacts leading to many very stimulating encounters, which have profoundly influenced my intellectual, spiritual, and cultural maturation throughout my lifetime.

My personal reflections on geography now will be presented in four sections:

1. looking for a workable and pragmatically fertile doctrine in (human) geography;
2. my education and training in geography and my experiences with geography in practice;
3. crises and doubts as preconditions for the conception of new ideas in answer to new challenges;
4. the ecology of my thinking on geography and planning.

Search for a workable and pragmatically fertile doctrine in (human) geography

The existence of university departments of geography does not prove that geography is a science or an art. The academic status of geography may depend on those few professors and staff-members who produced research papers welcomed by their critics; the continuation of those university departments depends more, however, on society's demand for young professors either as school teachers or as handy professionals for finding imaginative solutions to practical problems.

School geography is, of course, very different from academic geography, the former being directed to world orientation by teaching students to recognize typical physical forms and social ways of life (*genres de vie*) in reality, the latter directed toward critical evaluation of those forms and ways of life. From the early 1930s in the Netherlands the national market for school geographers became saturated and young geographers had to find jobs by steering their discipline toward solving practical sociospatial problems. The analytical skills of their discipline, however, fell short of the challenge. Their conceptual typologies served to describe concrete geographical structures, but not to solve sociospatial problems. In order to survive and to succeed in their jobs they had to borrow concepts, methods, and theories from other disciplines like economics or sociology; in doing so they moved away from what was then considered as the core of geography. Academic human geographers did not want to get involved in what they considered to be politics and did not meet the new challenge. In the Netherlands, however, with its dense and fast growing population, its expanding cities and towns, and its colonial empire not yet fully under control, professors in human geography gradually began to remold disciplinary theory and methodology in close cooperation with their young disciples who were engaged in practical problem-solving jobs.

At first this remolding seemed to involve mainly the incorporation of economic and sociological concepts and reasoning into human geography, e.g., the concepts and theories of locational and social change, but soon a full blown switch occurred away from sociospatial *structure* as the keystone of human geography theory and methodology to sociospatial *processes*. The contributions of modern human geography to the elucidation of those processes—like urbanization and its counterpoint ruralization, regionalization (in quite another sense than the construction of a conceptual typology of geographical regions), migration and mobilization (e.g., motorization), diffusion and sociospatial innovation—have been astonishing both in the Netherlands and elsewhere, e.g., Sweden and Poland.

How to relate a theory and methodology of geographical processes to those already conceived and tested concerning physical forms and ways of life? Geographical processes occur in space and time, and materialize cumulative effects by investments in infrastructure and buildings and by social institutions, behaviour patterns, and concrete value systems. Geographical processes also remold and change, even annihilate, the effects of previous geographical processes. The well-known 'structuring forces underlying physical forms and social ways of life', are exactly the processes which geographers employ when they are asked to study and elucidate problems, and using the results, to manipulate and change problem situations.

Academic geographers, with some exceptions, appeared to be eager and capable of integrating a more dynamic approach to the analysis of change.

Many, however, were very reluctant to investigate processes, and thus a major conceptual, theoretical, and methodological cleavage arose between those concerned with problem-solving and academics. Some of the former felt themselves betrayed by their discipline and joined other disciplines (economics or sociology or political sciences). Others created something like a new discipline (e.g., in the Netherlands 'planology', and dedicated themselves to critical reflection on the theory and praxis of practical problem-solving. From the point of view of geography as a science or an art this development has to be deplored. Unfortunately for the discipline, geography thus has voluntarily yielded an important and promising new pragmatic part of its terrain to scientists of other (old or new) disciplines.

The movement away from deterministic (causal) to probabilistic (stochastic) approaches to the understanding of geographical processes has meant that now these processes have become more amenable to being influenced by planning. Geographical processes in reality are subject to political processes or, in any case, are subject to the official or hidden policy of interest groups and political bodies. This means, in fact, that geographical processes are implicated in the molding of contemporary history. The interwovenness of human geography and history is also manifest in another and very pragmatic way: geographers who dedicate themselves to solving social problems handle both disciplines not retrospectively as many academics do, but prospectively, with their eyes on the future of the concrete community that puts their problems before them. They are also conscious that the present adult generations will be fully accountable for the potencies and qualities of that future. Geographers engaged in practical problem-solving have become aware that they are far more occupied with 'action-oriented geography' than with 'applied geography' (a term loaded with the mental reserves by academic geographers who failed to support their colleagues involved in practical problem-solving), and that scientific activities have to do fundamentally with social ethics – they must be other-directed, not self-directed.[1]

These considerations have convinced me that modern geography needs an explicitly articulated scientific doctrine. This conviction has evolved during the experiences of my lifetime. I started about thirty years ago with a public statement in a meeting of a Dutch professional society of geographers. 'Geography,' I said, 'has to be human or social geography or it has not to be at all.' At that time I was laughed at and blamed as well, quite understandably; but now I still consider this as a fundamental and primary part of a workable and pragmatically fertile doctrine in geography.

The second part of my doctrine concerns the central focus of the discipline. It is not world orientation because world orientation is not other-directed but self-directed. The central focus is ecological – the relation between human beings and their total environment, which consists of three inseparable components: the natural component; the built-up and arranged component; and the social (socio-cultural) component. These components are open to change by geographical processes even when planning addresses itself to one or two of them only. It is the merit of modern urban, rural, and regional planning to have elaborated this insight. American planners particularly (only few of them being geographers), have clarified this part of the scientific doctrine in geography.

The third part of the doctrine has to do with the study of concrete communities in the context of the integral environment on and around their territory. Two essential processes should characterize community self-control – humanization and environmentalization. Humanization means providing every member of the community with responsibility for his personal choice of individual purposeful activities in support of the community's democratically defined objectives. Environmentalization means purposefully contribution to the goals of surrounding and embracing communities that have power to remove or mitigate the constraints to the community's own purposeful activities. This third part of the doctrine has been derived from my experiences in urban renewal and regional planning during the 1970s and the appropriate instruments have been borrowed from the American operational research scientist Russel Lincoln Ackoff.[2]

Geographical processes occur where concrete societies, with conscious discretion over their environment, transform constraints to their activities into basic conditions for action, and plan for the design and collective implementation of threshold conditions for new developments previously considered unfeasible. This may sound like a paradox, but it is not. Self-control plus humanization plus environmentalization, however, will bring these transformations within the community's reach, always and everywhere, unexpectedly.

With such a doctrine geography becomes a very workable art within the four fields of planning: urban planning, rural planning, regional planning, and urban and rural revitalization (social + physical rehabilitation). I have recently developed a general working model for planological actions on those four fields with the help of the above doctrine in geography. This planological action model places all building processes and community organization and development processes, mutually interwoven, under the guidance of learning and planning processes, properly integrated, directed by a common command in each case for that purpose voluntarily articulated by the community's members. This working model allows one to discern both the geographical processes behind the cumulative effects they generate and their creative and remolding power. All members of the community are thus urged to become accountable for a stewardship of their integral environment. In this way, action-oriented geography will become other-directed and, thereby, a valuable support to maintain academic geography in our universities, provided academic geographers prove sensitive to the challenge.

My education and training in geography and my experiences with geography in practice against the background of my life-line (Fig. 14.2)

I was a student in the geography department of the State University of Utrecht, 1935–41. Our year group of undergraduate students comprised twelve; one of them, Christiaan van Paassen, immediately became my best friend and alter ego, another, Nelly Brouwer, became my wife. We three studied for our B.A. and M.A. degrees as a group.

Our professor in human geography, Louis van Vuuren, was a self-made geographer with truly unorthodox scientific opinions and views on the field. He made his career in the former Dutch East Indies, first as a military pacification officer on Sumatra. Later he became a member of the colonial civil service as the first (and only) director of the governmental encyclopedic bureau preserving all the official reports of local administration and local government in the colony's districts written by the local officers on the occasion of transfer of their duties to their successors. The director's task was to write regional monographs on all parts of the large colony accounting for the government's impact on sociospatial development by investments in infrastructure and by economic and administrative measures. In both jobs van Vuuren experienced the fundamental importance of geographical processes.

As a professor he followed the French School of Vidal de la Blache and his successors where geography was closely allied to history. But he emphasized the role of geographical processes. His doctrine was: man is creative but nature limits the range of possibilities for remolding the environment to his wishes as to the process and pattern of his occupation of space and quest for more wealth. Man can improve his level of technology however, and thus is master over geographical processes in his environment.

This doctrine of my professor, although primitive and obviously molded by ethical liberalism and by the widely held belief in social and economic progress before the Second World War, had impressed me intensely, partly because it was consistent with the ideas and values I brought with me from home.

Professor van Vuuren played a prominent role in seeking and creating new jobs for his students who could not be employed as teachers in school geography. He initiated projects, run by the young geographers he had trained personally for field research, and executed under his personal supervision with the help of graduate students. Some projects concerned regional structural unemployment and aimed at sketching concrete solutions for that problem; other projects had to do with urban development and planning or with rural reconstruction programs. Professor van Vuuren managed to get those projects financed, however minimally, in the unfavorable economic conditions of the late 1930s, by Local Authorities, Chambers of Commerce and Industry, and the Ministry of Agriculture.

In these projects he made his colonial experiences productive and showed how geography could be used as a pragmatic instrument for social problem-solving on a territorial basis. Moreover, he diffused his personal enthusiasm to his disciples and students. Christiaan van Paassen and I belong to the last generation of students in geography he trained in practical problem-solving by using their growing knowledge and primitive conception of geographical processes. And using his personal relations and influence, Professor van Vuuren found permanent jobs for us as well.

I became employed by the Provincial Physical Planning Board of North-Holland (the province to which Amsterdam belongs), 1943–66. I assisted in preparing the first regional plans in this province which provided the larger cities and towns with possibilities for physical expansion in the surrounding municipalities, and in determining green belts between and around those 'region-cities'. We had to pioneer in that field of physical planning. I had to train myself to find answers to the challenges by learning from my own failures and mistakes, and after some years I developed a conceptual framework for this task and tested it in practice. I was instructed by older civil engineers/town planners how to produce and to screen municipal development plans to be brought before provincial government for approval. They had produced a clear guide book for such development plans, based on estimations of future industrial and population growth projections that could be 'translated' into adequate (clearly not excessive) extensions of residential and industrial estates and recreational facilities. This guide book had already been conceived in 1927 but it was not available in the library of the geography department where I had been educated. Besides these tasks I was secretary of the provincial committee on reorganizing municipal subdivision of the province, a job that lasted more than twelve years and broadened my views on geographical processes enormously. So, too, did other secretariat positions I had to perform in later years as well as a long series of special reports I was called upon to make, many of which had scarcely anything to do with physical planning. This resulted in a more or less independent position as a provincial civil servant in relation to my colleagues in engineering and planning, I got a department of my own staffed with more geographers and research-fellows from other disciplines. Together we produced hundreds of reports and major recommendations, most of them being approved by an independent committee of provincial politicians, chief officers of the provincial civil service, university professors, and other experts.

In 1960 I was appointed by the Minister of Education as a part-time 'docent' in applied geography and planology at the State University of Utrecht where Dr Christiaan van Paasen had been nominated as an assistant professor with the same teaching and research tasks. This job came in addition to my job in North-Holland. I had to teach graduate students in geography what I had learned from my experiences in practice. This appointment seemed to offer the fulfillment of my highest aspirations, and I accepted it eagerly.

In 1966 I became employed by SISWO (the Netherlands Interuniversity Center for Socioscientific Research) at Amsterdam as a coordinator for planological research programs. This was again a very challenging job that provided me with new contacts in most Dutch universities and with several branches of the national, provincial, and municipal civil services. In Amsterdam the horizons of my practice extended from the provincial to the national level, opportunities expanded for international encounters which I had already enjoyed on a small scale since my appointment at the State University of Utrecht (that had to be withdrawn in 1967 because of my change of main job).

Although I managed to administer a few new planological research programs initiated by small clusters of several university departments, my new job soon appeared to be an interim one. In 1968 I was nominated for a new chair in planology and demography at the State University of Gronigen. A new and formidable challenge waited me and I decided to make the best of it. In fact, I had learned much since 1960 in Utrecht and in particular since 1966 in Amsterdam about education and research in geography, planology, sociology, and other sciences and arts with relevance for the solving of practical social problems. I had learned still more about the problems between professors and emancipating staff members in university departments, the tide of university democratization just having risen in the Netherlands. I concluded that I had to start in Groningen fully afresh: I had to conceive a totally new curriculum and I had to choose and to develop my own program of student projects.

At first the design of my curriculum had been based on my own experiences in practice in the fields of urban and regional planning. I endeavored to introduce my students to a set of activities which a geographer faces in concrete problem-solving situations, emphasizing comprehensive prognostics and coherent plan implementation. However, my students obviously needed a theoretical framework as well and so the emphasis in my teaching of planology shifted from the theory of geographical process as the hard core of the discipline to the theory and methodology of planning processes.

Before this shift had been completed a new problem arose. We had meanwhile started research and experimental projects on urban renewal, a field in which I had scarcely any personal experience, and our tentative projects on urban renewal proved that my above-mentioned shift toward comprehensive prognostics and coherent plan implementation was inadequate. Participation of the people concerned in the planning process appeared to be indispensable and had to be emphasized in theory and practice as well. So I felt enforced to develop a more integral curriculum focusing on both the substance and procedures of planological actions as an integral expression of a community's dissatisfaction with their actual situation, and proceeding with new collective feelings of self-respect about having formulated the right question to be answered. In fact the education of social planologists had to be remolded for the second time within a decade. In the meantime computerization of information and data-banking had opened new paths for planning methodology and technics all over the world. A formal planning technocracy seemed underway to affirm the 1984 agony syndrome of 'Big Brother is Watching You'.

Confusion and uncertainty have become widespread in the turbulent developments of physical planning practice. This endangers the education of planners throughout the world. Planning literature is unanimous in seeking a new foothold. Suggestions for new approaches have been presented by some, but most of them did not succeed in reaching general support from other experts. In my opinion a balanced synthesis and integration of many new ideas promises more success than any new one-sided 'ism' as such. In such a synthesis and integration each expert can rediscover his own contributions.

I decided to begin research and experiments to find a workable synthesis and integration of new ideas, leaning on my long experiences in regional planning as well as on my recent experiences in urban renewal. My efforts aimed at the construction of a general integrative working model of planological actions in all fields. The model was completed last year and will be published soon. It should be regarded as a provisional model, still to be tested and revised, and not as a final remedy.

As a full professor I continued to work in close contact with physical planning in order to build reliable bridges between it and the concepts and doctrines of planological theory. I did so creatively, critically, but obligingly (not subserviently) trying to avoid both the Scylla of abstract reasoning and the Charybdis of pure fact-finding research only.

These considerations concerning my education and experiences as a geographer in social problem-solving in general and in physical planning in particular can gain more depth when put against my career profile (Fig. 14.2). The course of my career has two remarkable features, one being my long stay in the dynamic (Western) part of the Netherlands (1935–68), the other my predominant employment (1943–66) in the provincial civil service of North-Holland. The former feature is due to the early start of regional physical planning in that part of the country, the latter had been decided by my own free will (North-Holland being the province with all Dutch physical planning problems in a nutshell). My return in 1968 to my home area, Groningen, occurred by chance and was considered by some outsiders as a deliberate banishment from the very heart of the country; in fact I had not applied for the Groningen chair but once invited to it I could not refuse.

Three main components of my life are indicated on the right side: (i) broad education and training which never stopped but showed a revival in the SISWO period 1966–68; (ii) intensive activity in physical planning practice in the period 1941–62 and afterwards reduced to more incidental proportions; and (iii) activity in university teaching of planology in the 1960s (part-time) and 1970s (full-time). My research activities mainly have been connected with component (ii) and far less with component (iii) and my publications often have been influenced by component (i).

Further to the right. I indicated the periods of crises that had a strong impact on my career (marked in). I had to overcome them personally. The first crisis concerned my physical handicap at the age of six to seven the second crisis my start as a young geographer employed in the practice of social problem-solving being scarcely equipped, either mentally and intellectually for the task. The third crisis refers to the gradual reduction of my engagement in the North-Holland physical planning with my full consent, and the fourth crisis had to do with remolding the Groningen curriculum twice.

As counterparts to the crises, I indicated the great challenges I experienced with the letters a–d. All of them arose by the chance of very stimulating encounters with other people; (a) confrontation with human geography in general and with Professor van Vuuren and Christiaan van Paassen in particular; (b) practical work in North-Holland and the chief officers and politicians who molded my experiences and practical work: (c) encounters with the emancipating members of staff in many university departments; and (d) intensive contacts with a growing group of older students and disciples

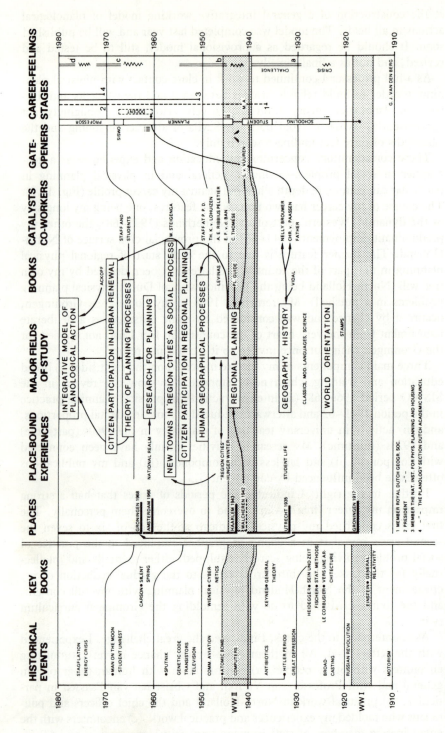

Fig. 14.2 Career profile: Gerrit Jan van den Berg

employed in several stimulating positions and willing to inform me about their experiences.

Activities abroad started in the early 1960s and since 1965 have grown to assume a substantial part of my time. Most of these activities have been induced by chance and, in any case, by initiatives of other people.

On the choice of my career, I consciously and voluntarily decided *not* to become a secondary school teacher in geography when offered that job (1942) and *not* to become a politician (around 1955) or a chief planner in the civil service (1962), and I refused to design and to take responsibility as a full professor for a master's course in quasi-unidisciplinary planology at Groningen State University (1969–70).

Crises and doubts as preconditions for the conception of new ideas in answer to new challenges

Social problem-solving can generate crises and doubts about methodology and doctrine. In fact no social problem is identical to any social problem solved before, so solutions are difficult to routinize. It is a great mistake to suppose that social problem-solving could be managed or routinized in the same way as data processing because then the specific integral character in time and space cannot be preserved. If managed in a routine way, the result is only a partial solution leaving the unsolved parts of a problem more insolvable than before. The keypoint is to find an approach which could guarantee an integral solution. Such an approach is difficult to find in general and not easy even case by case. Many experts in practical social problem-solving have acknowledged the so-called 'wicked' character both of many social problems and of their solutions. They do not reject the use of data processing and partial computerizing in their work but they think it wise to keep those methods subordinate to human reasoning and – what is more – to maximal participation of all the people concerned in the problem and to social bargaining within and between groups of participants in order to harmonize and integrate their contributions to the solutions of problems. The use of strategic choice methods is recommended to master feelings of individual uncertainty and group insecurity as to the outcome of the solution-seeking and decision-making process, but, in my view, this recommendation is only valuable in so far as the participants individually and as groups are able to apply such strategic choice methods. Alas, most of those methods have not been designed or remolded yet for use by the general public and many participants have not been sufficiently trained in their use.

Geographers and other experts who are involved in social problem-solving jobs therefore know periods of doubts and even crises during their career. Changes of job every five years or so apparently offer an escape; but in fact, doubts and crises cannot be overcome by escaping and suppressing those feelings and experiences but only by resolutely searching for and formulating the mistakes personally made in attitude, behavior, doctrine, and method-

	local level	national level
persons	○ my father (1942) ○ Professor Louis van Vuuren 1935–42 ○ Christiaan van Paassen 1935–68 ○ My former chief planners at the PPD C. Thomése C. E. and E. F. van den Ban C. E. 1943–51 ○ politician A. E. Ribbius Peletier M.A. in Law, Ph.D. 1946–58 ○ my staff members at the PPD 1952–62 ○ my students and staff 1969–now	○ Th.K. van Lohuizen C. E. chief planner of Amsterdam and professor at Delft Technical University 1943–59 ○ Professor Willem Steigenga, M.A. in Geography, Ph.D. professor in planology and demography, University of Amsterdam 1935–74
institutions	○ own family 1943–now ○ Provincial Civil Service Noord-Holland 1943–62 ○ SISWO 1966–68 ○ State University Groningen (at least 5 of the 12 faculties) 1968–now	○ The National Institute for Physical Planning and Housing 1950–now ○ The Planology Section of the Dutch Academic Council 1970–now ○ The Royal Dutch Geographical Society. Member since 1936. Incidental activities. President 1960–71
landscape	○ urban dweller from birth and ever since (urban environment)	● feel most at home in the coastal provinces (dunes, polders)
general currents of thought	○ religious socialism (unorthodox, i.e., liberal) ○ freemasonry	● problem solving by conditioning interdependent processes
political events	● democratization in self-reliance, self-help, self-control ● environmentalization both as constraint of and as an instrument for planned action	● remained free from military service ● fostering countervailing social and economic power of small initiatives aiming at self-determination, self-employment, and everyman's personal responsibility
technical innovations	● should at first be considered as hidden persuaders	● should be better controlled and cautiously selected
books	● I am far more a writer than a reader and stopped systematic let consciously. I am now reading very selectively according to subjects ● I think personal encounters and real dialogues far more inspiring	

Fig. 14.3 The ecological impacts on my thinking about geography and planning

○ IGU Congress, Amsterdam, 1938
○ International Federation of Housing and Planning Congresses 1962, 1963
○ Council of Europe, Colloquium on Preservation of Historical Buildings and Sites The Hague 1966, Split 1978
○ Sigtuna Seminar on Creativity and Science 1978
○ Most of the yearly Planertreffens of small delegations from Austria, Switzerland, Fed. Republic of Germany, Luxembourg, and the Netherlands. 1962 – now

○ don't feel happy in mountainous regions without spacious lakes

● syncretism of positive and constructive contributions into integrative and creative actions

● regional planning by selective closure of regions in order to enable the people concerned to master and provide for their own local and regional basic needs at first, by organising a better use of their common human and natural resources

● did much harm by their dehumanizing effects on human life
● far less important than promoting social innovations and (re-)activation of cultural values and creative capacities

alone random reading of scientific literature by 1-1-1962 very
and items I become interested in. I am no book reviewer either.
and productive.

ology. This implies painful learning for the expert involved and he has to do it consciously in order to open his heart and mind for his slumbering creativity. To really probe one's own intellectual capacity is as stimulating as it is seductive, and for me it proved an indispensable aid to self-confidence in periods of crisis. I learned to consider my own intuitive and reasonable intellectual doubts creatively as signals inviting purposeful encounters and interactions. At the Sigtuna seminar[3] I emphasized that such encounters and interactions were the most important preconditions for fertile creative scientific work for me as a geographer planner. As a matter of fact, to express one's creativity in terms of new ideas and concepts and methods demands relative loneliness because of the necessary self-criticism involved. And that requires much more rational and physical toil than creative inspiration.

Creativity has to be directed to specific goals, objectives, and targets in finding alternative approaches to the solution of social problems. Case by case, the expert discerns challenges which can direct his creativity, and doubts can be seen as signals for fresh discovery. Challenges must be selected, formulated, and defined very deliberately. The expert's ability to do that can obviously grow through experiences of handling his own doubts and crises; it is a product of his personal 'human growth' on the base of a continuous learning process.

Creativity demands a confrontation with challenges not met before. Thus, the role of crises and doubts as preconditions for the conception of new ideas is also bound up with the occurrences of new challenges in one's career as an expert in solving social problems. If those challenges do not occur sufficiently in one's job it may be wiser to consider resigning and look for another job with more promising challenges. Perhaps, this is a more valid argument for changing jobs every five years or so than the fear of surpassing the level of incompetence handling its forefelt crises.

The ecology of my thinking on geography and planning

Figure 14.3 sketches some of the most important environmental influences on my thinking on geography.

I have endeavored to systematize these impacts using the classification and categories suggested in the first initiation to write this essay. It remains simply to remark on the weight and personal significance of these levels.

While composing this sketch (Fig. 14.3) I became aware that landscape and technical innovations have had no specific impact on my career. The only impact of home landscape is that I remained a typical Dutch geographer/planner not very much inclined to be active in international science development. As for technical innovations – I do not trust them any more since the 1940s.

On books I had to take a radical decision in the early 1960s. Given my physical capacities, and need to use my time efficiently, I have decided not to read many books. Since the 1960s and the weakening of my memory capacity (perhaps a post-war effect of malnutrition during the hunger winter 1944–

45?), I have found other sources of inspiration more beneficial for personal human growth.

General currents of thought and political events have had a larger although a selective impact on my career. My orientation on problem-solving, on the base of the doctrine in geography I formulated in the introduction to this essay, feeds me with criteria for selecting those currents and events that appear to be remoldable for planned geographical action. Institutions have indeed had a great impact both as constraints and as supports. I tried to serve within them faithfully, not subserviently, and at the local and the national levels rather than on an international one.

People have had the most crucial impact of all. But again they were predominantly those encountered on the local and the national level. I did not encounter foreign people so often personally or by correspondence, but the names of Torsten Hägerstrand, Wolfgang Hartke, and Jean Gottman come to mind as scientists who are important and sympathetic to me.

Conclusion

These are some personal reflections by a geographer in his sixties who did not confine himself to the official academic traditions and development of his discipline. On the contrary, I felt obliged to cross its thresholds outside teaching school geography or doing academic research in search of new ways to make geography useful in solving social problems for the benefit of the people. I have always profited by favorable conditions but had to experience the travail of a pioneer and the loneliness which comes from being considered self-opinionated and troublesome.

Many research fellows have the same experiences. They are the indispensable price for the courage to pioneer. I do not consider myself as a real research fellow, and perhaps my remarks should not be taken that seriously!

Notes

1. Levinas, E. (1951) L'ontologie, est-elle fondamental? *Revue de Métaphysique et de Morale*, **56**, 88–98.
2. Ackoff, R. L. (1974) *Redesigning the Future: A System Approach to Societal Problems*. Wiley, New York.
3. Buttimer, A., Seamon, D. et al. (forthcoming) Creativity and Context. Report on a Symposium at Sigtuna, June 1978.

Selected readings

1957(a) Op zoek naar een kader (algemene inleiding) en De mensen en de nieuwe steden (pre-advies) (In search of a framework (general introduction) and People and new towns (proposals)), *Pre-adviezen inzake de vraag 'Nieuwe steden in Nederland?'* Publikatie LVII van het Ned. Instituut v. Volkshuisvesting en Stedebouw. Alphen aan den Rijn, 7–77.

(b) Structuur en functie van het Westen (Structure and function of the West), in *Regionale structuur en interregionale functies*. Zevende lustrum de Ver. van Utrechtse Geografische Studenten. Utrecht, 36–55.

1966(a) Motorisering en samenleving. Een voorlopige verkenning van de literatuur (Motorization and society. A provisional reconnaissance of literature), in *Ruimtelijke ordening in een gemotorseerde wereld*. Publicatie van het Ned. Instituut v. Ruimtelijke Ordening en Volkshuisvesting. Alphen aan den Rijn, 13–53.

(b) Regional planning in the Netherlands. Paper delivered at the Conference on Regional Planning in Dublin (19–21 May 1965). *Proceedings of the National Conference on Regional Planning*. Dublin, 36–44.

1967 Changing regional planning goals in a changing country. Paper read in the Town and Country Planning Summer School, Keele University (6 September 1966). London, 70–89.

1968 Inner cities and monuments in the light of the social sciences. Paper read at Symposium 'D'. The Hague (25 May 1967). Council of Europe: Council for Cultural Cooperation, Strasbourg. Also printed in Dutch in *Bouw*, **23**, 1486–90 and 1706–11.

1974 *Naar een theoretisch model van de stadsvernieuwing* (Toward a Theoretical Model of Urban Renewal). R. C. University, Leuven.

1978 A new approach to regional planning in Ireland's northwest region. Paper read at the Regional Studies Association Meeting (Irish Branch) in Sligo, Ireland, 2 May, Dublin.

1979 *Waartoe inspaakverlening bij streekplannen?* (*Why Participation in Regional Planning?*). Planologische Studiecentrum, universiteit de Groningen, Groningen.

1981 *'Voor ieder een plaats onder de zon?' Een inleiding tot de planologie* (*A Place under the Sun for Everyone*). Alphen aan den Rijn.

Chapter 15

Fig. 15.1 Wolfgang Hartke, Germany

Mention the term 'social geography' and thoughts turn to Frankfurt and München, to the founder of that very productive research institute of post-war Europe. Meet the person, and what life and passion one finds: a rebel since youth, keen social critic, and a tireless adventurer. Our conversations have always been in French and in many ways I associate him with that School. So do many French colleagues who cite his work to support their own efforts to render geography more applicable to practical problem-solving.

The story presented here is really a collage of impressions, mostly gleaned from the text of his interview and subsequent correspondence. I hope that it captures something of his dynamic personality and innovative spirit. To engage in dialogue with Wolfgang Hartke is to invite continuous surprise. One begins to appreciate why his influence on students has been so fruitful, and why indeed his name evokes that smile of recognition from colleagues in diverse lands.

An interview with Wolfgang Hartke[1]

Anne Buttimer: It is now more than twelve years since we first met here at München – remember how I came to consult you on the question of social geography?

Wolfgang Hartke: I'm pleased that you are not asking me the same question this time: you notice I have not answered it in the past twelve years. Shall we try again?

Buttimer: Not unless you wish. Now I am more interested in your own story – the story of your life and career.

Hartke: You begin with serious matters. I come from a liberal bourgeois family of the turn of the century. I was born in 1908 in the middle of the Rhineland, in Bonn, and went to grade school and began high school there. My father's side of the family come from the west, from Fürstenau, in the county of Bentheim near the Dutch border. My grandfather was a copper-smith and a *Bourgemestre*. My mother's family comes from the east, from the land of Hessen. When I was a teenager we crossed over toward the east, even-

225

tually settling at Potsdam, near Berlin. My father was President of the Higher Education Department in the Province of *Ost Preussen* (East Prussia) until 1933 and so we used to travel there during vacation times. In fact, my very first scientific steps in geography were taken in Prussia – a study on the new agrarian reform settlements.[2] My sister was a law student at Hamburg and Heidelberg, but in 1933 she was forced to leave her profession. She later became secretary to the famous anti-Nazi, Reverend Niemöller. I entered the University of Berlin in 1926.

Buttimer: Your father was a university scholar, then?

Hartke: Yes, a university scholar with training in classical philology, and a doctorate (*licenciat*) in theology – which indicates that he was one of those religious (Protestant) socialists, a group which belonged to the old Social Democratic Party.

Buttimer: There was a religious element in that movement?

Hartke: Surely. It was a kind of liberal religion, inclined toward social activism. Already this gave me a starting point for my personal horizons. Like most young people, however, what I wanted to be was a taxi driver, or a banker.

Buttimer: A banker?

Hartke: Exactly. I should have preferred to be a banker, or a medical doctor. Even at the age of 50, I could have started all over in medicine. But there was something else which I dreamed about during my fifth or sixth year in school. I wanted to be mayor of a big city. The reason? I was a very small person and I wanted to have many contacts with all the people of the city. I used to go on long walks alone when I was six years old and used to talk to people. You could say that I was already doing interviews in a childish way even at that age.

Buttimer: So it was the human contact and communication which were important in becoming a mayor, not power and authority?

Hartke: It was certainly the desire for human contact. My father had a strong personality and as a child I felt a lack of companionship. The most important adult in my life, however, was my mother, and she came from an intellectual family with pastors and professionals. In fact at that time the family owned a large boarding school for the education of young people. My mother's attitudes helped to shape the opinions which have guided me in later life.

Buttimer: Can you explain about this?

Hartke: I left home at an early age. I finished high school at seventeen with my diploma and good grades. This was exceptional – most boys do not get their high school diploma until they are 19 or 20. I felt I was extremely fortunate and had bought two years of freedom. It was a good middle-class family life and I had all the normal experiences of teenage life, first an unhappy love affair, then a misunderstanding with my father, which sometimes became quite serious. One day, during my first semester of university studies, he told me angrily 'If this house does not please you, you can just leave!' I left the next day. But my mother encouraged me to avail of a student exchange program which I had discovered, quite by fluke.

Buttimer: Your father, he did not expect that!

Hartke: No, naturally not. I've done the same thing several times in my life. I did it to my great Professor Mr Krebs in Berlin, a man to whom I owe a lot. I was one of the last students of Mr Penck, a naturalist and classical style geographer with an 'environmental–determinist' orientation. As a small insignificant student I was not supposed to have my own opinions. However, I had already picked my thesis topic – during my stay in Geneva which I'll describe shortly – without his permission or approval, and that led to difficulties. I must say, however, that Krebs was kind to me, and even favored me after the political difficulties in 1933. It was Krebs who, along with others, had me transferred to the Prussian Academy of Sciences, a politically safe place in the shelter of the Academy. He thereby gained a good worker for his atlas project.[3]

Buttimer: You were in charge of that famous *Atlas des Deutschen Lebensraumes*?

Hartke: Yes, and I worked on it for three years with very poor pay, beginning with 90 marks per month and improving later to 125 and even 150 marks. You see my father was not permitted to give us youngsters any allowance beyond that which was set by the Nazis. I lived with a family at Berlin, of course, otherwise I could never have managed to survive. Then one fine day, at the end of a purely academic discussion, a conflict of a political nature arose. I was not refusing any advice, not unjustified in what I said, but I simply disagreed, and felt a deep disappointment. So I left the Institute.

Buttimer: You cut out again! This is now in the mid-1920s – where did you go?

Hartke: I headed for Geneva without knowing a word of French. This was an extremely interesting period politically in Geneva and I flung myself wholeheartedly into this milieu. I went to the magnificent library at the *Bureau International de Travail* (ILO) and began eagerly to learn French. You see I was already imagining my return home to show off to my school professors how well I was speaking French

Buttimer: Another declaration of autonomy?

Hartke: In a sense that is how I felt. But during these years in Geneva, 1926–28, I also virtually abandoned any ideas of doing geography. At least I was not sure. I followed courses in geography with old Professor Chaix and the young Burky, and learned about the practical school of French studies. I became very keen on anthropology – in the broad sense of that term – and the human sciences. I wanted somehow to understand how people could be so damned hostile. Even in high school I was very interested in psychology –

you know this was the time of the great debates over Freud, Adler, and Jung.

Buttimer: You were interested in Jung?

Hartke: Yes, especially his ideas about mass psychology. But what seemed missing in all these treatises was any explicit attention to the spatial factor. I read LeBon's *Psychologie der Masses*, for example, which traces the progress of psychological evolution, but it spoke of internal and external forces in the shaping of psyche, without any reference to the environment outside. On the other hand, treatises in history and geography which I had found in universities did not really deal with man or the human sciences. The only reference to human thought or feeling was usually confined to a few pages on population tagged on at the end of the textbooks – not too different really from what you see today when we are getting back to simplistic sociometric and even biometric measures of people again – facts, without interpretation.

Buttimer: But did you find anything in geography which interested you at that time?

Harke: No doubt there were interesting things being written then but I remember being very disappointed. There were, of course, certain French authors . . .

Buttimer: You also read from the French tradition?

Hartke: Yes, in Geneva I read the French literature. There I also met one of the first real political economists, William Martin. You could say he was one of the first professors of political and social science of French nomenclature.

Buttimer: And Jean Brunhes?

Hartke: Jean Brunhes, Lucien Febvre, and Marc Bloch, of course. Jean Brunhes was a little too much of a mystic for me. His work, however, was fascinating. He had made the distinction between rich and poor regions, 'active' and 'passive' regions, and that was interesting. It's not that you can really call regions rich or poor – it is the people who are rich or poor. I cycled all around Geneva looking for rich and poor regions, asking people about their circumstances. Then for the first time I got arrested and put in a prison cell.

Buttimer: In France or Geneva?

Hartke: My first stay in prison was in France. I'd rather not talk about the second time. In prison, for the first two days I got plenty of time to think. It was a Saturday and I was given a piece of bread and cheese, a pot of coffee and some wine. On Monday they came with curiosity about why a German, without a valid passport, had left the free zone and was wandering about in France. All went well afterwards, but the memory of 'poor' and 'rich' regions kept haunting me. Before I left prison I knew that I'd never be a physical geographer. At the age of nineteen I was ready to return, as a true revolutionary, to study social problems.

Buttimer: You were already certain about the kind of geography you would do at the age of nineteen?

Hartke: Yes, that's right. I was 'independent' now and even if I had no money I had made many useful contacts in Geneva, for example, with politicians like Stresseman and Briand. Even though I knew little French, Briand welcomed me and introduced me to Streesseman, saying 'Here is a fellow who has some questions to ask – a compatriot of yours.' Stresseman received us cordially. What a great discussion we had! It was such a privilege to have known that great peacemaker, Briand. In the ILO library I became keenly interested in the workers. I read book reviews, reports, and anything I could find on the processes of migration, the assimilation problems of migrant workers, particularly in France, the land of immigrants *par excellence* during that period.

Buttimer: I notice you published a few items and maps on the problems of migration and of migrant assimilation questions in France.[4] You yourself were in a situation of exile, in a sense, but you were surrounded by friends . . .

Hartke: Yes, by all means. I came into contact with many students, and with one girl student in particular. One Christmas I decided not to go home because of 'flu and instead I was invited to spend the holiday season at an exclusive ski club near Saint Gervais. This girl really engaged my attention. It would have been a risk for me to stay around for a second semester . . . but her uncle, a kind and intelligent man, soon put me on the right track.

Buttimer: How was that?

Hartke: He was an intelligent and understanding person: two great gifts! He set up contacts for me with the Greek representative at the League of Nations, who was living in Paris at the time. I already had relatives in Paris so I went there. One of my professors, Thibaudet, finally got after me about my language abilities. 'Nonsense,' he says, 'to think you do not have a gift for language . . . on the contrary!' He helped me with the Romance languages first and English was at the bottom of the list. I read Spanish, Italian, and French, and even Provençal *patois*, geography became my second field of study.

Buttimer: Are we still in Paris now, or back in Berlin?

Hartke: In Berlin, and the year is 1928. My first course had been in classical geography and was taught by Alfred Rühl, an economist as well as a geographer, who impressed me deeply. He was a scholar educated in Penck's dynamic geomorphology, Davis's cyclical theory of morphological change – he had participated in the famous Transcontinental Expedition in 1911 with Penck and Davis – and then later turned to a kind of economic geography of the man-made landscape. He taught us, for instance, to think about the function of gold in the spatial organization of the world, to think about values, and the effect on us youngsters was enormous, 'infectious' one could say.

Buttimer: My impression is that Berlin must have really been a meeting ground of nations during the late 1920s, a place of excitement and diversity . . .

Hartke: Yes, indeed; we had Chinese, Japanese, people who undertook expeditions to Central Asia, people from the Balkan countries, and of course Bulgarians. Doctoral candidates used to come to Berlin and Professors Louis and Keyser (later in Cologne) had a major research project on the Balkans. This was one of Berlin's oldest traditions. There were the polar and oceanographic expeditions led by Wegener, Merz, Defant, and Wüst, and Mr Troll, formerly a biologist and a mountaineer, would return from his classical 'Humboldt'-style investigations in South America. These and others all brought in fresh ideas. At the same time Carl Schott was also a doctoral candidate at the Berlin Institute. Professor Hans Bobek arrived as an assistant professor from Innsbruck, bringing with him a lot of primarily historical–social interests and new ideas.

Buttimer: Was Krebs in charge?

Hartke: Yes, exactly. He had already begun on his comparative geography and his regional treatise on India. It was that project which turned me away from working further with him. Alfred Rühl impressed me deeply; even if his ideas were not welcomed by most of the 'good geographers' of the day, some young students were keen.

Buttimer: Can you tell me what the orthodox views were among those so-called 'good geographers' of that time?

Hartke: The 'official' geography at Berlin at that time was a completely classical type. Nature was given. Of course, one could be curious about the forces underlying the forms of nature. Space was also a 'given'. What mattered most was the form and pattern of facts. Earth was still the scene, the stage upon which history played itself out. All that was boring for me. I was not so much interested in mountains and what lay beyond the mountains – what I wanted to know was what kind of people were behind the mountains, why they behaved and lived in certain ways and not another, whether the allegedly 'free will' of people was limited by other spatial constraints. I was not willing to accept that the configurations of mountain and valley, river and plain were responsible for the way people lived and behaved. I was also a rebellious youth and was not ready to accept any 'explanations' associated with 'the good Lord'.

Buttimer: So what was controversial about the view propounded by Rühl?

Hartke: He claimed that general geography was a social science and not an environmental–deterministic 'natural' science. He wrote books on subjects like the 'economic spirit' of the Spanish and Oriental people in the Near East. He talked about gold not simply as a layer of mining industry, a production site, but as an economic principle, however fictional, and a principle of value. He was the first to introduce me to questions of values and motivations as forces in creating regional disparities all over the world. Rühl was the only German geographer who was invited to the first IGU Congress (in Paris) after the Second World War. He died tragically at Lugano, thoroughly dismayed by the Nazi story. He made one of the strongest influences on my thinking at the time.

Buttimer: Did his ideas resonate with your previous interest in psychology?

Hartke: Yes, of course. I was fascinated by the potential contribution of value theory to the development of geography. I continued to study theories of value and psychology – this was the time when Kurt Lewin was developing child psychology and introduced the notion of 'field theory'. Likewise I used to listen to Mr Köhler who told about his experiments with apes and behavior problems. I even sneaked in a remark on the utility of psychological insights for geography in a footnote to my article on the new agrarian reform villages in eastern Prussia in 1931.[5]

Buttimer: Kurt Lewin's *Field Theory* became quite important in America ...

Hartke: But that was much later. At that time people did not know him very well and he was rarely invited to speak. There were others, for example, the anthropologist Vierkandt, who fascinated me. Naturally I was curious because of my own family history and Geneva experiences. It was about this same time that I discovered the Dutch *sociaale geografie* – van Vuuren and Steinmetz – I know you are familiar with their work, and you can see why I was drawn to this kind of geography.

Buttimer: *Sociografie* or *sociaale geografie*?

Hartke: *Sociaale geografie*, of course. I soon met Demangeon, Dion, Chabot, Sorre, George, and later, of course, other Dutch colleagues like Keuning, de Vries Reilingh, Steigenga, de Vooys, van Paassen, and van den Berg. I found some real success in my work with the Dutch, and that was mostly because of good personal contacts. You know that generation of van Vuuren was as hard-nosed as myself, and so is the young generation. I was invited back to Holland after the war, at a time when very few Germans were welcome there. Mr Troll was, of course, but when I was invited to become an honorary member of the Dutch Geographical Society, it did not please everyone in Germany either! But that's a long story. What else would you like to know?

Buttimer: Frankly, I'd like to hear more about the Berlin years – you skipped a bit ...

Hartke: What was important about Berlin was that you had students and other young people from all over the world. I could begin to speculate about studies of the 'social landscape' without reference to the physical landscape. After this came the political judgement. After 1936 they made me wait for three years to graduate, but I was allowed to do assistant work. This was a period of NEP in politics. We thought 'we'll manage somehow ... things will work out'. It was possible to return to the university without joining party organizations or even becoming a member of the SA. There was good old Mr Behrmann, geographer and cartographer and famous traveler to New Guinea.

He was a totally just man and revered his emperor, William II. The departure of the emperor meant a great personal loss to Behrman. He understood my situation, as indeed did Carl Troll and some others. I was told about a position in Frankfurt which would be vacated in less than one year because its occupant would be moving up the ranks. 'In the meantime,' they asked, 'why not stay here and earn 90 marks a month?' I found a little room and began working, all the time reminding myself 'If it does not work, I shall quit the Institute, quit geography, and even abandon all thoughts of a university career.' At that time I had become quite enthusiastic about academic life because of the dynamic atmosphere at Berlin while I was assistant and my colleagues' insistence that I stay. However, if it was wealth or a materially secure life one wanted, then one might perhaps have to leave Germany altogether.

Buttimer: Did you ever think of leaving Germany?

Hartke: Oh yes. I approached the Rockefeller Foundation for help in order to leave the country. There I was in a situation where my father was already removed from office and I could not bring myself to join any of those political associations. I contemplated taking a position in Canada. It was much later – at the end of the Second World War – that I was offered a quasi-permanent position in Canada. I could not bring myself to take that step, I considered myself too old, so I stayed.

Buttimer: Your attachment to your native country was stronger than your desire to make a better career?

Hartke: Very early I recognized that I simply did not have the proper disposition to become an emigrant. I just did not have the right mentality for that. Despite all my revolutionary and radical feelings, I stayed. And this was largely because of personal contacts and friends. I had many warm relations with French colleagues too and they considered me as one of them.

Buttimer: So you stayed at Frankfurt?

Hartke: I was associated with the Institute in Frankfurt all the time between 1936 and 1952 and I kept in touch even between 1940 and 1946. At the end of 1946, I came back to Frankfurt. There was nobody left. All the professors had gone. Behrmann was still alive but he stayed at Berlin. At this bombed-out Institute there were two young girls and myself who went in with papers from the US military authorities. Mine did not even mention my situation as prisoner in Italy or France! I tried to pick up the pieces and start all over. Remember, I was not yet 'reinstated' as a 're-educated' civilian. My French colleagues had previously tried to reclaim me as a prisoner of war in France and to install me at Mainz at the new French-directed university there. We were tired of waiting, however, and so when I was liberated in October of 1946 I presented myself to the University of Frankfurt, which was in the American Occupation Zone, and luckily fell upon an old (pre-1933) Director of the Prussian Ministry of Cultural Affairs who knew my father. Right away he understood my situation, and he nominated me as Commissioned Director. From that moment on we were placed under the auspices of the Americans, and enjoyed the services of a fine university officer, Mr Hutchinson, the well-known humanist, from University of Chicago. In the midst of my difficulties, I finally wrote him a letter directly, bypassing the bureaucracy: 'Look here,' I wrote, 'here is a fine university officer who is unable to do anything because of the military authorities, because of the Germans, and also in my own personal case because of a Jewish returned migrant who is now the big censor here. He is unable to get anything done, even though he himself, as member of a prominent Jewish family, had profited from our personal help to leave the country and settle down in France.' Hutchinson put me in touch with Chauncy Harris and I became director of that abandoned institute of geography. Harris helped me to make many interesting contacts and we got on with the job without any irritations from the old guard. There were no books, no rooms, no food, no equipment. We organized food supplies for excursions and field trips. So I transformed everything into a kind of 'field' geography. We went out on excursions and field trips: I tried to provide training in field studies, research in action, interviews, field analyses of observed social problems as a foundation for theoretical reflection later on.

We went mostly on foot. There were few cars or trains around, but sometimes we got an American army truck with driver. This is how my applied social geography began.[6] Very soon we were asked to cooperate in reconstruction work and planning. Even the young recruits to the US Army Agrarian Offices, who were not very familiar with their new jobs, often received training from us. There were some French Occupation officers also from the other side of the Rhine, colleagues in history and geography who came to give a lecture now and then. At that time, you must know, there were frontiers between the French and US Occupation zones.

Buttimer: So your conceptions of social geography were formulated in a situation of action?

Hartke: In courses and excursions I only made a few remarks on underlying theoretical ideas. Later on, when I did articulate my theory, it was often not acceptable to the older colleagues who were, litttle by little, returning to their positions. If you want to know what the theory was, you have to look between the lines, in occasional short notes, in comments on articles, theses and texts, and dissertations which were done during those years.[7] More and more I turned to a behavior-oriented comprehension of spatial phenomena – processes behind the forms of landscape. I began to try cartographic methods not for the sake of representing facts, but to find out about processes, for example, the 'sociogeographic levées', 'social fallow', etc.

Buttimer: Your 'social fallow' idea was one of the first to prick the imagination of Anglo-geographers, I think. There was also that fine piece on the *Hütekinder* which impressed many.[8] As for your theoretical ideas, did you deliberately hold off expressing them until later? For example, I did not have to 'read between the lines' of your famous paper to the German National

Academy of 1960, or the other in *Erdkunde* in 1961: were you more convinced by then, or was that just a matter of timing?[9]

Hartke: Both, I think.

Buttimer: I notice that you returned to Frankfurt in the mid-1940s. Did you have much contact with the famous social theorists, Horkheimer and Adorno during that time?

Hartke: Sometimes my contacts with Horkheimer, Adorno, and the others were warm, friendly but not frequent, but I do not think they influenced my social geography to a great extent. Horkheimer was director at Frankfurt when they returned in the 1950s – intellectually radical Jews – to influence the direction of the university. After that initial period, one must say, it was finished. In 1952 I also went to München. Then came the political radicalism particularly in the Frankfurt School of the late 1950s. In fact, we did not have much scientific interaction. I had read their publications, and already in the 1930s was very impressed by Wittfogel's book and articles on geography, especially of hydraulic societies before 1933.[10] Later on I had some contact with Pollock on the problems of automation in a geographical context. When the Institute returned to Frankfurt, it did not make a deep impression on geography in Germany as a whole. I was the only professor of geography in West Germany who tried to include some fellows from the Frankfurt Social Science Institute in my staff at München. That was the period of multidisciplinary developments – long since over. But we were not very happy with them. Even before I left Frankfurt I felt there was something missing in the theories of these sociologists.

Buttimer: What was that?

Hartke: Space. Problems of space-boundedness – perhaps 'place and placelessness' – they took society as a given base for their surveys, interviews, and discussions about evolution. They even took the whole spatial anatomy of the administrative system for granted and seemed to pay little attention to time-boundedness of the individual's conditions of existence (e.g. Hägerstrand). Some would even accept the 'natural' unities of geography, just like those geographers who later simply accepted the statistical facts from sociologists and mapped them. One geographer, for example, mapped out such data to discover formal 'identities', and if he found any isomorphism between such 'natural areas' and the 'social areas', the explanation would be couched in terms of 'nature'. Simple-minded stuff. Later sociologists began to study 'social indicators' – a mode which developed later in Europe than in America. 'Indicators' for what? They were trying to elucidate questions related to 'quality of life', but not related to quality of spatial units of life process.

Buttimer: But you yourself were a radical – both intellectually and politically . . .

Hartke: Yes, and I lived in anxious anticipation for a repetition of the 'May 1968' events here. Those were famous years, between 1969 and 1971, even if they did not yield much result. But the debate went on: how to deal, in truly geographic fashion, with social problems? I'm still convinced that one must begin with mapping – to map the facts which have been derived from surveys, interviews, and most of all, from observations of people in action. Why did I insist on this? To prevent the classical geographers from denying its status as geography! I convinced them about my points by using their own language, their own methods. To show how spatial structures are a function of social structuration, and not determined by the physical conditions of soil or physical environment.

Buttimer: Did you find the cartographic language helpful in articulating your point, or were you simply 'fighting the oppressor with the oppressor's own tools'?

Hartke: No. It was necessary to change the very language itself. But many adopted simply the sociological language and many complain that they do not understand a thing any more: I am very upset about the simplifications which have been done, especially by our dear quantifiers.

Buttimer: You have been known as a great teacher, one who spent lots of time explaining orally with your students, encouraging them to take up their own areas of research – does it not delight you now to see how well they have done?

Hartke: Surely, there were good moments. Only now I think, the sociopolitical developments of the 1970s have made things awful. When one sees students, it seems too easy to offend them. Try to provoke them intentionally and they do not answer any more. There seems to be some kind of undefinable fear about the future or something. Perhaps it is only a small little group, always the same ones: they happen to be the brightest ones, most morally sensitive, and therefore vulnerable, when they react against 'normalcy'.

Buttimer: As you look back on your career, what do you consider as the most significant contribution of geography?

Hartke: There is so much research to be done on the spatial aspects of individual and group life, showing constraints, as Hägerstrand does, but especially showing the influences of power on the shaping of those spatial structures. Geography concerns space, place, frontier, and limits. Findings from its research in the field can lead to better theoretical understanding of process and likewise to better prognoses about the future. In geography there are always many 'black boxes', but the more one tries to make it practicably useful, the better geography can serve humanity. But sometimes I am not sure I'm a geographer any more . . .

Buttimer: You did not set out to be one, but you managed to fulfill some of your childhood aspirations: you did not need to become a taxi driver or mayor of a big city to come into contact with many people and gain many friends and disciples. As for the medical vocation, can you not also consider that your applied work was oriented toward a healing of society? This seems like a good note on which to bring our conversation to an end. I want to thank you for this interview and to congratulate you again on a fascinating and important career.

Notes

1. Edited and expanded text of an interview recorded at München in January 1979.
2. Zur Kulturgeographie der ländlichen Neusiedlung Ostpreussens, *Zeitschrift Ges. f. Erdkunde* (1933), 347–70.
3. Karte der Bevölkerungsverteilung in Mitteleuropa 1:300,000, *Atlas des Deutschen Lebensraumes* (1934), Bibl. Inst., Leipzig; Beispiele für die Wandlung der Kulturlandschaft, Karten 1:200,000. Oberrhein, Tiefebene, Oderbruch, *op. cit.* (1938).
4. Kulturgeographische Wandlungen in Nordostfrankreich, *Berliner Geographische Arb.*, **1** (1932), Engelhorn, Stuttgart; Die Ausländer in Nordfrankreich, *Petermanns Mittellungen* (1933), 6–9; Die Ausländer in Südostfrankreich, *ibid.* (1934), 52–54.
5. See Note 2 and Die ländliche Neusiedlung als geographisches Problem, *Erdkunde*, **1** (1947), 90–106.
6. Stadtbesichtigung. Ein Problem des Fremdenverkehres in kriegszerstörten Städten, *Die Erde*, **3–4** (1951–52), 258–71.
7. See Notes 2 and 5.
8. See Selected readings, 1956 (a) and (b).
9. Le problème essentiel de Géographie sociale dans le monde, *Synthèses*, **166** (1960), 393–407; Gedanken über die Bestimmung von Räumen gleichen sozialgeographischen Verhaltens. Troll-Festschrift, *Erdkunde*, **15** (1961), 426–36; Eine ländliche Kleinstadt im sozialen Umbruch der Gegenwart, *Raumforschung und Raumordnung*, **22** (1964), 126–36; Die sozialgeographische Differenzierung der Gemarkungen ländlicher Kleinstädte, *Geografiska Annaler*, **43** (1961), 105–13; see also Selected readings, 1960.
10. Wittfogel, K. H. *Wirtschaft und Gesellschaft Chinas. Versuch einer Wissenschaftlichen analyse einer grossen asiatischen Agrargesellschaft.* Leipzig, 1931; and *Oriental Despotism: A Comparative Study of Total Power.* Yale U. P., New Haven, 1957.

Selected readings

1938   Das Arbeits- und Wohnortgebiet im Rhein–Mainischen Lebensraum (Work and dwelling in the Rhine-Main living space), *Rhein–Mainische Forschungen*, **18**.

1939   Pendelwanderung und kulturgeographische Raumbildung im Rhein–Maingebiet (Commuters and their role in the sociogeographic formation of the Rhine–Main area), *Petermanns Geographische Mitteilungen*, **85**, 185–90.

1952   Die Zeitung als Funktion sozial-geographischer Verhältnisse im Rhein–Maingebiet (Newspapers in the Rhine–Main area – a sociogeographic perspective), *Rhein–Mainische Forschungen*, **32**.

1953   Die soziale Differenzierung der Agrarlandschaft im Rhein-Maingebiet (The social differentiation of the rural landscape in the Rhine–Main area), *Erdkunde*, **7**, 11–27.

1956(a) Die Hütekinder im Hohen Vogelsberg. Der geographische Charakter eines Sozialproblems (Children of the Hohen Vogelsberg, who tend animals – geographic features of a social problem), *Münchner Geographische Hefte*, **11**.

(b)   Die 'Sozialbrache' als Phänomen der geographischen Differenzierung der Landschaft (Social fallow as a phenomenon of the geographic differentiation of the landscape), *Erdkunde*, **10**, 257–69.

1960(a) *Denkschrift zur Lage der Geographie*. Verfasst im Auftrag der Deutschen Forschungsgemeinschaft, Wiesbaden (Memorandum regarding the present state of geography prepared for the German Research Society).

(b) Gedanken über die Bestimmung von Raumen gleichen sozialgeographischen Verhaltens (Thoughts/reflections on the definition of regions/space(s) of identical sociogeographic behavior), *Erdkunde*, **13**, 426–36.

1962 Die Bedeutung der geographischen Wissenschaft in der Gegenwart (The relevance of geographical science in the present day), *Tagungsbericht und wissenschaftliche Abhandlungen des 33*. Deutschen Geographentags in Köln 1961. Steiner Verlag, Wiesbaden, 113–31.

1963 *Das Land Frankreich als sozialgeographische Einheit* (*The State of France as a Sociogeographic Unit*). Diesterweg, Frankfurt, 2nd edn 1966, 3rd edn 1968.

Chapter 16

Fig. 16.1 Torsten Hägerstrand, Sweden

'Diorama' is perhaps the most satisfactory metaphor to capture Torsten Hägerstrand's view of the world. In many ways it also epitomizes the story of his life and the heuristic import of his written and spoken word. A school teacher's son from Småland, eager explorer of his boyhood Linnean milieu, Torsten still personifies well the call of his baptismal name, Stig.

What a challenge it must be to harmonize two such different vocations to life: Stig's ('pathway') and Torsten's ('the stone of Tor'). A large international body of scholars have come to appreciate the monumental achievements of Torsten; this essay should reveal how much there is also to the story of Stig.

It has been a privilege to work so closely with Torsten. My eyes have opened to realities hitherto unknown and I have been forced to question many of my former 'certainties'. For someone to engage so wholeheartedly in a dialogue process such as ours demands courage and energy. Energetic and courageous indeed is a scholar who sees no intrinsic conflict between art and science, between poetry and mathematics, who can be just as thrilled by the esthetics of landscape and music as by the rigor and power of scientific models. I have saved his essay to the end, for he has been most intimately involved with this project as a whole, and has therefore had the opportunity to seek common denominators between his own experiences and those of the other authors.

In search for the sources of concepts

It stands behind the experience that we are driven, together with everything else, from the past toward the future without a moment of time which does not vanish immediately. (Paul Tillich)

Preamble

My lifelong marriage to geography has exposed me to an enduring tension between attraction and repulsion from the moment when I was engulfed in the academic version of the discipline. Before that time the relation was close but unreflected, innocent and happy. Afterwards I have lived in a state of dissatisfaction with my chosen field of learning. This feeling also has formed the soil from which many sorts of ideas have sprung up. And how painful would my professional life have been if it had not provided room for the play of imagination? So, despite everything, I am happy with my choice. I have learned through the years that one can be an incredulous person and still in one's heart remain a true believer.

I have been interested in many things through my life – in fact too many to allow a thorough competence in anything particular. It is too late to repair this today. But I do not really regret it. I cannot see how I could have pursued my days otherwise. I am happy that I ended up in an academic discipline which had very few limiting fences in its structure and content. I probably could have felt at home in history or anthropology as well. But then I would have missed those contacts with the natural sciences that geography has offered.

Already at a superficial level my attraction to the fields just mentioned seems to hide a contradiction. I am not much interested in collections of facts about regions, periods, or cultures. My basic inclination is reductionistic. I am attracted to efforts to find out what is common to seemingly disparate phenomena. Now, of course, if one really looks for diversity, geography is a place to go to. So, this contradiction can be explained away. The deeper roots of the tension I have felt lie elsewhere. To exist is to be carried forward with time. Human consciousness feeds on what has happened and works on what is going to happen. A time-animal placed as an element in a pure spatial situation tells as little about the world as a violinist playing solo in front of the empty chairs of an orchestra would say about polyphonic music. I felt this dilemma very early and have been grappling with it ever since.

Today I maintain a world-picture in my mind which I feel to be sufficiently coherent and productive as a source of questions and insights for keeping me busy during my remaining days. It rests on that central part of the geographical tradition which tries to grasp phenomena where they appear as neighbors in the given world instead of separated out and removed from their situational ties as the dominating species of scientists prefer to do. But it radically departs from the tradition by assuming that people and things are processes, and that the essence of any geographic now (a landscape in its fullness, if you like) is not best understood in terms of its stable individuality but in terms of its double face of graveyard and cradle of creation. I have become convinced that there is a unifying power in this perspective and an ethical challenge which extends far beyond the limited issues one used to treat in traditional regional geography. As every geographer I love maps. But the great adventure of my scholarly life has been to try to transcend the map. I see, almost literally, the opulence of the world as a *moiré* of processes in conversation.

It is futile to argue about how much of a person's world-picture receives its shape from inborn tendencies and how much from sources along the life-path. In my own case I have some suggestions. I am going to expose them by recalling sections of my life which happened to become particularly sharply outlined in my memory as productive. In this perspective my years at high school, my professional preoccupations as a university employee, and my service in national and international organizations do not appear as well-springs of inspiration. They are more like busy transit halls in which to spend some useful time while the craft is away for repair and refuelling. I am going to devote only little attention to these periods and activities. I will instead turn the spotlight toward crucial realms which a biographer would never be able to trace in public records.

Childhood years

My mother was the daughter of a farmer who in his youth had wanted to study for the priesthood and later in life gave up farming to become a dealer in timber. This gave him time to read and talk and speculate. My father had generations of craftsmen behind him but uprooted in his thirties to become a school teacher. So, I was predestined to proceed to the academic ladder. The combination provided me with three enduring strands in my personality. My mother transferred to me a strong *consciousness of the flow of time*. The craftsman tradition lives on as an impulse *to touch* and *to shape*. The total environment of my childhood encouraged the development of these tendencies.

My home was hidden in the forest at a small river in the heart of Småland, situated just between a farming village and a little factory town. Our well and outhouse lay some short distance away, guarded by mighty spruce trees. I can easily transport myself back to the intense experience of dark and icy winter nights at the well, when the stars sparkled high above the flag pole, the ground smelled of rotten potato tops and new-cut firewood. A near and familiar world at the edge of infinity. All was deep silence except for the surge from the waterfall half a mile away. Time itself sounded from it, on and on without pulse or temperament, indifferent to my breathing and heartbeat.

We lived on the second floor in the school-house. We had restricted privacy – at least I felt so – as long as the children were there. Daily life had a peculiar rhythmicity with silent lecture hours and noisy 10 minute breaks. The children ran in and out in wooden shoes. Before I went to school myself I waited with mounting impatience as the day went by for the psalm to sound up through the floor at a quarter to four in the afternoon, marking the end of school:

Yet another day goes from our time and does not come back any more. . . .

This moment every weekday had a still wider significance for me than just a bar-line in the flow of time. The whole place – schoolroom, yard and garden – was suddenly my empire after the children had disappeared. I knew long before researchers in these later days have started to talk about it that human territories can be strongly time-dependent, periodic phenomena. Space and time were glued together in my small world.

When I think back on that time, I do not hear many speaking voices resound, but I can feel in my palms door-knobs and fences, spruce resin and strawberries, and in my nose the smell of attics and steam-engines. I never got to know much about what people had in their minds or how they viewed their circumstances. This is explainable, maybe, because in practice I left my home area for secondary school when I was between twelve and thirteen. But I knew a lot about their outward behaviors and their barns and backyards. In a strange way I came to be an outsider of society but an insider of landscapes.

I found my true satisfaction in bringing together unlike things and combining them into new mechanical constructions. I was not content with just understanding things. I wanted to proceed and remake them. I still own a strange little machine for pulverizing crayons – I tried oil-painting and had to produce the powder for the colours as best I could. It was easy to crush the crayons in a mortar, but that was too simple to be fun. When I look at the little machine today, I can see that it was very much inspired by the hand-driven knitting-machine with which a lady up the road made my underwear and stockings.

The universe of my childhood was not rich in words. I lived in a much more intimate contact with nature – snow, grass and trees, tools and malleable materials – than with people. I have never fully given up a primitive way of exploration and understanding. In order really to be convinced that I have a truth in front of me, I need to be able to touch and grasp the phenomenon in question with my hands, at least in principle. Today the important point is that I can give the phenomenon or model such a shape in my imagination that I could touch it if it materialized. I feel akin to chemists who build three-dimensional portraits of molecules with balls and pins. But I distrust sociologists and economists when they speak about 'preferences' and 'full information'. My world is 'matter-realistic'.

I do not exclude the human world from this perspective. I used to watch birds and observe people, children as well as grown-ups, in much the same way. I spent much of my time just watching the school-children's play from the porch and window.

My cost-free material was wood and tinplate. I picked up the wood from our heap of firewood (for making wheels) or from the boxes in which the exercise books for the school were sent from the paper mill. I procured the tinplate by cutting up spice tins and cake tins, at the price of a great deal of blood-shedding. I also saved remnants of worn-out gadgets and plundered the farmer's junk-holes in the wood for larger leftovers from threshers and harrows. Thus I adopted as a leading principle that things can be used for purposes other than what they were made for. My hero from twelve and a couple of years onwards was Thomas Alva Edison. I admired his imagination and manner of work immensely. I read the books about his life that I could find in the libraries around, and I composed a rather long essay about his life one of the first years in secondary school.

My father was one of the very first school teachers who seriously included 'home-area study' (*hembygdskunskap*) in the form that A. Goës and G. Sjöholm of Göteborg had presented it in books and courses. This approach to primary teaching was an application for Sweden of the German idea *Gesamtunterricht* (approx. 'wholeness or everything together education' – I wonder what it means that such simple ideas cannot be expressed properly in

English?). The momentum from my early exposure to this idea became so strong that it survived all the seven years in secondary school and gymnasium where the teaching of world geography was not very inspiring.

When I was eight or nine my father introduced the class to a bit of 'academic' geography. He explained the concept of scale and suggested that we draw a map of the classroom on the blackboard. The exercise must have been exciting, since I recall it so well. The new language suited me perfectly. Not that I had any difficulties in finding words when I spoke – it may have been easier at that time than now – but the wordless graph offered a comprehensive way of depicting my world which had room for all the various things that interested me. Perhaps map-making became so exciting because we dealt with my home and the space where I used to play in the evenings. My classmates could not possibly be equally attached to the various corners and objects.

About the same time I learned another language of a similar nature. My mother taught me musical notation when I was around seven, and she then made sure that I had at least half an hour of exercise every day at her little cabinet organ. My brother – six years older than I – studied to become an organist. We had the use of the key to the parish church and thereby access to its nineteenth-century pneumatic organ. Everybody who has had the opportunity to step into the forest of pipes in an organ will realize what a natural bridge between engineering and art this intrument could be for a boy. It was my duty to provide the air as bellows-blower. This was heavy work. My remuneration came in kind: I was granted the right to play the organ for 10 minutes per hour. My brother also took the trouble of teaching me the elements of harmony and counterpoint.

Thus I learned at an early date that whatever flows in time can also be translated into tangible shape. Perhaps those intimate links between experience and representation in both map-making and musical notation have led me later on to overestimate the power of non-verbal communication.

The purpose of the 'home-area study' was to make the child familiar with the nature, resources, history, and life of its locality in such a way that these things came to hang together in the mind and a caring attitude took shape. The emotional side was enhanced by song and art. This perspective took me firmly in its grasp. It posed no intellectual problem at the time. I was omnivorous, undisturbed by academic ideas about disciplines and specialization. My favorite reading between ten and twelve – these marvellous years of free exploration in woods and bookshelves – was a Danish–Swedish popular magazine with the provocative name: *Ljus* (**Light**). The Danes called it *Frem* (**Forwards**) which I think was less adequate. But both names tell something of the jaunty temper by which the editors – three extrovert professors at the University of Lund – presented nature and history, anthropology, psychology, language, business, and sports.

The home-area study included not only such technical aspects as map-making but also dealt with everyday life around us. An understanding of economic relationships and of the importance of natural resources came almost without effort. The rye from the fields around made only a short detour to the mill at the nearest waterfall before it was brought back home for bread-making. We had potatoes and vegetables in our own garden and had to do the planting and picking ourselves. Energy for heating and cooking came as cord-wood on sledges in the middle of winter, brought in by farmers from the neighborhood who used the dead season in farming for cutting and the snow and ice for easing transport. We had to split the big pieces into firewood by saw and axe ourselves. We had to carry buckets of water several times a day upstairs from the well. When electricity first came it was produced by the small flour-mills as a sideline. A mile away in the other direction lay a factory and foundry. They produced threshers and locomobiles for bigger farms. The location of my home between the village and the factory town gave me a political lesson which I have never forgotten.

Although my father had a salary in money, the tradition still survived from earlier periods among the village people to bring farm products to the school teacher's household – eggs, cream, cheesecakes, blood. These delicacies were welcome, of course. But the factory workers, who were the poorer at that time and had nothing to give, resented this traffic very much. Sometimes regular wars broke out between the two groups of children, particularly at Easter time. I saw many a tear over smashed eggs in the kitchen. To me this was all a very tangible experience of class conflict. I think that my delicate situation in the middle explains a lot of my relative isolation as a child and my disposition toward compromise later in life. My only close friend was Karl Nygren, son of the woman who swept the school and probably the poorest among the poor. He was my faithful assistant in experimentation and exploration. He was taken away by tuberculosis in his early teens. The funeral marked farewell to my childhood world.

Grundzüge der Länderkunde

After a year of military service I matriculated at the University of Lund in the autumn of 1937. Since I had just worked with my father on a new local history museum in my home parish I am convinced that I would have chosen Scandinavian ethnology if it had then existed at the university. I tried to approximate, and during the first year I came to divide my time between geography and the history of art.

I had not been in geography for very long before I became worried about most of the enterprise in the form I encountered it in instruction and reading. Some parts were fine, geomorphology, climatology, and field mapping, for example, as well as map-making in the laboratory. It was mostly great fun to be in charge of the weather station and report to the national network. I think I can still classify clouds, if need be. But how was everything held together intellectually? The answer is simple: it was not. Now forty years afterwards, I can quite well see as a possibility that senior geographers of the time might have owned integrated winds of regional data in their heads which made sense to them. But neither the specific interconnections of phenomena in chosen regions nor the general principles of location or territorial integration were made clear to the students. Lectures in regional geography were abominably boring. Geography appeared, not as a realm of ideas or a perspective on the world but as an endless array of encyclopedic data. We read Hettner, of course, in German – not his theoretical work, unfortunately, but his descriptive. One of my friends, who was a thorough student, worked over *Grundzüge der Länderkunde, Europa*, thirteen times in order to be able to memorize the characteristics of every little tiny piece in the mosaic of the continent for the examination. I hope it helped him later when buying wine and cheese abroad.

The dead-end of regional geography was not the only source of frustration. Obscurities concerning the relation between physical geography and human arrangements were still preached, perhaps more as a matter of routine than of belief.

Why did I not turn my back upon such a strange discipline? I do not think that possibility occurred to me. My job as a teaching assistant and librarian – which I was appointed to in the second year – rendered me free accommodation in the department building and a modest salary. My father had died in 1938. There was no family in the background to support me or to furnish security for bank loans. I could have moved out as a master in some secondary school. But one of the few things I was cocksure about during my early life was to avoid becoming a school teacher. I stayed on, but eagerly discussing the nature of the discipline with fellow students.

The intellectual problem with geography became really acute when I was to take on a research task of my own. Invited to a field station at Lake Sommen by Sven Björnsson, a geomorphologist, I had composed a couple of term papers about the retreat of settlement and land-use changes since the end of the nineteenth century. Around 1940 my professor, Helge Nelson, suggested that I continue in the area and write an in-depth regional study. The book was going to become one in a series of monographs dealing, piece by piece, with the southern part of Sweden. Every author was supposed to cover everything: physiography, population, land use and industry, settlement and transport, and perhaps also endeavor a sketch of the 'personality' of the area.

I set out to do the work, interrupted by long periods of military service. Finland was at war, Norway and Denmark occupied. After one or two summers of rather unproductive field work, spent locating farms with respect to land forms, I decided to focus on population development as the backbone of my study. There were good reasons for doing so. The second half of the last century had seen a heavy loss of people to America. Who moved, from where, why, what were the effects afterwards?

I noted very soon a strange thing. Almost no correspondence could be found between emigration and abandoned farmsteads of which there were great numbers. At this point I felt a need to have a more tangible idea of events than figures could provide. I decided to do something nobody had ever tried before: to trace the life from year to year of every human being within a selected area.

Ten thousand lives

The friendly vicar of Asby, Sigurd Högquist, gave me free access to the church archives of his parish. Several summers of 'population archeology' began. My intention was not to count anonymous numbers but to excavate the individual biographies of everybody who had breathed the air of Asby from 1840 and onwards up to 1940.

Because of the necessity to make continual cross-references between various registers, sometimes difficult handwritings and inklines across most names, reconstruction was a tedious job for only two eyes and hands. My fiancée and future wife Britt came to my rescue. In the autumn of 1945 we had finished the last volume. We had 10,000 lives in the files. By now I had discovered the manifold reasons why the correspondence I just mentioned was missing.

The Swedish church registers are a mixture of strict book-keeping and frank notes. They most of all remind of the notations in old family Bibles. While one is unravelling how a family moved, found its livelihood and branched out, one gets personally acquainted with people and begins to share their joy and sorrow. Life-line by life-line is interlaced over the area like the voices in an infinitely complex fugue, repeatedly the same but never wholly the same. In a more sophisticated way I was back in my childhood's preoccupation with viewing the school children play and fight from the window above.

We did not sit with these volumes all day long. We took off on bicycles regularly, talked to people, studied soil and vegetation, made drawings of buildings, paced the distance to wells, and picked cherries in overgrown gardens. In fact we visited every glade where somebody had tried to make a living. This unusual form of exploration in time and space produced a general world-picture in my mind which today perhaps one would call 'holographic'. It is not a special way of formulating problems. It is a special way of forming an image before any questions at all can be asked or answers sought. The arduous acquisition of this image is my single most essential experience as an adult geographer. Almost everything I have done since is somehow extrapolated from it. But at the time I did not really know how to handle and communicate what I knew.

In order to get something in print we decided to cut out just one aspect of all information the data contained. We mapped in detail decade by decade all individual migrations to and from Asby inside Sweden – men and women separately. I managed to condense the whole story in one little diagram which I am still rather proud of. I do not think that the whole social history literature contains anything similar even today. One should remember that by that time we had only heard rumors about computers. The exercise gave me my *licentiate* degree (in the Swedish system of the period approximately similar to a Ph.D.).

Teachers, friends, and books

In the impasse at which I found myself in the middle of the 1940s two persons came to play a crucial role for the future: my father-in-law, unknowingly, and Edgar Kant.

Edgar Kant, a student of J. G. Granö and Professor of Economic Geography at the University of Dorpat, came to our department in 1944 as a refugee. We began to interact the following year, since I was asked to assist him in Swedish matters until he could master this new language. Edgar Kant brought Europe to Lund: he had been a student in Hungary, Austria, Switzerland, Germany, Holland, and France. He spoke six languages and could read at least fifteen. He lectured during the first years about Steinmetz and Dutch sociography and he frequently referred to Vidal de la Blache, Jean Brunhes, Albert Demangeon – I think his real teacher at Sorbonne – and Max. Sorre. He introduced a climate of learned discourse at the department. He spent hours with Sven Godlund, Karl Erik Bergsten, myself and others when we worked in the map laboratory. It was a great advantage of the old-fashioned way of drawing maps and graphs oneself that one could produce and talk simultaneously. Kant could walk and discuss up and down the streets of Lund until hours beyond midnight. He also tried to place his younger colleagues in an international network. He frequently came over with addresses of people abroad who in his opinion should have this or that paper.

Seen in retrospect I recall two influences stronger than others. The first was Kant's version of social geography. In one of his first lecture series, when he still spoke in German, he presented his studies in Dorpat about how different social classes used urban space. He had brought in the daily life of ordinary people as a valid domain for geographic investigation. Since I had grown up exactly in a zone of conflict between farmers and industrial workers, I could easily see the point in making such studies. He also showed pictures of how social groups furnished their homes. This legitimated my old interest in material culture, developed when I explored my home area. The views were a refreshing contrast to the dominating approach. I think that this transcending of established boundaries paved the way in my mind for my later work on diffusion of innovation. But the idea to embark on these studies as such did not come from Kant.

The second impact grew out from the tremendous range of Edgar Kant's readings. Whenever one needed a reference he could give it, frequently far back in the nineteenth century. Then he used to quote somebody who had said that we did not need to dream up new problems and ideas today. We could just pick up again what last century had thrown in the wastepaper basket.

Among those authors Edgar Kant brought to my attention I remember in particular Johan von Thünen, J. G. Kohl (an early German central place theorist), Tord Palander, Walter Christaller, and – from a quite different stream of thinking – the famous Russian-American sociologist Pitirim Sorokin. Their respective works came up in our discussions as parts of the gradual re-evaluation of the relation between physical and human geography which went on during the later half of the 1940s.

Kant did not deny the importance of natural factors, but by holding forth the ideas of the early location theorists he said in his unobtrusive way that the human world tends toward a geographical structure by its own inherent forces. In this period of eye-opening I enjoyed to read, and show to others, Tord Palander's devastating criticism of Sten de Geer's statement that the American industrial belt coincided with the distribution of the sweet apple. Sorokin provided more food for thought in the chapter on the 'geographical school' in his work *Contemporary Sociological Theories* where he attacked much geographic reasoning as a kind of astrology. My father-in-law, John Lundberg, was a senior master at a gymnasium in Göteborg and an authority in science teaching in the Swedish schools. In 1945 or so I came across in his study a series of books of absorbing interest to me: Sir Arthur Eddington's *Space, Time and Gravitation* (the 1935 edition), *The Nature of the Physical World* (1931), *The Expanding Universe* (1933), and *New Pathways in Science* (1949).

Sir Arthur was a marvelous writer, well respected as a scholar and almost shockingly witty. His explanations and reflexions made a strong impression on my mind. I remember in particular from *The Nature of the Physical World* the story of an elephant sliding down a grassy hillside. When trying to summarize this event the physicist prefers to forget the poetry of the situation and states that the 'mass of the elephant is two tons', the 'slope of the hill is 60° and the 'softly yielding turf' is replaced by a coefficient of friction. So, 'by the time the serious application of exact science begins we are left with only pointer readings'.

I tried in my own work to compromise between this extreme form of reductionism and my need to grasp the shape of phenomena and not just their size and intensity. The concept of 'mean information field' (MIF in some American literature), introduced in my study of innovation diffusion, stands in the same relation to people's gossiping as Sir Arthur's coefficient of friction to the green turf which his elephant is enjoying.

I began to look for writings in the human and social fields which at least approximately corresponded to the image of science that Eddington painted. One of the first promising non-geographic books I came across was *Foundations of Sociology* (1939) by George Lundberg. This author won considerable fame in Sweden at that time, more than elsewhere, it seems. Sociology began to break away from its earlier home in moral philosophy at the Swedish universities. George Lundberg, of Swedish ancestry by the way, was one of the Americans who came over to lecture. I went to listen to him but did not really meet him personally until much later in Seattle.

By looking at the sentences I underlined in his book, I can easily reconstruct what I picked up with approval from him. Here is one example: 'The ends of science are the same in all fields namely, to arrive at verifiable generalizations as to the sequences of events' (p. 149). Note that geographers at that time hardly spoke of 'sequences of events'. They advocated this sterile idea of 'areal differentiation'. Today Lundberg's physicalism, behaviorism, positivism, or what you care to call it, may seem naive. But given the cackling in geography this new song sounded lovely in my ears. It also sounded promising with respect to the masses of unused population data I had in my cupboard. I was captivated.

I think the references given by Lundberg helped me to pick up some books which were not exactly new but nevertheless demonstrated a course to follow. Very impressive among those books was A. J. Lotka, *Elements of Physical Biology* (1925). He discussed at length for example, 'chess as a conventional model of the battlefield of life'.

It was also a rewarding experience to read J. L. Moreno, *Who Shall Survive? A New Approach to the Problem of Inter-human Relations* (1934). In this book Moreno is laying the foundations of sociometry. Being a doctor and psychiatrist he dealt only with the network structure of small groups of people, predominantly school classes. But given my background in empirical work on migration I readily envisioned a projection of his concepts on to the population map and converted it in my mind into a huge structure of social atoms extending over regions, nations, and continents. This vision, I think, is still not fully exploited in human geography.

Among the earlier literature I ought to mention also K. Lewin, *Principles of Topological Psychology* (1936). It is obvious that his concrete way of giving structure to the internal world would appeal very strongly to me.

In all these books I was looking for aid in reaching a perspective which Eddington in his witty way had described in his *Space, Time and Gravitation* (1935): 'If we had had two eyes of different sizes, we might have evolved a faculty for combining the points of view of the mammoth and the microbe.' I felt that the ability to join the micro and macro levels of understanding was a very central element in the task of giving life to geography.

In practical terms I took the advice given by Lundberg seriously. Apart from geography (including some soil science) and a little bit of the history of art, my earlier training included history, political science, and education, in other words many matters of fact but little theoretical consciousness. Now, I began a second round at the university including courses in statistics, numerical analysis, and sociology – without, I must confess, proceeding to final exams. I already had my job as assistant professor. My goal was to pick up ideas for my geography.

Clearly, these excursions mostly made me familiar with research techniques, and they did not help very much in dealing with the theoretical problems of geography. There emerged one exception, however. For numerical analysis I went to the class of my school-mate since secondary school, Carl-Erik Fröberg, who had just come back from a stay in the United States, and was now an associate professor in the physics department. He introduced me to the concept of random numbers – something mathematicians laughed at when they first heard about it – and handed over to me a thin pamphlet on the *Monte Carlo Method*, a proceedings report from a symposium held in 1949 by RAND, National Bureaux of Standards, and Oak Ridge National Laboratory. Ulam, von Neumann, and Fermi are given credit as originators. These ideas to combine sampling theory and numerical analysis helped me to see how I could apply Lotka's physical biology to my empirical insights into population processes.

In 1950 I still had to write and defend a thesis for the degree of Doctor of Philosophy according to the requirements in Sweden at that time (approx. equivalent to the German *Habilitation*). Despite years of deviation into population dynamics and migration, I had not liberated my thoughts from the original task of writing a general regional study of my given area. But I did not feel ready to do it yet. I was still looking for a perspective which could hold a study of that kind together.

Nelson had retired in 1947. The chair was soon divided into two. The new professor in human geography, David Hannerberg, had a strong historical orientation, it is true. But he was also a mathematician and gave his full support in a very generous way to the group around Edgar Kant.

Although technologies and artefacts were close to early interests of mine, I had not paid much attention to 'folk-life' research, because we had so little of it in Lund. Such work was concentrated in and around the Nordic Museum in Stockholm. But in 1944–45, when we were just married and I served in the military forces near the capital, Britt worked in this museum under Dr Sigfrid Svensson who had just (1942) published a book called *Bygd och yttervärld. Studier över förhållandet mellan nyheter och tradition* ('Bygd' and outer world. Studies in the relation between innovation and tradition).

When reading the book a few years later I found that my unique material on rural migration would permit a rather strict definition of the vague verbal concept 'outer world'. The surprising stability over time of directions and quantities of migrations – even deviations from a smooth distance decay – indicated that social communication had a definite geographical structure

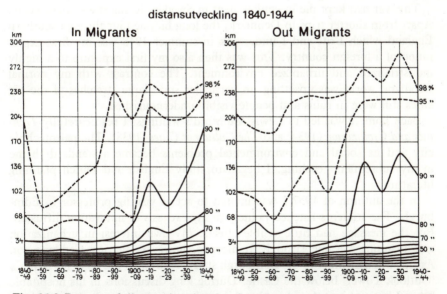

Fig. 16.2 Patterns of distance in migration flows between Asby and other Swedish communities, 1840–1944 (Hagerstrand, 1947)

which likewise changed only slowly. If so, this structure, as soon as it had become estimated in quantitative terms, would provide human geography with a connecting element with much the same role as climatology played for geomorphology and biogeography.

I enjoyed composing the latter half of my book on *Innovation Diffusion*. I had found a tool-kit by which I could construe things, I could make full use of my physicalistic thinking, I was able to portray a process over time and the whole emerged as a theme with variations. I also was confident that what I was doing would have some general validity. Fred Kniffen's papers, for example, told about similar processes in other parts of the world.

In order to be able to work full-time on my research I quit the job at the department in 1950 for three years and supported my family on a small grant, bank loans, and gifts to the children from Britt's aunt Ebba Lundberg, head-mistress of a girls' school in nearby Helsingborg and the most generous person I have known. The constrained economy was a problem. I was not able to pay back my debts fully until 1962.

The first steps abroad

My earlier references to readings may give the impression that my interests were exclusively directed toward theoretical and methodological problems and not toward matters of substance. This was not the case. Among several themes one in particular has been dominating: the growing interpenetration of society, technology, and landscape. Childhood memories, research on innovation diffusion, readings and experiences from travels abroad converge here.

The war had kept me in confinement in Sweden during my student years. Apart from shorter trips to Denmark, the long bicycle tour through southern England which Britt and I undertook in summer 1948 was my first exploration of a foreign country. This was then also my first exposure to a highly industrialized and urbanized environment. The contrast with my normal woodland world was upsetting, in particular since large stretches of the big cities still were in ruins and people visibly undernourished.

Back home I was led to Lewis Mumford's writings. I read carefully *Technics and Civilization* (1934) and *The Culture of Cities* (1938). The paleotechnic city and its social and environmental problems became a standard theme in my subsequent teaching. I began to study English and Dutch planning literature.

In the autumn of 1957 I was invited to act as visiting professor in Edinburgh. The students listened politely to my exegesis of innovation diffusion. But the message had little visible effects. My own interest in urbanization and planning became strongly reinforced, however. Patrick Geddes's son Arthur was a member of the Edinburgh staff. He brought me up to his father's 'Outlook Tower' below the Castle. During long discussions in his hospitable home I became introduced to the philosophy and ideology which formed the platform on which Mumford stood.

Two years later when I crossed the American continent by bus, train, and car to and from Seattle, I had the shocking experience that the same kind of problems from which England and Scotland were trying to heal themselves through planning were still building up in the world's wealthiest country: urban sprawl, technological erosion, formation of slums, devastation of forests. I could not understand the love of big size. I could not understand why these friendly people treated their land and their poor in such a careless way.

Our good friends, C. Arnold Anderson, sociologist and student of Sorokin, and his wife, Mary Jean Bowman, economist, who brought Britt and me by car through Kentucky, Tennessee, and Virginia, were tolerant but worried when I always wanted to stop and take pictures of heaps of car wrecks spread out on meadows along the road rather than views from the hill-tops.

Back in Sweden I held many lectures for planners and politicians about the destructive impact of the motor-car on urban and rural landscapes and public transport. The reaction was a shrug of the shoulders: 'We will never reach the American density of cars.' We did, at least almost, and a similar kind of urban destruction.

With *Man's Role in Changing the Face of the Earth* (1957) a book had finally appeared that treated Man and Nature in a manner which made sense to me in an age of expanding technology. I used large parts of the book as compulsory reading for graduate students all through the 1960s. The philosophy behind the symposium tied in well with my conception of geography as the study of processes. By this time I was fully committed to the thought that the transformation of society and landscape ought to be not only studied and understood but also guided by planning.

Ordinarius and planner

After the publication of *Innovation Diffusion* in 1953, David Hannerberg immediately arranged for me to come back to the department as associate professor. He also acted on my behalf – together with Karl Erik Bergsten in Lund and Gerd Enequist in Uppsala – to become his successor when he, four years later, moved to Stockholm in order to come closer to his historical archives.

My promotion to the university chair in 1957 brought me into a new and very different situation. I became an 'ordinarius' which meant safe income and full intellectual independence. That was good. But as a civil servant, appointed by the King in Council, I got entangled in time-consuming administrative tasks and sometimes burdensome responsibilities for teaching, employees, and students. The time for adventure shrank drastically because of unavoidable tasks. Only sometimes, on early Saturday mornings, I could hide away among the stacks of the university library where I had spent so much of my time before, browsing among books and journals, using them as if they were my old junk-holes in the wood, in other words as sources for my own constructions. Now I felt obliged to concentrate on the problem children of the discipline and department. One of these was, of course, regional geography. Another was the labor market for our graduates.

Whatever my feelings were with respect to regional geography, I could not neglect it in teaching. A limit had to be set to the splitting up of geography into diverging branches. Fortunately, my former teacher in map-making and geomorphology, Karl Erik Bergsten, had now come back from Göteborg to hold the chair in physical geography. He was strongly in favor of the unity of geography. This was of particular importance at that time, because geography was threatened by a total split in the schools. One part was supposed to be treated with science, the other with social studies.

We tried to implement a unifying program for the first-year students at least, by taking them out to make an intense field study in a local area in Sweden and in addition, giving each of them a country somewhere on the globe about which he or she had to acquire an all-round knowledge by library work and seminar discussion.

But what about readings? I tried to get away from the layer cake arrangement of regional texts and find more clearly integrated studies to hold up for the students as models for their thinking. A worthwhile book I came across was Robert S. Platt, *Latin America: Country-Sides and United Regions* (1952). I was attracted by Platt's insisting on the value of microgeography. Later on I became still more fond of William R. Mead, *An Economic Geography of the Scandinavian States and Finland* (1957). Again the reader is taken back and forth between local samples and broad overviews. The book has rare literary qualities. And there are diagrams showing the working rhythm of the year round in Finland and Norway! On somewhat different grounds I welcomed as student reading Jean Gottmann, *Virginia in the Mid-Century* (1955), a book very much dealing with the transformation of a region.

During the 1960s and a few years of the 1970s – while the 'quantitative wave' in America passed by – my time was almost fully occupied with matters related to teaching and to regional policy and planning. Although I could not produce much original work these years, I had the stimulation of a growing interaction with the international community of geographers. Bill Bunge, Peter Gould, Chauncy Harris, Walter Isard, Dick Morrill, John Nystuen, Ed. Ullman, to mention just a few, helped me to remain in touch with the American scene. Michael Chisholm, Richard Chorley, John Cole, Peter Haggett, Peter Hall, and Michael Wise, among many others, made me feel at home in Britain. Walter Christaller visited Lund several times. Christiaan van Paassen came up from Holland and Antoni Kuklinski from Poland.

My work in planning was by no means only an instance of social and environmental idealism. Other circumstances were perhaps more decisive at least in the beginning of the story. Since geography had an uncertain future on the higher levels in the school system we were several university geographers who felt it as our obligation to try to open a new labor market for our advanced students. Sven Godlund, then Associate professor in Stockholm, played the leading role in this respect. My position as full professor made it possible for me to recruit good students. This eventually became a loss for the department, because a whole generation of graduate students disappeared into the world of planning. But on the other hand, these persons were able to demonstrate to the administration how useful geographers could be in practical matters. Within a decade they had all reached central positions in the new planning hierarchy.

In my own case a second motivation was private. My old mother lived alone in Växjö, my former school town. I could easily stop to see her on my frequent travels by train to Stockholm. I became a 'meeting machine'. I would never have been able to afford such frequent visits if they could not have been included in the bargain. I clearly also appreciated the opportunity to go to concerts and exhibitions in the capital. I became involved in four major undertakings: the redivision of Sweden into new local government areas; the reorganization of the provincial government; the national land- and water-use plan; and the formulation of a national settlement strategy. Three leading ideals lay behind it all. One was to create as far as possible equal access to basic services, in particular education and health care. Another was to create robust local labor markets, also to absorb structural changes in a humane way. The third was to protect natural resources and create a healthy and attractive physical environment. In many ways present-day critics are right when they say that we tried to sweep up after the moves of a capitalistic industry, involved in international competition. At that time, however, this seemed to be the sensible thing to do.

My involvement with administrators and politicians did not add much to my conceptual development. But it gave valuable insights into how the transformation of localities and regions is bound up with events in society at large, and in effect was a kind of field study.

The central theme in my efforts in planning has been a constant fight against the negative influences on life and landscape of the functionally sectorized 'system society'. I suppose one can say that I have wanted to preserve or re-capture the home-area perspective of my childhood. That success has been limited is another matter.

After the middle of the 1960s research into regional development issues became institutionalized through the creation of ERU (the Expert Group for Regional Studies). I left the geography department in other hands and became a kind of full-time research director, floating somewhere between the government and the universities. In 1966 the new 'Bank of Sweden Tercentenary Foundation' decided to give Sven Godlund and me the largest research grant any social scientists in the country had ever received (1.2 million Swedish Kronor) for a joint project on the urbanization of Sweden. We could now assemble teams around us in a way that had not been possible before.

Graduate student once again

In 1971 I became invited to move over to a personal professorship, funded by the Social Science Research Council, and to lead a small group for development work in human geography, still at Lund University. It has been a blessing since then to be independent of an increasingly cumbersome university system as well as conditions set by the government machinery. Something like the years as graduate student came back, but now without the economic worries of that period. I am very conscious of my privileged position.

With my group of co-workers[1] I decided to make a second attempt to lift out into the open that old world-picture which had begun to take shape back in the 1940s. After all, I had learned some things in the meantime. Work at the national level had made me more aware than before about the need to look into the relation between the structures and arrangements of society at large and ordinary people's everyday lives. The problem of the mammoth and the microbe again. In addition a purely practical facilitator, the computer, had come into being. Geo-coding, suggested in a paper in 1955, was now implemented as a government responsibility.

This is not the place to make an exposé of the time-geographic outlook. Let me say only that publications which have come out so far, in particular in foreign languages, reflect only limited aspects of what I have in mind. The whole perspective is very broad and general. For myself at least, I have managed to unite within one frame of thought all the various subject matters I have dealt with earlier: settlement, migration, social communication, diffusion, domain structure, and impact of technology. There is also the beginning of an answer to my old quest for the purpose and form of the 'regional' approach to geography and the placing of man in nature.

A source of great delight in these recent years has been the support and criticism given by friends who speak other languages, have roots in other cultural settings, and have studied other authors. Christiaan van Paassen, Allan Pred, Egon Matzner, Anne Buttimer, and Olavi Granö – to mention them in the time-order I got to know them more closely – have all opened doors in various new directions. A shift of my awareness of new perspectives has taken place which can only be compared to what my interaction with Edgar Kant entailed back in the 1940s.

As far back as I can remember my dominating stance has been that of observer rather than participant. I have felt very much at home with the ruling scientific attitude to approach matters from the outside. That is why I have preferred to take the 'corporeality' of society as vantage point and focus open opportunities and constraints while avoiding attempts to explain human purposes and behaviors.

The weakness of this position was well taken by Chr. van Paassen in conversation and writing.[2] However, I did not clearly see the risks of this kind of one-sidedness until Anne Buttimer frankly told me that the world I depicted reminded her of a *dance macabre*. My time-geography diagrams, for example, to her eyes, seemed to omit those crucial dimensions of human temporality which seemed so essential to her view of life – images and perceptions of time on the one hand, and bio-ecological rhythms on the other. In my system of concepts – just as in the objective scientific stance in general – there lay hidden a blueprint of a world which neglected and ran over the more important part of human existence: the internal realms of experience and meaning. I had to admit that I was trying to walk on only one leg.

We agreed to try to initiate a dialogue on this urgent theme of a bridge

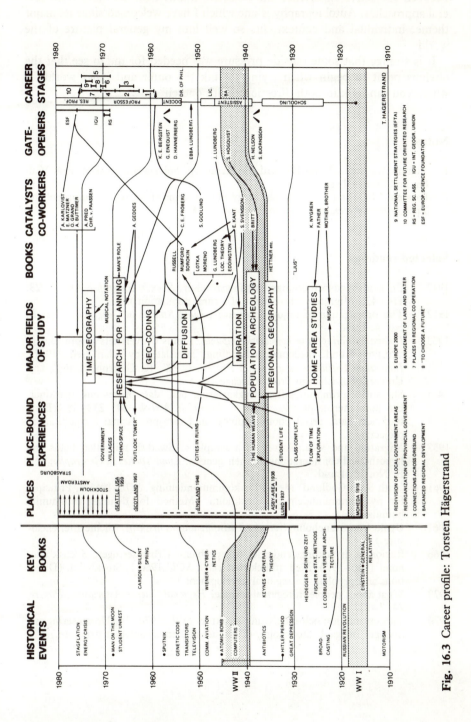

Fig. 16.3 Career profile: Torsten Hägerstrand

255

between subjective experience and objective knowledge. We have tried several approaches. Autobiography is one which I have welcomed since its major theme, individual and context, fits so well into my general picture of the world.

It is strange that some outside pressure was needed for me to see that the obvious point to begin when trying to link the outer and inner worlds is where you can look in both directions: with oneself.

Notes

1. Tommy Carlstein, Kajsa Ellegård, Bo Lenntorp, Solveig Mårtensson, Erik Wallin, and Sture Öberg.
2. Paassen, Chr. van (1976) Human geography in terms of existential anthropology, *Tijdschrift voor economische en sociale geografie*, **6**.

Selected readings

1947   En landsbygdsbefolknings flyttningsrörelser, *Svensk Geografisk Årsbok*, **23**, 114–42. (Migratory movements of a rural population.)
1952   The propagation of innovation waves, *Lund Studies in Geography*, Ser. B, No. 4, Lund.
1953   *Innovationsförloppet ur korologisk synpunkt*. Lunds Universitets geografiska institution. Lund, akademisk avhandling. Translated by Allan Pred, (1967). *Innovation Diffusion as a Spatial Process*. C. W. K. Gleerup, 'Lund', and University of Chicago Press, Chicago.
1955   Statistiska primäruppgifter, flygkartering och 'data-processing' maskiner: Ett kombineringsprojekt, *Svensk Geografisk Årsbok*, **31**, 233–55. (Basic statistical data, aerial photo-mapping, and data-processing: a combinatory project.)
1957   Migration and area. Survey of a sample of Swedish migration fields and hypothetical considerations on their genesis. *Lund Studies in Geography*, Ser. B, No. 13, Lund.
1961a (with Sven Godlund) Metod för kommunindelning. *SOU* 1961:9, Principer för ny kommunindelning, Stockholm. (Method for defining community boundaries.)
1961b Utsikt från Svaneholm, *Svenska Turistföreningens årsskrift*. (Skåne after enclosure.)
1966   Regionala utvecklingstendenser och problem. Urbaniseringen. Svensk ekonomi 1966–1970. Med utblick mot 1880. *SOU* 1966:1, Stockholm. (Directions and problems in regional development.)
1970   What about people in regional science? *Regional Science Association, Papers*, **24**, 7–21.
1970   Tidsanvändning och omgivningsstruktur. *SOU* 1970: 14, Urbaniseringen i Sverige. Bilaga 4, Stockholm. (Time use and environmental structure.)

Appendix A.1

Participants in the international dialogue project

COUNTRY	GEOGRAPHERS							OTHERS						
	Autobiogr. interview	Recorded discussion	Autobiogr. essay	Internat. workshop	Written response	Other	Persons involved	Autobiogr. interview	Recorded discussion	Autobiogr. essay	Internat. workshop	Written response	Other	Persons involved
Australia	•						•							
Austria	•		•		•	•	••	•	••		•			∴
Canada						•	•							
Denmark											∴	•		::
Eire					•	•	•	••	•				•	∴
Finland		•	•	::	••		∴∙				•			•
France	•	::∙	•		•	•	::::∙							
Germany (W.)	••		•	••	∴	••	∴∙				•	•		••
Iceland	•						•							
India											•			•
Israel		•				•	•	•						•
Japan					•	•	••				•	•		•
Luxembourg											•		•	•
Netherlands	•	•	•	••	••		∴				•			•
Norway	••		•	••	••	•	::		•		•	•		••
Poland	•						•					'		
Portugal		•					•							
S. Africa	•		•			•	••							
Sweden	∴	::	::	:::	:::	∴	::::∙	::::	∴∙	•	::::∙	:::	∴	▓
United Kingdom	••		••		••		::							
United States	::	:::::∙	••	∴	••	∴	:::::∙	:::	•		•		•	::∙
Yugoslavia											•			•

Appendix A.1 Participants in the international dialogue project (diagram by Torsten Hägerstrand)

Appendix A.2

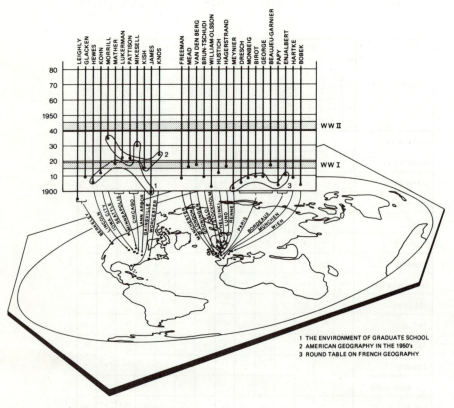

Appendix A.2 Essayists and discussants

Alphabetical list of contributors to this volume

Ch. 9 Bataillon, Claude (n. 1931), Prof. de Géographie, Institut de Géographie, Université de Toulouse, Toulouse, France.

Ch. 10 Beaujeu-Garnier, Jacqueline (n. 1917), Prof. de. Géographie. Université de Paris-I, 191, rue Saint Jacques, 75005 Paris, France.

Ch. 9 Birot, Pierre (n. 1909), Prof. de Géograhie, Université de Paris-IV, Laboratoire de Géographie Physique de CNRS, Paris, France.

Ch. 12 Bobek, Hans (n. 1903), Emer. Prof., Mahlerstrasse 4, A-1010 Vienna 1, Austria.

Ch. 3 Brun-Tschudi, Aadel (1909–1980).

Ch. 9 Dresch, Jean (n. 1905), Prof. de Géographie, Université de Paris-VII, Paris, France.

Ch. 9 Enjalbert, Henri (n. 1910), Prof. de Géographie, Université de Bordeaux-III, Bordeaux, France.

Ch. 7 Freeman, T. Walter (n. 1908), Prof. of Geography, 1 Thurston Close, Abingdon OX14 5RD, England.

Ch. 9 George, Pierre (n. 1909), Prof. de Géographie, Université de Paris-I, Paris, France.

Ch. 2 Glacken, Clarence (n. 1909), Prof. of Geography, University of California, Berkeley, California 94720, USA.

Ch. 15 Hartke, Wolfgang (n. 1908), Emer. Prof., Tengstrasse 3, 8 München 40, West Germany.

Ch. 16 Hägerstrand, Torsten (n. 1916) Prof. of Geography, University of Lund, Box 716, 220 07 Lund, Sweden.

Ch. 5 Hewes, Leslie (n. 1906), Prof. of Geography, University of Nebraska, Lincoln, Nebraska 68588, USA.

Ch. 8 Hustich, Ilmari (1911–1982).

Ch. 5 James, Preston E. (n. 1899), Emer, Prof., 379 Villa Drive South, Atlantis, Florida 33462, USA.

Ch. 13 Kish, George (n. 1914), Prof. of Geography, University of Michigan, 1028 LSA Building, Ann Arbor, Michigan 48109, USA.

Ch. 5 Kohn, Clyde (n. 1911), Prof. of Geography, University of Iowa, Iowa City, Iowa 52242, USA.

Ch. 13 Knos, Duane (n. 1924), Prof. of Geography, Clark University, Worcester, Massachusetts 01610, USA.

Ch. 6 Leighly, John (n. 1895), Prof. of Geography, University of California, Berkeley, California 94720, USA.

Ch. 13 Lukermann, Fred (n. 1921), Prof. of Geography, University of Minnesota, 414 Social Science Building, Minneaplis, Minnesota 55455, USA.

Ch. 5 Mather, Cotton E. (n. 1918), Prof. of Geography, University of Minnesota, 414 Social Science Building, Minneapolis, Minnesota 55455, USA.

Ch. 4 Mead, William R. (n. 1911), Prof. of Geography, University College, London, England.

Ch. 9 Meynier, André (n. 1901), Prof. de Géographie, Université de Rennes, Rennes, France.

Ch. 5 Mikesell, Marvin (n. 1930), Prof. of Geography, University of Chicago, 5828 South University Avenue, Chicago, Illinois 60637, USA.

Ch. 9 Monbeig, Pierre (n. 1908), Prof. de Géographie, Centre National des Recherches Scientifiques, Paris, France.

Ch. 13 Morrill, Richard (n. 1934), Prof. of Geography, University of Washington, Seattle, Washington 98195, USA.

Ch. 9 Papy, Louis (n. 1903), Doyen de la Faculté des Lettres, Université de Bordeaux, Bordeaux, France.

Ch. 13 Pattison, William (n. 1921), Prof. of Geography, University of Chicago, 5828 South University Avenue, Chicago, Illinois 60637, USA.

Ch. 14 van den Berg, G. J. (n. 1916), Prof., Geographical Institute, Department of Planology and Demography, State University of Groningen, The Netherlands.

Ch. 11 William-Olsson, William (n. 1902), Prof. of Geography, Stavgårdsgatan 12, S-161 37 Stockholm, Sweden.

Highlights of the decades (1900–80) in nine countries

Austria–Germany: Professor Elisabeth Lichtenberger, İnstitut für Geographie, A–1010 WIEN, Universitätstrasse 7, Austria. Professor Dietrich Bartels, Geographisches Institut des Universität Kiel, Olshausenstrasse 40–60, D–2300 KIEL, West Germany.

Finland: Professor Olavi Granö, Department of Geography, University of Turku, Turku, Finland.

France: Professor Paul Claval, Institut de Géographie, Université de Paris IV, 75005 Paris, France.

Great Britain: Professor Ronald Johnston, Department of Geography, University of Sheffield, Sheffield, England S10 2TN.

The Netherlands: Professor Christian van Paassen, Subfaculteit des Sociale Geografie, Universiteit van Amsterdam, 1011 NH Amsterdam, The Netherlands.

Norway: Professor Hallstein Myklebost, Department of Geography, University of Oslo, Blindern 2, Oslo, Norway.

Sweden: Professor Staffan Helmfrid, President, Stockholms Universitet, S–10691 Stockholm, Sweden.

United States: Professor Marvin Mikesell, Department of Geography, University of Chicago, Chicago, Illinois 60637, USA.

Appendix B

Highlights of the decades (1900–80) in nine countries

Pre–1900

1870 – Norway
First program in geography at University of Oslo (1871). Strong links with history and natural history. Environmental determinism a dominant philosophy and service to the needs of school geography emphasized.

1880 Finland
Geography begins to acquire formal structure 1881. Two rival societies: (1) Society for the Geography of Finland (SGF), an elitist, interdisciplinary 'Academy', promoting research related to national identity, with leaders such as Palmén (zoology), Norrlin (botany), and Ignatius (statistics, author of *Geography of Finland*, 1881), and (2) the Geographical Association (GA), comprised of teachers and amateur geographers led by Hult (plant geography) who strives to establish geography as a discipline with an autonomous place in the university along German lines. First department in Helsinki (1890) chaired by Hult, emphasizing teacher training. Geography assigned place as a natural science, geomorphology dominant. First Ph.D. (Rosberg on Finnish deltas, 1895), SGF publishes in *Fennia* series and completes *Atlas of Finland* (1899). GA publishes work of Hult and disciples emphasizing teaching interests.

1880 Sweden
Pre-academic background in geographical and topographical survey since 1628; Linnaean nature research and exploration since eighteenth century. Strong German influence (Ritter) on school textbooks, and encouragement for 'home area' study. Swedish Society for Anthropology and Geography (SSAG) founded 1877. Arctic exploration (Nordenskjöld) Northeast Passage, 1800. First Swedes, e.g., Hedin and others, take Ph.D. at German universities. First department at Lund (1894) chaired (1897) by von Schwerin, historian and political scientist. First geographical journal, *Ymer*, founded 1880.

1890 France
Beginning of modern geography in France. Foundation of *Annales de Géographie* (1891). Vidal de la Blache appointed to Sorbonne (1898). Strong influence of German geography (Ritter and Ratzel).

1900

1900 The Netherlands
First chair of geography at Amsterdam (1877) held by a classicist, Kan, who advocates comparative, political–economic macro geography (as alternative to Hettner's *Länderkunde*) and separation of physical and human branches. His successor, the ethnologist Steinmetz (1907), initiates an entirely new course by relating human geography to sociology, establishing the Amsterdam School of 'sociography', which became the cradle of social sciences in the Netherlands. At Utrecht second chair of human geography held by Niermeyer (1908), disciple of Kan, who cultivates *la géographie humaine*; chair in physical geography held by a German, Oestreich, who emphasizes geomorphology.

261

1900 Finland

Geography firmly established as popular discipline following Rosberg's appointment as professor (1901) and J. G. Granö as first assistant (1902) in cartography. Ex-botanist Leiviskä gets Ph.D. for geomorphological studies of western Finland and Granö for studies on glaciation of Central Asia. GA flourishes and publishes revised and expanded edition of the *Atlas of Finland* (Palmén, ed. 1911).

1900 Sweden

Chairs in geography set up in Uppsala (1901) chaired (1904–06) by Ahlenius, student of Ratzel, in Göteborg (1901) with emphasis on commercial geography and ethnography, and at the Stockholm School of Economics (1909) with an orientation toward economic geography and resources. Geography well established in school curricula. Research specializes in anthropology, glacial morphology (e.g., Hamberg, Uppsala) and climatic change, history of maps, and expeditions to Antarctica, South America, and Asia. No basic philosophical commitment except Darwinism. Close links to geology and anthropology. Ahlenius initiates topographical survey of Sweden, 6 vols. to be completed (1924) by Sjögren; Sten de Geer introduces dot-map of population (1908).

1900 France

Developments of French School of geography, with emphasis on regional geography: Vidal de la Blache sets up the models with the *Tableau de la géographie de la France* (1903); Demangeon and Blanchard give the first monographs (1906, 1907); Lucien Gallois puts more emphasis on physical setting (1908). Geomorphology develops along lines coming from Davis (USA) and from Germany. First textbooks: physical geography (de Martonne 1909), human geography (Brunhes 1910). The *excursion interuniversitaire*, a field trip open to all professors teaching in the universites, and to graduate students (1904), promotes collegial contacts and professional cohesion.

1900 Great Britain

First publication of Mackinder's heartland thesis and Herbertson's ideas of natural regions. Institutional dominance of the Royal Geographical Society with its continued emphasis on exploration and discovery; Scott reached the South Pole. Establishment of departments of geography in a number of universities. Foundation of *The Geographical Teacher* (later *Geography*) as the journal of the Geographical Association.

1900 United States

Founding of the Association of American Geographers (AAG) in 1904 and first program of graduate studies at Chicago (1903). Dominant role of physiography and prominence of ex-geologists. Rivalry of Davis (Harvard) and Salisbury (Chicago) for leadership of the profession. Beginning of 'environmentalism' and attraction of students trained initially in history and other non-physical science fields.

1910

1900 Germany–Austria

Geography firmly established at universities, full range of periodicals and associations, extremely prolific period of research. Enduring influence of Ratzel (1844–1904) in deterministic conception of human society within the framework set by 'nature', later modified in Vidal de la Blache's possibilistic model. Axiomatic principles of site and situation emphasize the individuality of regions, later resumed by Haushofer's 'geopolitics'. Concepts of migration and diffusion of people and material objects strongly influencing historical and social sciences, an approach later on extended in the theoretical work of Torsten Hägerstrand (Sweden). Important theory-building in physical geography. Penck, founder of glacial morphology, quaternary research, morphometry, and preliminary discussions of the earth's 'carrying capacity'. Karst research by Grund and Cvijic. Landscape proclaimed as the research object of geography (1906) by Otto Schlüter (1872–81). First phase develops the historical and morphological perspective. Hugo Hassinger (1877–1952) develops cultural-historical urban geography emphasizing sequent occupance; he co-initiates the movement for historical preservation of monuments and buildings in Vienna.

1910 The Netherlands

At Rotterdam (1914) economic geography becomes a requirement for students in economics. *Tidjschrift voor Economische Geographie* (TEG) founded in 1910 – the first of its kind in the world. Steinmetz (1913) defines human geography as sociography, empirical counterparts of speculative German sociology, advocates 'grass roots' research, and publishes volume on European nationalities (1920).

1910 Finland

Controversy over nature of geography. Sederholm (Chairman, SGF) maintains that geography is the sum of different sciences and not a science on its own. Rosberg and Granö, representing disciplinary interests, claim that geography is an independent and monistic discipline. Leiviskä continues research on glacial forms in Finland and Granö on geomorphology of Central Asia. Ph.D.'s by Hänninen on drumlins of northern Finland, and Hildén on anthropological studies in Central Asia. National congresses sponsored by GA and Helsinki geography department. Granö professor at Tartu, Estonia (1919–22).

1910 Sweden

Pioneering work in thematic mapping covering entire country, e.g., population, land forms, arable land, quantitative approach to distributions of phenomena. Settlement studies in association with archeology and inventory of rural house and farm types (Erixon). Nelson (1913) publishes pioneering study of the historical geography of a mining area. *Geografiska Annaler* founded 1919.

1910 Norway

Werenskiold (1925–53) interprets physical geography as an exact science. Much activity directed toward exploration of polar regions.

1910 France

Opening of new research lines in urban geography (Blanchard 1913), political geography (Vallaux 1911, Siegfried 1913). In regional geography, growing interest in the role of cities and industry (Vidal de la Blache 1913, 1917). First provincial geographical journal (*Revue de géographie alpine* 1913). First World War stops research for practically ten years.

1910 Great Britain

Honours schools of geography begin to established (Liverpool, Aberystwyth 1917).

1910 United States

Geography seen increasingly as enterprise devoted to demonstration of physical-environmental influence on human affairs. Growing importance of Semple after publication (1911) of her translation and elaboration of Ratzel's *Anthropogeographie*. Yet physiography remains most important professional commitment. *Annals* of the AAG launched in 1911 and *Geographical Review* in 1919. Ph.D.'s earned by Jones, Parkins, Visher, Sauer, Colby, and Platt at Chicago. Fennemann (1919) argues that regional study is the unifying core of geography.

1920

1918 Germany–Austria

Core emphases in geography rest on geomorphology and on settlement geography. Both develop typologies and structural categorization of material objects based on *gestalt* perception. Berlin leads institutionally throughout the German-speaking world, producing most of the researchers of the interwar period. Important theoretical works in geomorphology, e.g., models of slopes, peneplain and valley development, glacial erosion, and other themes. In human geography main foci of research are (a) historical cultural research on landscape, some of it interdisciplinary, e.g., using archeological techniques, study of soils, mapping of plant association, pollen analysis, and focusing mostly on medieval settlements and field systems and research into their abandonment; (b) agricultural geography, marked especially by Waibel's research into *Agrarformation*. Landscape study now takes on a distinctly functional approach. (c) Landmark study of Innsbruck (Bobek 1927) sets up model for generations of city monographs and stimulates research into the urban fringe. The importance of Christaller's (1933) 'Central Places of Southern Germany' is not recognized. Extensive (12-volume) series of handbooks on general and regional geography, edited by Klute, illustrates the encyclopedic range of knowledge involved in the teaching of geography. Hettner's *Länderkundliches Schema*, is closely followed in many handbooks and texts, while also evoking some negative criticism.

1920 The Netherlands

Institutionalization (by law) of physical and human geography as two different university curricula, opening the possibility for Ph.D. dissertations (1912). 'Social geography', compromise between sociography and *géographie humaine*, now regarded as social science. First Ph.D. dissertations, e.g., ter Veen on the Haarlemmeer, 1925, who succeeds Steinmetz (1932). Van Vuuren assumes chair at Utrecht (1927) and continues French orientation. Boerman (Rotterdam, 1923–) begins to emphasize survey and planning.

1920 Finland

Second chair of geography established at Helsinki (Leiviskä 1921; Granö 1923–26), new chair at Helsinki School of Economics (Hildén 1921–52), Turku University (Granö 1926–44). Radical change in Granö's thinking at Tartu: moving away from Davis and Penck schools via Hettner, Passarge, and Banse to his own inductive approach to landscape, starting from environmental perception and proceeding to a definition of geographical regions (*Reine Geographie* 1929). Critique of Granö (e.g., Leiviskä). Granö sets stamp on Estonian regional and urban geography (Kant was one of his leading pupils). Ph.D.'s by Tolvanen and Heklaakoski on the evolution of Finnish lakes. Amalgamation of SGF and GA. Nationalistic aims preserved. Third edition of *Atlas of Finland* (Granö, ed. 1925–29).

1920 Sweden

S. de Geer introduces a chorological paradigm (1923) in one of the first essays of a theoretical nature, and he also pioneers in urban geography. Anna Kristoffersson explores the historical development of rural landscape, settlement pattern, and economy in a selected area – an issue which becomes central for subsequent reseach in Swedish geography. Chair established at Stockholm (1929). Now five professors of geography in Sweden. *Svensk Geografisk Årsbok* initiated in 1925. Last and greatest exploration by Sven Hedin in Central Asia 1927–33.

1920 France

Vidal de la Blache's posthumous *Principes de géographie humaine* (1922) and Lucien Febvre's *La terre et l'évolution humaine* (1922), proclaim the unique character of the French School. Emphasis on regional monographs in France, and broader syntheses on foreign countries: first volume of *La géographie universelle* (1927–48), *Les îles britanniques* (1927) by Demangeon, is a model of regional description. In physical geography, emphasis rests on geomorphology, along Davisian lines (Baulig, *Le plateau central de la France*, 1928). French geographers begin to work abroad, e.g., South America (Denis 1920), in North America (Gautier 1923, 1927) and in South-East Asia (Robequain 1929).

1920 Great Britain

Publications by Fleure and Daryll Forde linking human geography with anthropology. Little expansion in universities.

1920 United States
Criticism of 'environmentalism', especially by Sauer. Alternate orientation of the 'Morphology of Landscape' (1925). Profession turns to empirical, inductive studies of small areas. Unit-area mapping, concern for correlation rather than causation. Beginning of separation of physical and human geography. New graduate program started at Clark (1921), Michigan (1923), and Wisconsin (1928). Midwestern Field Conferences (1923–) provide setting for dialogue on professional objectives.

1930

1930 The Netherlands
Many doctoral dissertations both in Amsterdam, e.g., Kruyt's (1933) on secularization in the Netherlands, den Hollander's (1933) on poor whites in the American South, Hofstee's (1938) regional rural sociology, and in Utrecht, e.g., Keuning's (1933) regional monograph, de Vooys' (familiar with Sorokin (1929) and with American rural sociology) study of rural labor markets, all demonstrate predominantly rural focus. Growing concern for applied geography. Ter Veen initiates research on Zuiderzee polders, and van Heek writes first study on Wieringermeer (1929). Economic–technological provincial institutes promoting development engage geographers (e.g., Hofstee at Groningen). First sociographer appointed to Board of Works and Buildings in Amsterdam (Dijkhuis, 1932). A civil engineer, van Lohuizen, conducts survey for Amsterdam Extension Plan with economist Delfgaauw, author of first Ph.D. dissertation in land economics (1932). Van Vuuren promotes research on rural marginal areas for Chamber of Commerce, (Maas study, 1935), and pioneers in applied urban geography with studies of Utrecht (1938), Zwolle (1939), and Flushing (1941); growing national interest in physical and economic planning. IGU Congress in Amsterdam (1938). Cool reception for Christaller's theory. Repudiation of sociography by van Vuuren and his disciples. First Ph.D. dissertation on geographical theory by van der Valk on Kapp, disciple of Ritter (1939). Political geography (ter Veen, 1931, 1946) attacks German geopolitics.

1930 Finland
Auer succeeds Granö in Helsinki, introduces (Swedish) methods of micro-fossil and shoreline displacement, refining these methods in expeditions to Patagonia and Tierra del Fuego. Together with his students, Auer develops concept of urban hinterland, which is continued by his brother-in-law, Ajo, along mathematical lines. Tanner, a quaternary geologist, succeeds Rosberg and strengthens geomorphology, organizes expeditions to Labrador, accompanied by Hustich (plant geography). At Turku, Granö continues work on landscapes of Finland and Mongolia, and in collaboration with students, research into settlement and urban geography in Finland. Smeds awarded Ph.D. for French-type regional thesis. GA publishes *Handbook on the Geography of Finland* (ed., Hildén 1936).

1930 Sweden
Specialization in research and beginning split between physical and human geography which Nelson tries to overcome. Arctic exploration and research on climatic change continue (Ahlmann); major Swedish contributions to meteorology and oceanography. Greater Stockholm Project analyses urban development and inner differentiation of the city (Ahlmann *et al.* 1934). William-Olsson pioneers in urban social geography, using unexplored sources of quantitative data on socio-demographic patterns. Nelson researches areas of Swedish settlement in North America; Frödin (Uppsala) studies semi-permanent settlements and transhumance. First national gathering of Swedish geographers (1933): debate over nature and scope of the discipline.

1930 Norway
Isachsen (Professor in Oslo, 1931–69), strongly influenced by the French School introduces fields of research on urban geography, hinterlands, and traffic flows; also

valuable work in historical geography, dealing with the particular needs of marginal areas. Isachsen pursues research in geology and geomorphology but younger colleagues opt for either human or physical geography, with occasional excursions into regional geography. Mid-1930s witness a growing trend toward quantification through the influence of urban surveys, e.g., those of New York (Haig 1927) and Stockholm (Ahlmann *et al.* 1934). Case studies in the field, based on personal observations and interviews, with walking boots, map and note-book, however, continue to be the conventional mode.

1930 France

Continuing emphasis on regional monographs in France, syntheses on foreign countries in *La géographie universelle* (de Martonne, *L'Europe centrale*, 1930; Baulig, *L'Amérique septentrionale*, 1934). Increasing interest in South America and on colonial Empire: Middle East (Weulersse 1934, 1936), South-East Asia (Gourou 1936). First textbook on human geography is published (Lavedan 1936); main opening is in field studies and rural geography following Demangeon's studies of settlement, agrarian structures are analyzed (Bloch 1931; Roupnel 1932; Dion 1932). Renewed interest in political geography (Ancel 1936–38) and a curiosity for social geography (Cholley 1931). In gemomorphology, studies on climatic conditions develop (de Martonne 1935, 1940).

1930 Great Britain

First Land Utilization Survey – for the whole of Great Britain – organized by Dudley Stamp. Pioneering work in urban geography (Dickinson), historical geography (Darby), the use of statistics in climatology (Crowe), and geomorphology (Wooldridge and Linton). Foundation (1933) of the Institute of British Geographers. Major focus on regional geography.

1930 United States

Development of topical specialities (urban, economic, political, etc.). Contrast of Midwest (contemporary) and Berkeley (historical) interests. James (1934) produces first comprehensive textbook; keyed to Passarges's *Landschaftsgürtel*, the Köppen climatic classification, and 'sequent occupance'. Numerous articles on agricultural regions published in *Economic Geography*. Geographers employed by TVA promote the cause of applied or non-academic geography. Hartshorne responds to Leighly's critique of regional geography and Sauer's 'Morphology' with 'The Nature of Geography' (1939).

1940

1940 The Netherlands

National Institute for Social Scientific Research (ISONEVO) founded in 1940, proposes monographs on rural municipalities, e.g., Groenman (1947). National and provincial services for physical planning (since 1943), and National Census Bureau engage leading geographers. Keuning (1941) defines economic-geographic regionalization of the Netherlands; geographers participate in variety of post-war ISONEVO sponsored conferences and projects on societal planning (1947), population increase (1949). Research on population, location of industry (Winsemius, 1945, 1949). on functional classification of cities (Keuning, 1948), application of Christaller's method of settlement policies in Ijsselmeer-polder (Takes, 1948). Growing interest in social agrarian history (Slicher van Bath). *Geographisch Tijdschrift* (GT) founded in 1948. New department at Groningen (Keuning, 1948–), 'Brain drain' of sociographers to new departments of sociology. Hofstee, chairman at Wageningen, initiates 'differential sociology' (à la Steinmetz) in well-known school of rural sociology. 'Changing of the guard' in Amsterdam and Utrecht where de Vries Reilingh (1950–71) and de Vooys (1949–73), respectively, determine the course for about twenty years.

1940 Finland

Gradual post-war increase in student numbers and new stimulus for research. Tanner to Sweden and Auer to Argentina, the latter returns to become professor of geography and later geology. Hustich makes expeditions to Canada, Granö returns to Helsinki and proposes new edition of the *Atlas of Finland*, more close to traditions of GA. Tuominen introduces Christaller's ideas to Finland (Ph.D., 1949). Smeds professor at Swedish School of Economics in Helsinki and later at University of Helsinki (1950). Granö edits second edition of *Handbook on the Geography of Finland* (1952).

1940 Sweden

Scientific 'rearmament' of Swedish universities, geography chairs doubled (i.e., with chairs in both physical and human geography) in some institutes. By 1950 there are nine professors of geography in Sweden and the subject reaches its best position in school curricula. Good recruitment for higher education. Edgar Kant arrives from Estonia, introduces Christaller and European social geography to Lund, revitalizing the tradition of population geography and first experiments in model construction. Hannerberg formulates fundamental questions and introduces new mathematical methods for the analysis of rural landscape evolution and the origins of rural settlements. Dahl advocates more economic thinking in economic geography. Significant developments in physical geography including river studies (Hjulström), glacial morphology (Mannerfelt) tephrochronology and vulcanism (Thoraninson). *Lund Studies in Geography* initiated in 1949. William-Olsson argues for regional geographic understanding as goal for the discipline.

1940 Norway

Post-war reconstruction strengthens the need for regional planning and applied geography. Geographers become active in various kinds of government service with increase in student numbers. International contacts intensify, e.g., with the vigorous Swedish school (e.g., Kant, William-Olsson, Hannerberg, and Hägerstrand) and even with Britain, USA, and France. Topical research interests in population growth, food production, resource conservation and underdevelopment begin to burgeon.

1940 France

Wartime prevents field work, research interests focus on improving the structure to the discipline: Cholley's *Guide de l'étudiant en géographie* (1942), Sorre's *Fondements de la géographie humaine* (1940). Ecological and sociological perspectives are opened; Deffontaines tries in vain with the *Revue de géographie humaine et d'ethnologie*, to develop anthropological perspectives. Pierre George, then a Marxist, begins to orient research in human geography, along economic lines. The influence of German physical geography is very strong and explains the shift from Davisian to climatic geomorphology, with men like Tricart and Birot. *Les pays tropicaux* (1946) of Pierre Gourou gives unity to research on tropical geography and explains the development of the French school of zonal geography: climate becomes the main factor of explanation both in the physical and in the human field.

1940 Great Britain

Substantial role played by geographers in the war effort, especially in intelligence activities; development of air-photo interpretation and publication of the Admiralty Handbooks. Series of Royal Commissions preparing for post-war planning legislation. Regional geography still in the ascendant.

1940 United States

Research and training interrupted by the Second World War. Rapid expansion of graduate and undergraduate programs after the war. Proliferation of topical and regional specialities. New textbooks (e.g., Finch and Trewartha 1936) offer separate, systematic treatment of physical and cultural 'elements' of geography. Pervasive influence (except among Sauer's students) of Hartshorne's assessment of the character and objectives of the discipline. Preoccupation with the independence of geography. AAG becomes open rather than elite society. The Marshall Plan and other post-war expressions of internationalism encourage foreign-area studies.

1950

1945 Germany Austria

Political subdivision of the German-speaking area deeply influences the institutional context of academic research. In the Western countries a functional perspective on landscape prevails, 'external factors' cited as explanatory. Development of climatic geomorphology (Büdel), ecology of high mountain areas (Troll) and social geography, already anticipated by Busch-Zantner's *Faktorenlehre* (1937), and developed by Hans Bobek and Wolfgang Hartke. Dichotomy develops between macrolevel theory formulation (Bobek) and extremely detailed microanalysis wherever landscape observations could provide 'geographical indicators' that gave rise to a study of specific phenomena, mapping of individual houses and parcels, air-photo interpretation, interviews. Independent developments in marginal fields, e.g., population biology (Kinzl). Discovery of principles of spatial order in economy, e.g., early ideas of Christaller and Loesch (Otremba).

1950 The Netherlands

Trend setting by Utrecht and Amsterdam continues, applied geography grows, new directions opened. Utrecht promotes small-scale field research focusing on socio-economic structures rather than relations with physical environment. Historical and socio-demographic research (de Vooys) competes with sociology (van Heek 1956; Hofstee 1954). Field work in 'underdeveloped' regions in Spain and Greece (de Vooys 1959) and first world regional text is published (de Vooys and Tamsma 1959). Keuning's 'Mosaic of Functions' provides model of cross-sectional historical geography. New chair of geography at Nijmegen (Cools, disciple of van Vuuren, 1958–). At Amsterdam greater interaction with sociology leads to first textbook on research methodology (Kruyer, *Observation and Argumentation*, 1959). De Vries Reilingh (1951, 1964) proposes focus on territoriality and culture, illustrating with rural studies, as distinctly geographical style *vis-à-vis* competing claims of Hofstee's rural sociology, de Vooys' 'materialistic' and existential geography. His assistant Heinemeyer, opens up 'urban sociography' (à la Park/Burgess 1923; Hatt/Reiss 1951) in neighborhood studies with his students, later moving to politically-oriented development research in Morocco. Demographic research continues, regional texts published. Van Doorn *et al.* set up *The Sociological Gids* (1953) and launch an attack on the theoretical foundations of sociography (1958). ISONEVO (later re-named SISWO), promotes research on basic and non-basic activities (van Lohuizen, 1954), functional classification of cities (Steigenga 1955), population distribution and social integration of residential neighborhoods (ISONEVO 1955). Trend setting study on urbanization by van den Berg (1957), who introduces 'citizen participation' in North Holland. More explicitly spatial–situational paradigm for urban geography, ideographic in character, is proposed in a detailed study on the Haarlemmeer (Kouwe, Wissink, and van Paassen, 1955). Two major theoretical contributions of the fifties: van Paassen's phenomenologically-inspired work on the origins of geographical thought in Ancient Greece (1957) and de Jong's analysis of the chorological tradition (1955). First detailed historical geography of land reclamation in Zeeland-Flanders (Gottschalk 1956, 1958).

1950 Finland

Increase in student numbers. Period of maturation prior to following decade of expansion. At Helsinki Aario succeeds Granö, who has now moved toward human geography, and completes fourth edition of the *Atlas of Finland* (1960). Hustich professor at Swedish School of Economics turns to economic, social, and political interests, plant geography taking second place. Ajo moves to Helsinki University, heralding the 'quantitative reformation'. Ph.D.'s on regional topics (Varjo and Olavi Granö). By the end of the decade two new departments of geography established at Oulu and Turku School of Economics.

1950 Sweden
Atlas över Sverige (Swedish National Atlas) engaging numerous geographers for twenty-five years (M. Lundquist, Enequist *et al.*) Enequist explores distribution patterns in Swedish social geography. 'Quantitative revolution' and theory formulation in human geography, exploring statistical methods in central place studies and the definition of functional regions, migration studies, measures of centrality, and the application of all these in sectoral, national, and regional planning. Godlund works on Sweden's first national road plan. Fundamentally new approach to analysis of geographical processes by Hägerstrand, introducing Monte Carlo simulation and spatial models as tools of research. Beginning steps in computer cartography. Hannerberg's school of historical (rural) geography promotes extensive field work and micro-analysis of villages based on field evidence, old maps, and manuscripts (Helmfrid *et al.*).

1950 Norway
Changes associated with vibrant economic growth stimulates research in the linked phenomena of urbanization, migration, commuting and rural depopulation, in land-use changes and rural depopulation. Problems associated with urban expansion and agriculture, with recreation and conservation, exercise continuing appeal.

1950 France
Rapid increase of enrollments in departments of geography. New job opportunities are opened by French regional and planning policy; a movement for applied geography develops (Phlipponeau 1960), but it is unable to change the academic curricula, since both traditionalists and Marxists refuse innovation. In human geography, there is a growing interest in population studies, industrial geography, problems of capital formation and circulation (Labasse 1955), but rural landscape is still the main focus of research. Historical geography is more and more practiced by historians. Strong development of field work in North Africa, Western and Central Africa, South America and Mediterranean countries (Birot and Dresch 1953, 1956), but little contact with German- and English-speaking countries.

1950 Great Britain
Continued strength of regional geography – many new texts (some stimulated by wartime experiences) and publication of Wooldridge and East's *Spirit and Purpose of Geography* advocating regional synthesis as the major geographical focus. Beginning of substantial shift to sytematic studies; geomorphology leading the way. Continuing debate on environmental determinism, possibilism, and probabilism. Major contributions by geographers (1951 on, for two decades) to the annual regional handbook produced for the British Association conferences (several edited by geographers).

1950 United States
Era of consolidation and unprecedented academic prosperity. Rapid growth of established departments, creation of new departments (and entire new universities). 'Official' assessment of topical fields presented in *American Geography: Inventory and Prospect* (1954). Controversial character of Whittlesey's chapter on 'The Regional Concept and Regional Method', *Man's Role in Changing the Face of the Earth* (1956) offers additional perspective. Growing importance of urban geography and first signs in articles by Brush, Berry and Garrison, and Warntz of coming 'Quantitative Reformation'.

1960

1960 Germany – Austria
Qualitative and 'process-oriented' approaches disengage themselves from any connection with the landscape. 'Actual morphodynamics' begins to measure morphological

processes; social geography studies behavior of specific social groups, developing concepts of *Daseingrundfunktionen*, consumption, education, leisure activities. Research into centrality booms. Institutional growth and rapidly increasing student numbers. By 1965 two separate main paradigms evident: geographical geo-ecology (geomorphology remaining important) and modern behavioral social geography developing alongside the older landscape-oriented human geography. Incipient critique of geography's methodology and role within society, influenced by neo-Marxist stances of Frankfurt School continuing until 1975.

1960 The Netherlands

Exceptional growth in numbers of students, scholars, publications. University 'democratization'. Free University (Chair, Heslinga 1962–) adopts Utrecht style. Interests in 'underdevelopment' and applied geography. Chair in historical geography at Amsterdam (Gottschalk, 1962–74). Sociographers assume chairs in sociology, applied geographers move to 'planology'. Sociography now gets chair in Faculty of Social and Political Science (Kruyer 1961), and de Vries Relingh 'geographizes' the Steinmetz position. Socio-demographic research flourishes (ter Heide 1965; Buissink 1970); controversy between van Heek and Hofstee over impact of religion on demographic behavior (1963). Rural geography persists, e.g., Constandse's on Ijsselmeer-polders (1960), but urban and applied geography become wellsprings for innovation. *Urban Core and Inner City* (Amsterdam 1966) attracts wide range of international participants. Interdisciplinary interest in organizational structure of the city (Kouwe 1966) and city region (SISWO 1966; KNAG 1968). First ecological study on the urban core of Amsterdam and typology of residents on basis of attitudes (Heinemeyer, van Hulten, and de Vries Reilingh 1968). Van Paassen formulates a spatial-cum-ecological paradigm as geography's response to the challenge of interdisciplinary research (1962), and a systems approach to cities develops (Hoekveld 1964; Lambooy 1969). First modern urban thesis (Wissink 1962), review of urban ecology (Heinemeyer 1963) and empirical work on Youth and Leisure (Heinemeyer 1964). Political geography follows the line of Deutsch (Heinemeyer 1968) and later The Rokkan School. Heslinga's Ph.D. dissertation on the Irish Border (1962) has a strong cultural–geographical focus.

1960 Finland

Rapid expansion of graduate programs and growth of new departments, e.g., 'regional science' at Tampere University, the Vaasa School of Economics at Turku, new chairs at Turku and Oulu Universities. Departments begin their own publications and at Turku and Oulu their own geographical societies. Doctoral theses in human geography, especially in Helsinki. Quantitative methods now widely used, particularly in human geography. Physical geography, which overlaps with quaternary geology, now begins to grow in importance and separates itself from human geography. Regional geography forgotten. Applied geography acquires its own curriculum and becomes increasingly important in planning. Leadership of GA now almost entirely in hand of geographers. English supersedes German as language of publication.

1960 Sweden

XIX International Geographical Congress in Stockholm with special symposia in various centres in Sweden, intensifies contacts between Swedish geography and the 'new geography' of America as well as with the more historically oriented traditions of Europe. Geography reaches its fullest development and success in universities and society within a limited number of institutes. 'Modern geography', emphasizing spatial analysis, dominates, but paradigmatic pluralism is maintained. Experimental work on fluvial processes at Uppsala (Sundborg). Methodological expansion, computer applications, model development, and theory building in various fields supported through a series of governmentally sponsored projects for preparing regional development policy. Studies on the location of industry, schools, regional hospitals, revision of local government boundaries, reorganization of provincial government boards (Godlund, Hägerstrand, Törnquist, Bylund *et al.*). Widening job market when public planning administration is established in different sectors. Physical geographers employed in environmental protection administration. Big numbers of students and research positions in low and middle levels. Model development gradually more mathematically oriented (Olsson). Chair established at Umeå (1965) with specialization in planning for peripheral areas. School reforms weaken geography, splitting it in upper secondary level between natural science and social studies, thus undermining its disciplinary identity. Reduction of geography in teacher training programs. Debate between William-Olsson and Hannerberg on definition of human geography, the former maintaining a holistic view, the latter defining it as a set of special disciplines. Physical geography increasingly oriented toward process analysis and links with earth science.

1960 Norway
Quantitative revolution makes itself felt in the mid-1960s particularly due to Helvig (Ph.D. Chicago, 1964) who moves from the Institute of Transport Economy to the Geography department of Oslo (1966) and later to Bergen (1971). Non-academic employment opportunities open for geographers, and a significant increase in the number of these produced.

1960 France
The best results in regional studies now come from tropical countries (Sautter 1966; Pélissier 1966; Gallais 1967). In spite of its inadequacy for industrial and development societies, the classical view of geography is upheld by the majority of colleagues. Influences from economics, from some Marxist groups and from Anglo-American 'new geography' bring dissenting conceptions in economic and social geography (Claval 1963, 1968), in regional studies (Brunet 1968), and in physical geography (Bertrand 1968). The 1968 student movement supports reorientation toward quantitative methods, but its effects on new trends are often more dubious. CNRS gives new possibilities with the building of big geographical laboratories (geomorphology in Caen 1964; tropical geography in Bordeaux 1968), but nothing is done for human and social geography or for new methodologies. Applied geography develops mainly through regional atlases (Beaujeu-Garnier 1961).

1960 Great Britain
Rapid growth in student numbers and in the number and size of university geography departments; establishment of the polytechnics, many with geography departments. Expansion of the IBG and the creation of strong systematic study groups within it – British Geomorphological Research Group; Urban Geography Study Group; Population Geography Study Group; Quantitative Methods Study Group. The 'quantitative and theoretical revolutions' and the impact of transatlantic contacts; the Madingley lectures and resulting books edited by Chorley and Haggett (*Frontiers in Geographical Teaching; Models in Geography*); spatial analysis (Haggett) and logical positivism (Harvey) dominate human and physical geography. Responsibility for funding research taken over by Social Science Research Council (human geography) and Natural Environment Research Council (physical geography).

1960 United States
Decade of the 'Quantitative Reformation' and various 'counter-reformations'. Rapid diffusion from centers of innovation (Seattle, Evanston, Chicago, Iowa City, and even distant Lund). Proliferation of central-place studies and emergence of Christaller, Lösch, Weber, von Thünen, and Hägerstrand as leaders of a 'new geography'. But persistence of cultural and historical geography as alternative orientations. Beginning of interest in environmental perception and behavior. Growing commitment of geographers to function as social scientists. Social and anti-war activism among students and younger professors fosters disenchantment with 'mechanistic scientism'.

1970

1970 Germany – Austria

Institutional growth, doubling the number of university institutes, each one striving to publish its own periodical; considerable loss of research efficiency due to uncoordinated policies within the discipline and *Länder*. Adoption of quantitative-analytical methods and some Anglo-American perspectives, e.g., diffusion models, perception research, 'radical', and socially-engaged points of view. New curricula for school geography with emancipatory aims eventually has repercussions at level of research programs (*Raumwissenschaftliches Curriculum-Forschungsprojekt*). Broad spectrum of individual research aims and virtually no discussion of general questions regarding discipline as a unity. Undisputed pluralism in scientific basis of research theory (Hermeneutic, neo-Positivistic, neo-Marxist approaches simultaneously). Convergence of research aims with practitioners of other disciplines, e.g., biology and economics, in particular fields like environmental research and development policies, which eventually move to special institutions. Revival of regional geography with somewhat different interests. Divergence of trends between FRG and GDR in large part attributable to different political systems:

FRG	GDR
Research abroad predominant. Career norms fixed, subsidized by the *Deutsche Forschungsgemeinschaft* (German Research Foundation).	Research almost exclusively at home. Commissioned research organized via the *Akademie der Wissenschaften*.

1970 The Netherlands

Generation of the 1940s takes over in most institutions. Diversification, theoretical reflection, and beginnings of critical geography. New impetus in economic geography at Groningen (Lukkes 1974–), Nijmegen (Wever 1977–), and Amsterdam (Lambooy 1973, 1981). Cooperation in industrial geography (Jansen, de Smidt, and Wever 1974, 1979). Labor-market studies flourish at Utrecht under chairmanship of de Smidt (Ph.D. 1975), as does urban geography ('Randstadt' project, Ottens 1976; van Ginkel 1980). At Amsterdam urban work takes a socio-ecological orientation developing typologies based on degrees of urbanization (van Engelsdorp Gastelaars, Ostendorf 1980). Influence of Hägerstrand's time geography in the analysis of outdoor activities (SATURA I, II). Modeling of urban systems at Free University (Hoekveld) and at Nijmegen (Buursink 1972, 1979). New developments in the geography of 'underdevelopment' at Utrecht (Hinderink), Nijmegen (Kleinpenning), Free University (de Bruyne), and at Amsterdam (Heinemeyer *et al.*). Political geography continues at Amsterdam (van der Wusten 1977; Heinemeyer 1981). Studies of ethnic minorities and segregation (van Amersfoort 1974, 1980) also at Amsterdam and Rotterdam (Drewe and Mik 1975). Geography of cognition, with an emphasis on education, develops at the Free University (van Westrhenen, Dijkink). First chairs in research methodology at Utrecht (Hauer 1975–) and Free University (Dielemann 1978–) and first dissertation with mathematical–statistical applications (van der Knaap 1978; Dieleman 1978). Planology produces impressive work on Utrecht conurbation (Steigenga 1970–73) and first introductory textbook (van den Berg 1981). A decade of textbooks written mainly in Utrecht and Free University. Renaissance of Critical, Phenomenological and Marxist thought yields special conferences (Nijmegen 1973; KNAG 1976) and some ground-breaking articles (Jansen and van Paassen 1976) and critical exchange of viewpoints around phenomenology (Jansen and Dijkink, GT 1980 and 1981). Historical geography at Amsterdam culminates in three detailed studies of stormfloods and fluvial inundations (Gottschalk, 1971, 1975, 1977).

1970 Finland

End of expansion. Only new department at Joensuu. Geographers active in many tasks outside universities. Ph.D.'s awarded in human and physical geography. Interest in quantitative methods peaks in human geography but continues to expand in physical geography which is now relatively more important and embraces areas other than quaternary geology. Criticism of quantitative geography only on the level of principle; theoretical, epistemological, and historical questions discussed. Little concrete work in humanistic, perceptual, and behavioristic geography. Science policy spreads to geography. A time of nascent re-identification.

1970 Sweden
First half of decade still sees many governmentally commissioned projects, e.g., national plan for land and water use, national settlement strategy, second half fewer. Popular reaction against planning models of the 1960s, and an anti-intellectual movement within universities. New job market for geographers who have specialized in rural landscape evolution because of growing interest in land use and landscape protection. Interest in local history and historical geography but lack of trained researchers. Process-thinking takes shape in 'time-geography' and fundamentally new theoretical approaches suggested (Hägerstrand). Rapid technical development in computer cartography and remote sensing (Hoppe, Wastenson, Rapp.). Falling student numbers, stagnation in numbers of academic staff. Impact of Marxist, Critical, and Phenomenological thought (Olsson, Harvey), reappraisal of 'humanistic' and holistic visions of geography. Closer contact with behavioral sciences and with social anthropology.

1970 Norway
Aadel Brun Tschudi is appointed as Professor of Geography at Oslo (1972) bringing an emphasis on non-European countries; development geography becomes established as a field which excites both interest and enthusiasm. Late 1970s witness a growing reaction against the quantitative school and renewed interest in studies at the individual level and in the less easily quantifiable aspects of human behavior and welfare. Physical geography dominated by geomorphology and continuing emphasis on study of deglaciation. Societal problems arising from conflicting land claims, e.g., those associated with extensive hydro-electric power construction schemes, stimulate considerable research interest.

1970 France
Crisis and disillusionment. With economic crisis and the end of the student boom recruitment of teachers and scholars stops in 1973. Job opportunities for graduates in the secondary school system decline and curricula are not well adapted to alternatives. Despite the poor institutional situation of the universities, this is a time of hard work and of innovation. Old isolationism is dead. Strong influence of Anglo-American 'new geography', but French geographers are more interested in social and political processes and in regional structures than with radical views. Marxist theory is too old to get the majority of young geographers excited (Lacoste, 1976, is the main exception). Systems theory and Russian views on global physical geography are influential in the field of environmental studies. There is strong diversification of research in the economic field, with groups working on commercial structures, transportation, industrial location, regional development – and naturally on urban and rural problems. The development of critical views on the evolution of geography is partly the result of the action of P. Pinchemel. Historical and cultural geography, illustrated in the 1950s by Roger Dion and X. de Planhol, is now developing again under the aegis of the latter.

1970 Great Britain
Dominance of systematic studies (particularly urban geography and geomorphology but growth of others – biogeography; pedology – and some revivals – political geography). Decline of positivist dominance in human geography; increasing interest in philosophical and methodological questions. Major impact of Marxist appproaches (Harvey – *Social Justice and the City*); a commitment to human welfare (Smith – *Human Geography: A Welfare Approach*). Greater links with other social sciences and the investigation of common philosophical problems (e.g., Gregory – *Ideology, Science and Human Geography*). Strong interest in 'applied' topics in all fields partly as research

273

financing became difficult and economic crisis deepened and partly as outside contract funds became more available and acceptable. Growth of a number of specialist journals, many of them interdisciplinary (*Environment and Planning A, Regional Studies, Progress in Physical Geography, Progress in Human Geography, Journal of Biogeography*). Growing interest in environmental issues and resource management, argued by some as a means of reintegrating physical and human geography following decline of the regional paradigm. Major developments in spatial statistics, systems analysis, data manipulation, and historical geography. Growing involvement, especially by physical geographers, in the field of remote sensing.

1970 United States
Reassertion of pluralism. Most departments maintain commitments to multiple methods and objectives. Continued interest in model-building and quantification complemented by humanistic and behavioral studies. Vigorous development of historical geography. Public interest in 'ecology' encourages revival or reassertion of geography's 'earth science' orientation. Proliferation of philosophical, mainly epistemological writing. National spirit of neo-isolationism after the Vietnam war inhibits foreign-area studies. But national interest in historical preservation encourages local studies and 'topophilia'. 'Environmental impact assessments' present opportunity for 1930-style regional geography. 'Open meeting' policy of the AAG encourages a wide range of disparate and complementary interests. Curtailment of opportunities for academic employment fosters intense interest in 'applied geography'.

1980

1980 Sweden
Committee set up to prepare next reform in upper secondary school level, proposes re-establishment of geography in curriculum. Seventeen chairs of geography (inc. remote sensing) now in Sweden. Problems of recruitment continue. Philosophical introspection, reformulations, methodological uncertainty, and stagnation in empirical and applied work. Substantial projects, externally financed, in applied geography, e.g., remote sensing, data banks on population and real estate information. Physical geographers continue Arctic research (*Ymer* 1980) in collaboration with other natural scientists.

Selective Bibliography

Aay, H. (1972) A re-examination: geography – the science of space, *Monadnock*, **46**, 20–31.

Ackerman, E. A. (1945) Geographic training, wartime research, and immediate professional objectives, *Annals Assoc. Am. Geogr.* **35**, 121–43.

Ackerman, E. A. (1953) Geography as a fundamental research discipline, *Department of Geography Research Paper*, University of Chicago.

Ackerman, E. A. (1963) Where is a research frontier?, *Annals Assoc. Am. Geogr.* **53**, 429–440.

Adorno, Th. W. (1973) *The Jargon of Authenticity*. Routledge and Kegan Paul, London.

Adorno, Th. W. *et al.* (1950) *The Authoritarian Personality: Studies in Prejudice*. Harper and Row, New York.

Ajo, R. (1954) New aspects of geographic and social patterns of net migration rate, *Svensk Geografisk Årsbok*, Lund, Sweden.

Amadeo, D. and R. G. Golledge (1975) *An Introduction to Scientific Reasoning in Geography*. Wiley, New York

Appelbaum, W. (1940) How to measure the value of a trading area, *Chain Store Age*, 92–4, 111–14.

Appelbaum, W. and S. Cohen *et al.* (1961) Store location and development studies, *Economic Geography* (special publication), Worcester, Massachusetts.

Appelbaum, W. and R. F. Spears (1951) How to measure a trading area, *Chain Store Age*, 149–54.

Arendt, H. (1958) *The Human Condition*. University of Chicago Press, Chicago.

Arendt, H. (1968) *Men in Dark Times*. Harcourt Brace and World, New York.

Association of American Geographers (1979) *Annals* (special issue), vol. 69.

Ayer, A. J. (1959) *Logical Positivism*. Free Press, Glencoe, Illinois.

Baker, O. E., R. Borsodi and M. L. Wilson (1939) *Agriculture in Modern Life*. Harper and Bros, New York.

Barrows, H. H. (1923) Geography as human ecology, *Annals Assoc. Am. Geogr.*, **13**, 1–14.

Bartels, D. (1968) *Zum Wissenschaftstheoretischen Grundlegung einer Geographie des Menschen*. Steiner, Wiesbaden.

Bastié, J. (1958) La population de la région parisienne, *Annales de Géographie*, **67**, 12–58.

Baulig, H. (1950) William Morris Davis: master of method, *Annals Assoc. Am. Geogr.*, **40**, 188–95.

Baum, G. (1975) *Journeys*. Paulist Press, New York.

Beaujeu-Garnier, J. (1971) *La géographie: méthodes et perspectives* (Perspectives and Methods of Geography). Masson, Paris.

Beck, H. (1973) *Geographie, Europäische Entwicklung*. Alber Verlag, Freiburg, München.

Benjamin, W. (1969, 1978) *Illuminations*. Shocken Books, New York.

Berdoulay, V. (1974) *The Emergence of the French School of Geography 1870–1914*. Ph.D. dissertation, Department of Geography, University of California, Berkeley, California.

Berdoulay, V. (1978) The Vidal–Durkheim debate, in Ley, D. and M. Samuels (eds) *Humanistic Geography: Prospects and Problems*. Maroufa Press, Chicago pp. 77–90.

Berdoulay, V. (1980) La métaphore organiciste. Contribution à l'étude du langage métaphorique en géographie. Oral presentation to the IGU Commission on the History of Geographic Thought, Kyoto.

Berelson B. and G. A. Steiner (1964) *Human Behavior: An Inventory of Scientific Findings*. Harcourt Brace and

World, New York.

Bergmann, G. (1957) *The Philosophy of Science.* University of Wisconsin Press, Madison, Wisconsin.

Bernal, J. (1939) *The Social Function of Science.* Routledge and Kegan Paul, London.

Bernstein, R. J. (1976) *The Restructuring of Social and Political Theory.* Methuen, London.

Berr, H. (1911) *La synthèse en histoire. Son rapport avec la synthèse générale.* A. Michel, Paris.

Berry, B. J. L. (1964) Approaches to regional analysis: a synthesis, *Annals, Assoc. Am. Geogr.*, **54**, 2–11.

Berry, B. J. L. and W. L. Garrison (1958a) Functional bases of the central place hierarchy, *Economic Geography*, **34**, 145–54

Berry, B. J. L. and W. L. Garrison (1958b) Alternate explanations of urban rank size relationships, *Annals Assoc. Am. Geogr.*, **48**, 83–99.

Berry, B. J. L. and A. Pred (1961) *Central Place Studies: A Bibliography of Theory and Applications.* Regional Science Research Institute, Philadelphia.

Bertalanffy, L. Von (1951) General systems theory: a new approach to the unity of science, *Human Biology*, **23**, 303–61.

Bertalanffy, L. Von (1968) *General Systems Theory.* Braziller, New York.

Bettelheim, B. and M. Janowitz (1950) *Dynamics of Prejudice: A Psychological and Sociological Study of Veterans.* Harper & Row, New York.

Bettleheim, M. (1977) *The Uses of Enchantment.* Vintage Books, New York.

Bingham, M. T. (1959) A moment of drama, *The Radcliffe Quarterly*, (February), 13–17.

Birot, P. (1950) *Le Portugal: Etude de géographie regionale.* Armand Colin, Paris.

Birot, P. (1959) *Précis de géographie physique générale.* Armand Colin, Paris. Translated (1966) *General Physical Geography.* Harrap, London.

Birot, P. (1960) *Le cycle d'érosion sous les différents climats.* Universidado de Brasil, Rio de Janeiro. Translated (1968) *The Cycle of Erosion in Different Climates.* University of California Press, Berkeley, California.

Birot, P. and J. Dresch (1953, 1956) *La Méditerranée et le Moyen-Orient.* Presses Universitaires de France, Paris.

Blache, J. (1934) *L'homme et la montagne.* Gallimard, Paris.

Black, M. (1962) *Models and Metaphors.* Cornell University Press, Ithaca, New York.

Blanchard, R. (1925) *Les Alpes françaises.* Armand Colin, Paris.

Blanchard, R. (1958) *Les Alpes et leur destin.* Fayard, Paris.

Blaut, J. (1961) Space and process, *The Professional Geographer*, **13**, 1–7.

Blaut, J (1962) Object and relationship, *The Professional Geographer*, **14**, 1–7.

Bloch, M. (1931) *Les caractères originaux de l'histoire rurale française.* Les Belles Lettres, Paris. Translated (1966) *French Rural History. An Essay on Its Basic Characteristics.* University of California Press, Berkeley, California.

Blume, S. S., ed. (1977) *Perspectives in the Sociology of Science.* Wiley, New York.

Blüthgen, J. (1941) Entwicklung, Stand und Aufgaben der Geographie in Schweden. Sammelreferat. *Zeitschrift für Erdkund*, **9**, Heft 3/4.

Blythe, R. (1969) *Akenfield.* Penguin Books, London.

Blythe, R. (1979) *The View in Winter. Reflections on Old Age.* Penguin Books, London.

Bobek, H. (1959) Die Hauptstufend der Gesellschaftsund Wirtschaftsentfaltung in geographischer Sicht, *Die Erde: Zeitschrifte der Gesellschaft für Erdkunde zu Berlin*, **90**, 259–98. Translated in Wagner and Mikesell, (eds,) *op. cit.* (1962), 218–47.

Boorstin, D. J. (1953) *The Genius of American Politicis.* University of Chicago Press, Chicago.

Boorstin, D. J. (1958) *The Americans: The Colonial Experience.* Random House, New York.

Borsodi, R. (1933, rep. 1972) *Flight from the City – An Experiment in Creative Living on the Land.* Harper and Row, New York.

Boulding, K. (1956a) *The Image – Knowledge in Life and Society.* University of Michigan, Ann Arbor.

Boulding, K. (1956b) General systems theory – the skeleton of science, *Man-*

agement Science, **2**, 197–208.

Boulding, K. (1980) Science: our common heritage, Science, **207**, 831–6.

Bourdieu, P. (1978) Outlines of a Theory of Practice. Cambridge University Press, Cambridge.

Boutroux, E. (1874) De la contingence des lois de la nature. G. Baillière, Paris.

Boutroux, E. (1908) Science et méthode. Flammarion, Paris.

Bowden, M. J. (1959) Changes in Land Use in Jefferson County, Nebraska 1857–1957. University of Nebraska, Lincoln, Nebraska.

Bowden, M. J. (1980) The cognitive renaissance in American geography: the intellectual history of a movement, Organon, **14**, 199–204.

Bowers, L. (1943) The Country Life Movement in America. 1900–1920. Kennikot Press, Port Washington.

Bowman, I. (1921) The New World. Problems in Political Geography, World Book Company, Yonkers-on-Hudson and Chicago.

Braudel, F. (1949) La Méditerranée et le monde méditerranéen à l'époque de Philippe II. Armand Colin, Paris.

Breese, G. W. (1949) The Daytime Population of the Central Business District of Chicago with Particular Reference to the Factor of Transportation. University Press, of Chicago, Chicago.

Broc, N. (1974) L'établissement de la géographie en France, Annales de Géographie, **85**, 545–65.

Brodbeck, M. N. (1968) Readings in the Philosophy of the Social Sciences. Macmillan, London.

Brown, R. H. (1943) Mirror for Americans. American Geographical Society, New York.

Brunhes, J. (1902) L'irrigation, ses conditions géographiques, ses modes et son organisation dans la péninsule Ibérique et dans l'Afrique du Nord. C. Naud, Paris.

Brunhes, J. (1910, 1934) La géographie humaine. Essai de classification positive. Alcan, Paris.

Brunhes, J. (1913) Du caractère propre et du caractère complexe des faits de géographie humaine, Annales de Géographie, **22**, 1–40.

Bruss, E. P. W. (1972) Autobiography: The Changing Structure of a Literary Act. Ph.D. dissertation, University of Michigan.

Bunge, W. (1962) Theoretical Geography, Lund Studies in Geography, Series C, No. 1, Lund.

Bunge, W. (1971) Fitzgerald, Geography of a Revolution. Schenkman Pub., Cambridge, Massachusetts.

Bunge, W. (1979) Fred K. Schaeffer and the science of geography, Annals Assoc. Am. Geogr., **69**, No. 1, 128–32.

Burke, K. (1969) A Grammar of Motives. University of California Press, Berkeley, California.

Burton, I. (1963) The quantitative revolution and theoretical geography, Canadian Geographer, **7**, 151–62.

Buttimer, A. (1971) Society and Milieu in the French Geographic Tradition. Rand McNally, Chicago.

Buttimer, A. (1974) Values in geography, Association of American Geographers Resource Paper, **24**, Washington DC.

Buttimer, A. (1976) Grasping the Dynamism of Lifeworld, Annals. Assoc. Am. Geogr., **66**, 277–92.

Buttimer, A. (1978) On people, paradigms, and 'progress' in geography, Rapporter och Notiser, **47**, Kulturgeografiska institutionen, Lunds Universitet. Reprinted in Stoddart, D. (ed.) (1981) Geography, Ideology and Social Concern. Basil Blackwell, Oxford.

Buttimer, A. (1979) Reason, rationality, and human creativity, Geografiska Annaler, **61**B, 43–9.

Buttimer, A. and T. Hägerstrand (1980) Invitation to dialogue, DIA Paper No. 1, Lund.

Buttimer, A. and D. Seamon, eds. (1980) The Human Experience of Space and Place. Croom Helm, London.

Buttimer, A., Seamon, D., et al. (forthcoming) Creativity and Context. Report on a Symposium at Sigtuna, June 1978.

Buttimer, A. 1982 "Musing on Helicon: Root Metaphors in Geography" Geografiska Annaler, 64, 2, 89–96.

Cassirer, E. (1955) Philosophy of Symbolic Forms. Yale University Press, New Haven, Connecticut.

Castells, M. (1977) The Urban Question. Edward Arnold, London.

Cavell, S. (1980) The Claim of Reason. Harvard University Press, Cambridge, Massachusetts.

Chabot, G. and J. Beaujeu-Garnier (1963) *Traité de géographie urbaine*. A. Colin, Paris.

Chargaff, E. (1978) *Heraclitean Fire*. The Rockefeller University Press, New York.

Chase, S. (1954) How language shapes our thoughts, *Harper's Magazine* (April), 97–106.

Chicago Plan Commission (1943) *Master Plan of Residential Land Use of Chicago*. Chicago Plan Commission, Chicago.

Cholley, A. (1942) *Guide de l'étudiant en géographie*. Presses Universitaires de France, Paris.

Cholley, A. (1948) Géographie et sociologie, *Cahiers internationaux de sociologie*, 5, 3–20.

Chorley, R. J. and P. Haggett (1967) *Models in Geography*. Methuen, London.

Christaller, W. (1935) *Die zentralen Orte in Süddeutschland*. G. Fischer, Jena.

Christaller, W. (1968) Wie ich zu der Theorie der zentralen Orte gekommen bin, *Geographische Zeitschrift*, 56, 88–101.

Clark, T. (1973) *Prophets and Patrons. The French University System and the Emergence of the Social Sciences*. Harvard University Press, Cambridge, Massachusetts.

Claval, P. (1964) *Essai sur l'évolution de la géographie humaine*. Les Belles-Lettres, Paris.

Claval, P. (1972) *La pensée géographique*. S.E.D.E.S. Paris.

Claval, P. (1981) Methodology and geography, *Progress in Human Geography*, 5, 97–104.

Colby, C. C. (1933) Centrifugal and centripetal forces in urban geography, *Annals Assoc. Am. Geogr.*, 23, 1–20.

Coleman, J. S. (1980) The structure of society and the nature of social research, *Knowledge: Creation, Diffusion, Utilisation*, 1, 333–50.

Dacey, M. F. (1945) The geometry of central place theory, *Geografiska Annaler*, 47B, 111–24.

Dahl, S. (1973) Kulturgeografi (Human Geography), Report to the Social Science Research Council, Stockholm.

Dalenius, T. *et al.*, eds (1970) *Scientists at Work*, Almquist & Wiksell, Stockholm.

Dallmayr, F. R. and T. A. MacCarthy (1974) *Understanding and Social Enquiry*. Notre Dame University Press, Notre Dame, Indiana.

Darby, H. C. (1936) *An Historical Geography of England before A.D. 1800*. Cambridge University Press, Cambridge.

Darby, H. C. (1947) *The Theory and Practice of Geography*. Inaugural Address, University Press of Liverpool.

Davies, W. K. D. (1972) *The Conceptual Revolution in Geography*. Rowman and Littlefield, Totowa, New Jersey.

Davis, W. M. (1906) An inductive study of the content of geography, *Bulletin of the American Geographical Society*, 38, 67–84.

Davis, W. M. (1924) The progress of geography in the United States, *Annals Assoc. Am. Geogr.*, 14, 159–215.

Deasy, G. F. (1947) Training, professional work, and military experience of geographers, *The Professional Geographer*, 6, 1–15.

Deasy, G. F. (1948) War-time changes in occupation of geographers, *The Professional Geographer*, 7, 33–41.

de Dainville, A. (1941) *La géographie des humanistes*. Argence, Paris.

de Dainville, A. (1964) *Le langage des géographes*. A. et J. Picard, Paris.

Deffontainnes, P. (1933) *L'homme et la forêt*. Gallimard, Paris.

Deffontainnes, P. (1948) Défense et illustration de la géographie humaine, *La Revue de Géographie Humaine et d'Ethnologie*, I, 5–13.

de Geer, S. (1923) On the definition method, and classification of geography, *Geografiska Annaler*, 5, 1–37.

de Jong, G. (1955) *Het karakter van de geografische totaliteit*. J. B. Wolters, Groningen. Translated (1962) *Chorological Differentiation as the Fundamental Principle of Geography*, J. B. Wolters, Groningen.

de la Rué, A. (1935) *L'homme et les îles*. Gallimard, Paris.

Demangeon, A. (1927) La géographie de l'habitat rural, *Annales de Géographie*, 36, 1–23, 97–114.

Demangeon, A. (1942) *Problèmes de géographie humaine*. Colin, Paris.

de Martonne, E. (1925, rep. 1950, 1955) *Traité de géographie physique*. Academie des Sciences et la Société de Geographie de Paris, Paris.

Detienne, M. (1973) *Les maitres de vérité dans la Grèce Archäique*. F. Maspero, Paris.

Dickinson, R. E. (1969) *The Makers of Modern Geography*. Praeger, New York.

Dilthey, W. (1913) (1967) *Gesammelte Schriften*, 14 vols. Vandenhoeck & Ruprecht, Göttingen. Vols. I–XII, reissued (1958) by B. G. Tuebner, Stuttgart.

Dion, R. (1934) *Le val de Loire*. Arrault, Tours.

Duncan, J. S. (1980) The superorganic in American cultural geography, *Annals Assoc. Am. Geogr.*, **70**, 181–98.

Duncan, O. D. (1959) Human ecology and population studies, *The Study of Population* (Eds Hauser and Duncan). Chicago, 678–716.

Durkheim, E. (1897–98) Morphologie sociale, *L'Année Sociologique*, **2**, 520–521.

Durkheim, E. (1898–99) Review of *Anthropogeographie* Vol. I, *L'Année Sociologique*, **3**, 550–8.

Elzinga, A. (1980) Models in the theory of science: critique of the convergence thesis, in Baark, E. *et al.* (eds), *Technological Change and Cultural Impact in Asia and Europe: A Critical Review of the Western Theoretical Heritage*. Swedish Committee for Future Studies (SALFO), Stockholm, pp. 37–69.

Enequist, G. (1941) Bygd som geografisk term, *Svensk Geografisk Årsbok*, **17** 7–21.

Eriksson, G. (1978) Kartläggarna, *Acta Universitatis Umensis*, **15**, Umeå, Sweden.

Fallaci, O. (1976) *Interviews with History*. M. Joseph, London. Translated by J. Shepley.

Febvre, L. (1922) *La terre et l'évolution humaine*. Renaissance du livre, Paris. Translated 3rd ed. (1950) *A Geographical Introduction to History*. Routledge and Kegan Paul, London.

Festinger, L (1957) *A Theory of Cognitive Dissonance*. Harper & Row, New York.

Finch, V. C. (1927) Progress in the field of mapping detailed geographic interrelationships, *Annals. Assoc. Am. Geogr.*, **17**, 26–7.

Finch, V. C. (1937) A greater appreciation of maps, *Business Education World*, **18**, 1–5.

Firey, W. (1946) *Land Use in Central Boston*. Harvard University Press, Cambridge, Massachusetts.

Firey, W. (1960) *Man, Mind and Land*. Free Press, Glencoe, Illinois.

Fleck, L. (1935) *Entstehung und Entwicklung einer Wissenschaftsche Tatsache*. Benno Schwabe & Co. Verlags Buchhandlung, Basel. Translated (1980) by University of Chicago Press, Chicago.

Fleure, H. J. (1918) *Human Geography in Western Europe: A Study in Appreciation*. Williams Norgate, London.

Foucault, M. (1969) *L'archéologie du savoir*. Gallimard, Paris.

Foucault, M. (1971) *The Order of Things: An Archeology of the Human Sciences*. Pantheon Books, New York.

Foucault, M. (1972) Discourse on Language, *The Archeology of Knowledge*. Harper, New York.

Foucault, M. (1977) *Language, Counter-Memory, Practice*. Cornell University Press, Ithaca, New York.

Foucault, M. (1980) *Power and Knowledge*. Translated by C. Gordon, Harvester Press, Brighton, Sussex.

Frazier, F. (1957) *The Negro in the United States*. Rev. ed., Macmillan, New York.

Freeman, T. W. (1977) *Isolationism in Early American Geography*. Paper delivered at IGU Commission on the History of Geographical Thought, Edinburgh.

Fromm, E. (1941) (1968) *Escape from Freedom*. Rinehart, New York.

Frye, N. (1957) *The Anatomy of Criticism. Four Essays*. Princeton University Press, Princeton, New Jersey.

Gadamer, H. -G. (1965) *Wahrheit und Methode*. 2nd edn, J. C. B. Mohr, Tübingen. Translated (1975) *Truth and Method*. Seabury Press, New York.

Gale, S. and G. Olsson, eds (1979) *Philosophy in Geography*. Reidel, Dordrecht.

Garraty, J. A. (1970) *Interpreting American History. Conversations with Historians*. 2 vols. Macmillan Company, New York and Collier-Macmillan, London.

Garrison, W. L. (1959) Spatial structure of the economy: I, *Annals Assoc. Am. Geogr.*, **49**, 232–9; Part II, *Ibid.*, **49**

(1959) 471–82. Part III, *Ibid.*, **50** (1960) 232–9.

Garrison, W. L. (1960) Notes on the simulation of urban growth and development, *Discussion Paper*, 34, Department of Geography, University of Washington.

Garrison, W. L. (1979) Playing with ideas, *Annals Assoc. Am. Geogr.*, **69**, 118–20.

Garrison, W. and D. F. Marble, eds (1967) Quantitative Geography, Parts I and II, *Northwestern Studies in Geography*, 13 and 14.

Geografiska Annaler (1961) *Morphogenesis of the Agrarian Landscape*. Report of the Vadstena Symposium and the XIX International Congress, No. 1–2, 1960.

George, P. (1951) *Etude géographique de la population du monde*. Presses Universitaires de France, Paris.

Gottmann, J. (1946) French geography in wartime, *Geographical Review*, **36**, 80–91.

Gottmann, J. (1947) De la méthode d'analyse en géographie humaine, *Annales de Géographie*, **56**, 1–12.

Gottmann, J. (1951) Geography and International Relations, *World Politics* **3**, 153–73.

Gouldner, A. W. (1954) *Patterns of Industrial Bureaucracy*. Free Press, Glencoe.

Granö, J. G. (1929) Reine Geographie. Eine methodologische Studie beleuchtet mit Beispielen aus Finnland und Estland, *Acta Geographica*, **2**, 2, Helsinki.

Gregory, D. (1978) *Ideology, Science and Human Geography*. Hutchinson, London.

Grob, G. N. and G. A. Billias (1967, 3rd edn 1978) Interpretations of American History, *Patterns and Perspectives*, Vol. II. Free Press, New York and Collier-Macmillan, London.

Gunn, J. V. (1977) Autobiography and the narrative experience of temporality as depth, *Soundings*, **40**, 194–209.

Habermas, J. (1968) *Knowledge and Human Interests*. Beacon Press, Boston.

Habermas, J. (1979) *Communication and the Evolution of Society*. Beacon Press, Boston.

Hägerstrand, T. (1952) The propagation of innovation waves, *Lund Studies in Geography*, Series B, No. 4.

Hägerstrand, T. (1953) *Innovationsförloppet ur korologisk synpunkt*. C. W. K. Gleerup, Lund. Translation by Allan Pred.

Haggett, P. (1975) *Geography: A Modern Synthesis*. Harper & Row, New York.

Hard, G. (1973) *Die Geographie, eine wissenschaftstheoretische Einführung*. Walter de Gruyter, Berlin.

Harris, C. D. (1943) A functional classification of cities in the United States, *Geographical Review*, **33**, 86–99.

Harris, C. D. (1979) Geography at Chicago in the 1930's and the 1940's, *Annals Assoc. Am. Geogr.*, **69**, 21–32.

Harris, C. D. and J. D. Fellman (1960, 1971, 1980) *Annotated World List of Selected Current Geographical Serials*. University of Chicago.

Harris, C. D. and E. L. Ullman (1945) The nature of cities, *Annals Am. Academy Political Social Sciences*, **242**, 7–17.

Hart, J. F. (1979) The 1950's, *Annals Assoc. Am. Geogr.*, **69**, 109–14.

Hartmann, G. W. and J. C. Hook (1956) Substandard urban housing in the United States: a quantitative analysis, *Economic Geography*, **32**, 95–114.

Hartshorne, R. (1935) Research developments in political geography, *American Pol. Sc. Rev.*, **29**, 784–804, 943–66.

Hartshorne, R. (1939) *The Nature of Geography – A Critical Survey of Current Thought in the Light of the Past*. Association of American Geographers, Lancaster, Pennsylvania.

Hartshorne, R. (1950) The functional approach in political geography, *Annals Assoc. Am. Geogr.*, **40**, 95–130.

Hartshorne, R. (1954) Political geography, in James, P. E. and C. Jones (eds.) *Inventory and Prospect*, Syracuse, pp. 167–225.

Hartshorne, R. (1955) 'Exceptionalism in geography' re-examined, *Annals Assoc. Am. Geogr.*, **45**, 205–44.

Hartshorne, R. (1959) *Perspective of the Nature of Geography*. Rand McNally, Chicago.

Hartz, L. (1955) *The Liberal Tradition in America*. Hartcourt, Brace, New York.

Harvey, D. (1969) *Explanation in Geography*. St. Martin's Press, New York.

Hauser, P. M. and O. D. Duncan, eds (1959) *The Study of Population: An Inventory and Appraisal.* University of Chicago Press, Chicago.

Hedar, S. (1941) Geografi och historia. *Historisk Tidskrift 1941.*

Heidegger, M. (1971) *Letter on Humanism.* Translated and printed in Languilli, N. (ed.), *The Existentialist Tradition. Selected Writings.* Anchor Books, New York, pp. 205–48.

Heisenberg, W. (1969) *Der Teil und das Ganze.* R. Piper Verlag, München.

Hélias, P. J. (1975) *Le cheval d'orgueuil. Mémoires d'un Breton du pays bigouden.* Plon, Paris.

Helmfrid, S. (1976/77) Hundra år svensk geografi. En snabbskiss i anledning av Svenska Sällskapets för Antropologi och Geografi hundraårsjubileum. *Ymer, Årsbok 1976/77.*

Henrikson, A. K. (1975) The map as an 'idea': the role of cartographic imagery during the Second World War, *The American Cartographer*, 2, 19–53.

Hérodote (1975) Questions on geography, *Hérodote.* Reprinted in Foucault, *Power and Knowledge*, 63–7.

Hettner, A. (1927) *Die Geographie: ihre Geschichte, ihr Wesen, und ihre Methoden.* F. Hirt, Breslau.

Hettner, A. (1960) Drei autobiographische Skizzen. Herausgegeben von E. Plewe, *Heidelberger Geographische Arbeiten*, H. 6, 41–80.

Higbee, E. (1960) *The Squeeze: Cities without Space.* Morrow, New York.

Higham, J. (1959) The cult of the 'American Consensus': homogenizing our history, *Commentary*, 27, 93–100.

Higham, J. (1980) **The matrix of specialization**, *Bulletin of the American Academy of Art and Sciences*, 33, 9–29.

Hofstadter, R. (1979) *Gödel, Escher, Bach.* Harvester Press, Hussocks, Sussex.

Hoover, E. (1948) *The Location of Economic Activity.* McGraw–Hill, New York.

Horowitz, I. L. (1969) *Sociological Self-Images. A Collective Portrait.* Sage Publications, Beverly Hills, California.

Hougan, J. (1975) *Decadence: Radical Nostalgia, Narcissism, and Decline in the Seventies.* William Morrow, New York.

Humboldt, A. Von (1845–62) *Kosmos. Entwurf einer physischen Weltbeschreibung.* Cotta, Stuttgart, Tübingen. Translated by E. Sabine as *Cosmos: Sketch of a Physical Description of the Universe*, 4 vols.

L'Information Géographique (1957) *La géographie française au milieu du XX siècle.* J. B. Baillière et Fils, Paris.

Isard, W. (1956) *Location and Space-Economy.* MIT Press, Cambridge, Massachusetts.

Jackson, W. A. (1958) Whither political geography?, *Annals Assoc. Am. Geogr.*, 48, 178–83.

Jackson, W. A., ed. (1964) *Politics and Geographic Relationships.* Prentice Hall, Englewood Cliffs, New Jersey.

Jacoby, R. (1975) *Social Amnesia: A Critique of Conformist Psychology from Adler to Laing.* Beacon Press, Boston.

Jakobson, R. (1960) Linguistics and poetics, in Thomas A. Seboek (ed.) *Style in Language.* Technology Press, MIT, Cambridge, Massachusetts.

James, P. E. (1926) Some geographic relations in Trinidad, *Scottish Geographical Magazine*, 42, 84–93.

James, P. E. (1952) Toward a further understanding of the regional concept, *Annals Assoc. Am. Geogr.*, 42, 195–222.

James, P. E. and C. F. Jones, eds (1954) *American Geography: Inventory and Prospect.* Syracuse University Press, Syracuse, New York.

James, P. E. and G. Martin (1979) *The Association of American Geographers, Seventy-Five Years, 1904–1979.* Association of American Geographers, Washington, DC.

James, P. E. and E. C. Mather (1977) The role of periodic field conference in the development of geographic ideas in the United States, *The Geographical Review*, 67, 446–61.

James, W. (1907, 1955) *Pragmatism.* New American Library, New York.

Johnston, R. J. (1979) *Geography and Geographers.* Edward Arnold, London.

Jones, S. B. (1943) Field geography and post-war political problems, *Geographical Review*, 33, 446–56.

Jones, S. B. (1954) A unified field theory of political geography, *Annals Assoc. Am. Geogr.*, 44, 111–23.

Journaux, A. *et al.* eds (1966) *Géog-*

raphie générale. Gallimard, Paris.

Jung, C. J. (1957) *The Undiscovered Self*. Little Brown, Boston.

Kant, E. (1953) Migrationernas klassifikation och problematik, *Svensk Geografisk Årsbok*, **29**, 180–209. Reprinted in Wagner and Mikesell eds, op. cit., 342–54.

Kates, R. W. (1962) Hazard and choice perception in flood plain management, *Research Paper*, No. 78, Department of Geography, University of Chicago.

Kazin, A. (1979) The self as history: reflections on autobiography, in Pachter, M. (ed.), *Telling Lives*. New Republic Books, Washington DC, pp. 75–89.

King, L. (1961) A Multivariate analysis of the spacing of urban settlements in the United States, *Annals. Assoc. Am. Geogr.*, **51**, 222–33.

King, L. (1979) The seventies: disillusionment and consolidation, *Annals, Assoc. Am. Geogr.*, special issue, **69**, 155–6.

Kirk, W. (1951) Historical geography and the concept of the behavioural environment, *Indian Geographical Journal*, Silver Jubilee Volume, 152–60.

Kirk, R. (1954) *The Conservative Mind*. Faber and Faber, Chicago.

Kluckhohn, C. (1944) *Mirror for Man*, McGraw Hill, New York.

Kohn, C. (1979) The reconstituted Association of American Geographers 1949–1963, in James and Martin (eds) (1979), pp. 113–147.

Kollmorgen, W. (1954) And deliver us from big dams, *Land Economics*, **30**, 333–46.

Kollmorgen, W. and R. W. Harrison (1947) Drainage reclamation in the coastal marshlands' of the Mississippi River Delta, *The Louisiana Historical Quarterly*, **30**, 1–57.

Kouwenhoven, J. A. (1961) *The Beer Can by the Highway: Essays on What's American about America*. Doubleday, Garden City, New York.

Krebs, N. (1923) Natur- und Kulturlandschaft, *Zeit. d. Ges. für Erdkunde zu Berlin*, 81–95.

Kuhn, T. S. (1962) *The Structure of Scientific Revolutions*. University of Chicago Press, Chicago.

Labasse, J. (1955) *Les capitaux et la région. Etude géographique*. Colin, Paris.

Labasse, J. (1966) *L'organisation de l'espace*. Hermann, Paris.

Lakatos, I. (1974) History of science and its rational reconstruction, in Elkana, Y. (ed.), *The Interaction Between Science and Philosophy*. Free Press, London, pp. 195–204.

Landes, D. (1980) The creation of knowledge and technique: today's task and yesterday's experience, *Daedalus*, **4**, 111–20.

Langer, S. K. (1953) *Feeling and Form*. Scribners, New York.

Langer, S. K. (1957) *Philosophy in a New Key*. Harvard University Press, Cambridge, Massachusetts.

Lasch, C. (1978) (1979) *The Culture of Narcissism: American Life in an Age of Diminishing Expectations*. Norton, New York.

Lautensach, H. (1952) Otto Schlüter's Bedeutung für die methodische Entwicklung der Geographie, *Pet. Mitt.*, **96**.

Leach, E. (1970) *Claude Lévi-Strauss*. Viking Press, New York.

Leighly, J. (1937) Some comments on contemporary geographic methods, *Annals Assoc. Am. Georg.*, **27**, 125–41.

Leighly, J. (1938) Methodologic controversy in nineteenth century German geography, *Annals Assoc. Am. Geogr.*, **28**, 238–58.

Le Lannou, M. (1948) La vocation actuelle de la géographie humaine, *Les Etudes Rhodadiennes*, **4**, 272–80.

Le Lannou, M. (1959) *La géographie humaine*, Flammarion, Paris.

Lévi-Strauss C. (1966) *The Savage Mind*. Weidenfeld & Nicholson, London.

Lewin, K. (1952) *Field Theory in Social Science*. Tavistock, London.

Lilley, S. (1953) Cause and effect in the history of science, *Centaurus*, **3**, 58–72.

Lichtenberger, E. (1979) "The Impact of Political Systems upon Geography: The Case of the Federal Republic of Germany and the German Republic" *The Professional Geographer*, 31, 201–211.

Livingstone, D. N. and R. T. Harrison (1981) Meaning through metaphor: analogy as epistemology, *Annals Assoc. Am. Geogr.*, **71**, 95–107.

Løffler, E. (1911) *Min selvbiographi. En geographs levnadsløb*. Lehmann & Stages

Boghandel, Copenhagen.

Lord, A. (1971) *The Singer of Tales*. Atheneum Press, New York.

Lösch, A. (1954) *The Economics of Location*. Yale University Press, New Haven, Connecticut.

Lowenthal, D. (1961) Geography, experience, and imagination, *Annals Assoc. Am. Geogr.*, **51**, 241–60.

Lukermann, F. (1965) The 'Calcul des Probabilités' and the Ecole Française de Géographie, *Canadian Geographer*, **9**, 128–37.

Lynch, K. (1960) *The Image of the City*. MIT Press, Cambridge, Massachusetts.

MacIntyre, A. (1981) *After Virtue*. Notre Dame University Press, Notre Dame, Indiana.

Maini, S. M. and B. Nordbeck (1975) Critical moments, the creative process and research motivation, *International Social Science Journal*, **25**, 190–203.

Malin, J. C. (1947) *Grasslands of North America*. Lawrence, Kansas.

Malin, J. C. (1955) *The Contriving Brain and the Skillful Hand in the United States: Something about History and the Philosophy of History*. Lawrence, Kansas.

Mannheim, K. (1946) *Ideology and Utopia: An Introduction to the Sociology of Knowledge*. Hartcourt Brace and World, New York.

Marsh, G. P. (1864) *Man and Nature, or Physical Geography as Modified by Human Action*. C. Scribner, New York.

Mayer, H. M. (1942) Patterns and recent trends of Chicago's outlying business centers, *Journal of Land and Public Utility Economics*, **18**, 4–16.

Mayer, H. M. (1954) Geographers in City and Regional Planning, *The Professional Geographer*, **6**, 7–12.

Mayer, H. M. (1955) *Chicago: City of Decisions* (ed. Chauncy D. Harris) The Geographic Society of Chicago, Chicago.

Mayer, H. M. and C. Kohn, eds (1959) *Readings in Urban Geography*. University of Chicago Press, Chicago.

McCarty, H. H. (1940) *The Geographic Basis of American Economic Life*. Harper & Bros, New York.

McCarty, H. H., J. C. Hook and D. Knos (1956) *The Measurement of Association in Industrial Geography*. Department of Geography, State University of Iowa, Iowa City.

McDonald, J. R. (1964) Current controversy in French geography, *The Professional Geographer*, **16**, 20–23.

Mead, M. (1942) *And Keep Your Powder Dry*. W. Morrow and Co., New York.

Mendelsohn, E. *et al.* (1977) *The Social Production of Scientific Knowledge*. Sociology of Sciences Yearbook. Reidel, Dordrecht.

Merleau-Ponty, M. (1973) *The Prose of the World*. Northwestern University Press, Evanston, Illinois.

Merton, R. K. *et al.* eds (1959) *Sociology Today: Problems and Prospects*. Basic Books, New York.

Michelet, J. (1833, 1947) *Tableau de la France*. Les Belles Lettres, Paris.

Mikesell, M. W. (1968) Friedrich Ratzel, *International Encyclopedia of the Social Sciences*, vol. 13, New York, pp. 327–9.

Mill, H. R. (1951) *An Autobiography* with an introduction by L. Dudley Stamp. Longmans, Green, London, New York.

Mills, Ch. W. (1941) (1956) *The Power Elite*. Oxford University Press, New York.

Mink, L. O. (1968) *Philosophical Analysis and History*. Harper and Row, New York.

Moreno, J. L. (1951) *Sociometry, Experimental Method and the Science of Society*. Beacon House, Boston.

Morgan, G. (1980) Paradigms, metaphors, and puzzle solving in organization theory, *Administrative Science Quarterly*, **25**, 605–22.

Morse, P. M. and G. E. Kimball (1951) *Methods of Operations Research*. Wiley New York.

Mumford, L. (1934, 1971) *Technics and Civilization*. Hartcourt-Brace and World Inc., New York.

Murphey, R. (1954) The city as a center of change: Western Europe and China, *Annals Assoc. Am. Geogr.*, **44**, 349–62.

Myrdal, G. (1944) *An American Dilemma*. Harper & Row, New York.

Nakayama, S. (1972) Externalist approaches of Japanese historians of science, *Japanese Studies in the History of Science*, **11**, 5–6.

National Academy of Sciences–National Research Council (1965) *The Science of*

Geography. Publication 1277. Washington, DC.

Needham, J. (1964) Science and society in East and West, *Centaurus*, **10**.

Needham, J. (1969) *The Grand Titration. Science and Society in East and West*. Allen & Unwin, London.

Needham, J. N. (1954–78) *Science and Civilization in China*, 5 vols. Cambridge University Press, Cambridge.

Nelson, H. (1913) Hembygdsundervisning i folkhögskolan, *Svenska Folkhögskolans Årsbok*.

Nelson, H. (1916) Geografien som vetenskap, Inaugural Lecture, University of Lund (unpublished manuscript, Lund University: Department of Geography).

Nisbet, R. (1953) *Quest for Community*. Oxford University Press, New York.

Norborg, K. (1962) **Proceedings of the IGU Symposium in Urban Geography, Lund 1960**, *Lund Studies in Geography*, Series B, Vol. 24.

Odén, B. (1975) Det historiskt-kritiska genombrottet, *Scandia*, **41**, 5–29.

Oden, B. (1981) Överföring av värderingar genom forskarutbildning, in Hanson, B. *et al.* (eds), *Festschrift till Hampus Lyttkens*, Doxa Press, Lund, pp. 197–209.

Olsson, G. (1979) Social science and human action or on hitting your head against the ceiling of language, in Gale, S. and G. Olsson (eds), *Philosophy in Geography*. Reidel, Dordrecht, pp. 287–308.

Overbeck, H. (1954) Die Entwicklung der Anthropogeographie, *Blätter für deutsche Landesgeschichte*, **91**.

Palmer, R. E. (1969) *Hermeneutics*. Northwestern University Press, Evanston, Illinois.

Papers and Proceedings of the Regional Science Association (from 1954 on).

Parker, R. (1978) *Men of Dunwich: The Story of a Vanished Town*. Collins, London.

Parry, M. (1971) *The Making of Homeric Verse*. Oxford University Press, London and New York.

Pattison, W. (1964) The four traditions in geograpy, *Journal of Geography*, **63**, 211–16.

Peet, R. (1977) The development of radical geography in the United States, *Progress in Human Geography*, **1**, 240–63.

Pepper, S. C. (1942) *World Hypotheses*. University of California Press, Berkeley and Los Angeles.

Perry, R. B. (1949) *Characteristically American*. Knopf, New York.

Philbrick, A. (1957) Principles of areal functional organization in regional human geography, *Economic Geography*, **33**, 299–336.

Platt, R. S. (1928) A detail of regional geography: Ellison Bay community as an industrial organism, *Annals Assoc. Am. Geogr.*, **18**, 81–126.

Platt, R. S. (1935) Field approach to regions, *Annals Assoc. Am. Geogr.*, **25**, 153–74.

Platt, R. S. (1942) *Latin America: Country-sides and United Regions*. McGraw-Hill, New York.

Platt, R. S. (1952) The rise of cultural geograhy in America, *Proceedings of the Seventeenth International Geographical Congress*, Washington, DC.

Platt, R. S. (1957) A review of regional geography, *Annals Assoc. Am. Geogr.*, **47**, 187–90.

Platt, R. S. (1959) Field study in American geography: the development of theory and method exemplified by selections, *Research Paper* No. 61, Department of Geography, University of Chicago.

Poincaré, H. (1913) *The Foundations of Science: Science and Hypothesis; The Value of Science; Science and Method*. Science Press, New York.

Polanyi, M. (1960) *Personal Knowledge. Towards a Post-Critical Philosophy*. University of Chicago Press, Chicago.

Popper, K. R. (1961) *The Poverty of Historicism*. Routledge & Kegan Paul, London.

Potter, D. M. (1954) *People of Plenty: Economic Abundance and the American Character*. University of Chicago Press, Chicago.

Pribram, K. H. (1980) The role of analogy in transcending limits in the brain sciences, *Daedalus* (Spring), 19–38.

Rabinow, P. and W. M. Sullivan (1979) *Interpretive Social Science*. University of California Press, Berkeley, California.

Ratzel, F. (1898–99) Le sol, la société et l'Etat, *L'Année Sociologique*, **3**, 1–14.

Ratzel, F. (1911) *Glücksinseln und Träume*. George Reimer, Berlin.

Ratzel, F. (1966) *Jugenderinnerungen* Kösel, München.

Ravetz, J. R. (1971) *Scientific Knowledge and its Social Problems*. Oxford University Press, London and New York.

Redfield, R. (1948) The art of social science, *American Journal of Sociology*, **54**, 1–20.

Richter, H. (1959) Geografins historia i Sverige intill år 1800. Lychnosbibliotek. Studier och källskrifter utgivna av *Lärdomshistoriska Samfundet*, 17:1. Uppsala.

Ricoeur, P. (1975) *La métaphore vive*. Seuil, Paris. Translated (1977) as *The Rule of Metaphor*. University of Toronto Press, Toronto.

Riesman, D. (1950) *The Lonely Crowd: A Study of the Changing American Character*. Yale University Press, New Haven.

Ritter, C. (1862) *Allgemeine Erdkunde. Vorlesungen an der Universität zu Berlin*. Translated as *Comparative Geography*. William Blackwood and Sons, Edinburgh and London.

Rochefort, R. (1961) *Le travail en Sicile. Etude de géographie sociale*. Presses Universitaires de France.

Rorty, R. (1979) From epistemology to hermeneutics, in *Philosophy and the Mirror of Nature*. Princeton University Press, Princeton, New Jersey, pp. 315–94.

Rose, A. M. (1969) Varieties in sociological imagination, *American Sociological Review*, **34**, 623–30.

Rose, J. K. (1951) Geography in practice in the Federal Government, Washington, in Taylor, G. (ed.), *Geography in the Twentieth Century*. Methuen, London, pp. 566–86.

Saarinen, T. F. (1966) Perception of the drought hazard on the Great Plains, *Research Paper*, No. 106, Department of Geography, University of Chicago.

Sack, R. (1980) Conceptions of geographic space, *Progress in Human Geography*, **4**, No. 3, 315–45.

Sack, R. (1980) *Conceptions of Space in Social Thought*. Cambridge University Press, Cambridge.

Sauer, C. O. (1925) The morphology of landscape, *University of California Publications in Geography*, **2**, 19–53.

Sauer, C. O. (1941) Foreword to historical geography, *Annals Assoc. Am. Geogr.*, **31**, 1–24.

Sauer, C. O. (1956) The education of a geographer, *Annals Assoc. Am. Geogr.*, **46**, 287–99.

Schaeffer, K. (1953) Exceptionalism in geography, *Annals Assoc. Am. Geogr.*, **43**, 226–49.

Schiff, M. (1970) Some theoretical aspects of attitudes and perception, *Natural Hazards Working Paper*, No. 15, Toronto.

Schlanger, J. E. (1971) *Les métaphores de l'organisme*. Vrin, Paris.

Schlüter, O. (1952–58) *Die Siedlungsräume Mitteleuropas in frühgeschichtlicher Zeit*. Forschungen zur deutschen Landeskunden, Band 63, Remagen.

Schmieder, O. (1972) Lebenserinnerungen und Tagebuchblätter eines Geographen, *Schriften des Geographisches Institut der Universität Kiel*, Band 40.

Schouw, J. F. (1925) Selvbiografi, in C. Christensen and J. F. Schouw, *Botanisk Tidsskrift*, **38**, 3–4.

Schrag, C. O. (1980) *Radical Reflection and the Origin of the Human Sciences*. Purdue University Press, West Lafayette, Indiana.

Schramke, W. (1975) Zur Paradigmengeschichte der Geographie und ihrer Didaktik, *Geographische Hochschulmanuskripte*, **2**, Göttingen.

Schultz, H. D. (1980) *Die deutschsprachige Geographie von 1800 bis 1870. Ein Beitrag zur Geschichte ihrer Methodologie*. Geographischen Institut der Frein Universität Berlin. Abhandlungen des Geographisches Institut, Anthropogeographie, Band 29, Berlin

Schur, E. (1976) *The Awareness Trap: Self-Absorbation Instead of Social Change*. Quadrangle Press, New York.

Schwarz, G. (1961) *Allgemeine Siedlungsgeographie*. Walter de Gruyter, Berlin.

Schweitzer, A. (1949) *Goethe*. Adam and Charles Black, London.

Sellars, J. (1964) *Science, Perception and Reality*. Routledge and Kegan Paul, London.

Sestini, A. (1947) Le fasi regressive nello sviluppo del paesaggio antropogeogra-

285

fico, *Revista Geografia Italiana*, **54**, 153–71. Translated and reprinted in Wagner and Mikesell, op. cit., 479–490.

Shibles, W. A. (1971) *Metaphor: An Annotated Bibliography and History*. Whitewater, Wisconsin.

Siegfried, A. (1952) *Mes souvenirs de la Troisième République: mon père et son temps, Jules Siegfried, 1836–1922*. Presses Universitaires de France, Paris.

Simiand, F. (1906–09) Review of regional monographs by Demangeon, Blanchard, Vallaux, Vaucher, and Sion, *L'Année Sociologique*, **11**, 725–52.

Simon, H. A. (1956) Rational choice and the structure of the environment, *Psychological Review*, **63**, 129–238.

Simon, H. A. (1957) *Models of Man*. Wiley, New York.

Simon, H. A. (1960) *The New Science of Management Decisions*. Harper & Row, New York.

Sion, J. (1934) L'art de la description chez Vidal de la Blache, *Mélanges Vianey*. Presses Françaises, Paris, 479–87.

Smith, N. (1979) Geography, science, and post-positivist modes of explanation, *Progress in Human Geography*, **3**, 356–83.

Sorre, M. (1913) *Les Pyrénées méditerranéennes; Essai de géographie biologique*. Colin, Paris. The Andorra vignette is contained in this study.

Sorre, M. (1933) Complêxes pathogènes et géographie médicale, *Annales de Géographie*, **42**, 1–18.

Sorre, M. (1943, '48, '51) *Les fondements de la géographie humaine*, 3 vols, Colin, Paris.

Sorre, M. (1953) Le rôle de l'explication historique en géographie humaine, in Bussac, G. de (ed.), *Mélanges géographiques offerts à Philippe Arbos*, 2 vols, pp. 19–22.

Sorre, M. (1961) *L'homme sur la terre. Traité de géographie humaine*. Hachette, Paris.

Spencer, J. E. (1979) A geographer west of the Sierra Nevada, *Annals Assoc. Am. Geogr.*, **69**, 46–52.

Sprout, H. and M. Sprout (1965) *The Ecological Perspective on Human Affairs*. Princeton University Press, New Jersey.

Stewart, J. Q. (1947) Empirical mathematical rules concerning the distribution and equilibrium of population, *Geo-graphical Review*, **37**, 461–85.

Stoddart, D. R. (1966) Darwin's impact on geography, *Annals Assoc. Am. Geogr.*, **56**, 683–98.

Stone, K. H. (1979) Geography's wartime service, *Annals Assoc. Am. Geogr.*, **69**, 89–96.

Stouffer, S. A. (1940) Intervening opportunities: a theory relating mobility and distance, *American Sociological Review*, **5**, 845–67.

Ström, E. T. (1958) Svensk kulturgeografisk forskning med särskild hänsyn till tiden efter 1940, *Gothia*, **8**, Göteborg.

Szanton, D. L. (1980) The humanities and the social sciences: a symposium items, *Social Science Research Council*, vol. 34, No. 3/4, 54–57.

Taaffe, E. J. (1962) *The Air Passenger Hinterland of Chicago*. University of Chicago, Chicago.

Taaffe, E. J. (1979) Geography in the sixties in the Chicago area, *Annals Assoc. Am. Geogr.*, **69**, 133–8.

Tansley A. G. (1935) The use and abuse of vegetational concepts and terms, *Ecology*, **16**, 287–307.

Taylor, G., ed. (1951) *Geography in the Twentieth Century*. Methuen, London.

Taylor, G. R. (1958) *Journeyman Taylor. The Education of a Scientist*. Robert Hale, London.

Teich, M. and R. Young, eds (1973) *Changing Perspectives in the History of Science*. Heinemann, London.

Theodorson, G. A. (1958) *Studies in Human Ecology*. Row, Peterson, Evanston, Illinois.

Thomale, E. (1942) Sozialgeographie, eine disziplingeschichtliche Untersuchung, *Marburger Geographische Schriften*, **53**.

Thomas, E. N. (1960) *Maps of Residuals from Regression: Their Characteristics and Uses in Geographic Research*. Department of Geography, State University of Iowa, Iowa.

Thomas, W. L., ed. (1956) *Man's Role in Changing the Face of the Earth*. University of Chicago Press, Chicago, Illinois.

Tobler, W. (1961) *Map Transformation of Geographic Space*. University of Washington, Seattle. Unpublished dissertation.

Toynbee, A. (1975) Narrative history: The narrator's problem, *Clio*, **4**, 299–316.

Trewartha, G. (1953) A case of population geography, *Annals Assoc. Am. Geogr.*, **43**, 91–7.

Troll, C. (1947) Die geographische Wissenschaft in Deutschland in den Jahren 1933 bis 1945, *Erdkunde*, **1**, 1, 1–48.

Ullman, E. L. (1940–41) A theory of location for cities, *American Journal of Sociology*, **46**, 853–64.

Ullman, E. L. (1949a) *United States Railroads Classified According to Capacity and Relative Importance* (map). New York.

Ullman, E. L. (1949b) The railroad pattern of the United States, *Geographical Review*, **39**, 254–5.

Ullman, E. L. (1953) Human geography and area research, *Annals Assoc. Am. Geogr.*, **43**, 54–66.

Ullman, E. L. (1980) An autobiographical statement, in Boyce, R. R. (ed.), *Geography as Spatial Interaction*. University of Washington Press, Seattle and London.

UNESCO (1973) Autobiographical portraits, *International Social Science Journal*, **1**, 2, 13–168.

US House of Representatives (1966) Task force on federal flood control policy, a unified national program for managing flood plans. 9th Congress, 2nd session, *House Document* No. 465, Washington, DC.

Vallaux, C. (1925) *Les sciences géographiques*. Alcan, Paris.

Van Hise, C. R. (1910) *The Conservation of Natural Resources in the United States*. Macmillan, New York.

Vernant, J. R. (1962) *Les origines de la pensée grecque*. Presses Universitaires de France, Paris.

Vico, G. P. (1744) (1948) *The New Science of Giambattista Vico*, (Translated from the 3rd (1744) edition by T. G. Bergin) Fisch, M. H. (ed.), Cornell University Press.

Vidal de la Blache, P. M. (1898) La géographie politique. A propos des écrits de M. Frédéric Ratzel, *Annales de Géographie*, **7**, 97–111.

Vidal de la Blache, P. M. (1903a) La géographie humaine et ses rapports avec la géographie de la vie, *Revue de Synthèse Historique*, **7**, 219–40.

Vidal de la Blache, P. M. (1903b) *Tableau de la géographie de la France*, vol. I of E. Lavisse, *Histoire de France*, Hachette, Paris. Translated (1928) as *The Personality of France*, Christophers', London.

Vidal de la Blache, P. M. (1904) Rapports de la sociologie avec la géographie, *Revue International de Sociologie*, **12**, 309–13.

Vidal de la Blache, P. M. (1911) Sur la relativité des divisions régionales, *Athena*, **11**, 1–14.

Vidal de la Blache, P. M. (1913) Des caractères distinctifs de la géographie, *Annales de Géographie*, **22**, 290–9.

Vidal de la Blache, P. M. (1922) *Principes de géographie humaine*. (Ed. E. de Martonne) Colin, Paris. Translated (1926) as *Principles of Human Geography*, Henry Holt, New York.

Viereck, R. (1953) *The Shame and Glory of the Intellectuals*. Beacon Press, Boston.

Wagner, Ph. L. and M. W. Mikesell (1962) *Readings in Cultural Geography*. University of Chicago Press, Chicago.

Waibel, L. (1933) Was verstehen wirunter Landschaftskunde, *Geog. Anzeiger*, **34**, 197–207.

Walter, E. V. (1980–81) The places of experience, *The Philosophical Forum*, **12**, 159–81.

Warntz, W. (1964) *Geography Now and Then*. American Geographical Society, *Research Series*, **25**, New York.

Warntz, W. (1966) The topology of a socio-economic terrain and spatial flow, *Papers – Regional Sciences Association*, vol. 17, 47–61.

Watson, L. C. (1976) Understanding a life history as a subjective document: hermeneutical and phenomenological perspectives, *Ethos*, **4**, 95–131

Weber, A. (1909) Über den Standort der Industrien, 1. *Reine Theorie des Standorts*. Tübingen. Translated.

Weber, A. (1929) *Theory of the Location of Industries*. Russell and Russell, New York. Translation by C. J. Friedrich.

Weber, M. (1947) *The Theory of Social and Economic Organization*. Free Press, Glencoe, Illinois.

287

Weber, M. (1958) *The Protestant Ethic and the Rise of Capitalism.* Scribners', New York.

Weintraub, K. J. (1978) *The Value of the Individual: Self and Circumstance in Autobiography.* University of Chicago Press, Chicago.

Whitaker, J. R. (1954) The way lies open, *Annals Assoc. Am. Geogr.*, **44**, 231–44.

Whitaker, T. R. and E. A. Ackerman (1951) *American Resources, Their Management and Conservation.* Harcourt, Brace, New York.

White, G. F. (1966) Optimal flood damage management: retrospect, in Kneese, A. V. and S. C. Smith (eds), *Water Research.* Johns Hopkins Press, Baltimore.

White, G. F., ed. (1974) *Natural Hazards: Local, National, Global.* Oxford University Press, New York and London.

White, G. F. *et al.* (1958) Changes in urban occupance of flood plains in the United States, *Research Paper*, No. 57, University of Chicago, Dept. of Geography, Chicago.

White, H. (1973a) Foucault de-coded: notes from underground, *History and Theory*, **12**, 23–54.

White, H. (1973b) *Metahistory: The Historical Imagination in 19th Century Europe.* Johns Hopkins Press, Baltimore.

Whittlesey, D. W. (1929) Sequent occupance, *Annals Assoc. Am. Geogr.*, **19**, 162–5.

Whittlesey, D. (1935) Political geography: a complex aspect of geography, *Education*, **50**, 293–8..

Whittlesey, D. W. (1936) Major agricultural regions of the world, *Annals Assoc. Am. Geogr.*, **26**, 199–240.

Whittlesey, D. (1945) The horizon of geography, *Annals Assoc. Am. Geogr.*, **35**, 1–38.

Whittlesey, D. (1954) The regional concept and the regional method, in James, P. E. and C. Jones (eds), *Inventory and Prospect*, Syracuse, pp. 19–69.

Whorf, B. L. (1952) *Collected Papers on Metalinguistics.* United States Government Department of State.

Whorf, B. L. (1956) Languages and logic, in Carroll, J. B. (ed.), *Language,*

Thought, and Reality: The Selected Writings of Benjamin L. Whorf. MIT Press, Cambridge, Massachusetts.

Whyte, W. H. (1956) *The Organization Man.* Doubleday, Garden City, New York.

Wilson, L. S. (1948) Geographical training for the post war period, *Geographical Review*, **38**, 575–89.

Wirth, E. (1979) *Theoretische Geographie*, Teubner, Stuttgart.

Wittgenstein, L. (1922) *Tractatus Logico-Philosophicus.* Routledge & Kegan Paul, London.

Wittgenstein, L. (1960) *Philosophical Investigations.* Basil Blackwells, Oxford.

Wittgenstein, L. (1969) *On Certainty.* Basil Blackwells, Oxford.

Wolpert, J. (1966) Migration as an adjustment to environmental stress, *Journal of Social Issues*, **22**, 92–102.

Wright, J. K. (1947) Terrae Incognitae: the place of the imagination in geography, *Annals. Assoc. Am. Geogr*, **37**, 1–15.

Wright, J. K. (1952) *Geography in the Making: The American Geographical Society 1851–1951.* American Geographical Society, New York.

Wright, J. K. (1956) What's American about American geography? Lecture reprinted in Wright, J. K. (1966) *Human Nature in Geography. Fourteen Papers, 1925–65.* Harvard University Press, Cambridge, Massachusetts, pp. 124–39.

Wright, J. K. (1966) *Human Nature in Geography, Fourteen Papers, 1925–65.* Harvard University Press, Cambridge, Massachusetts.

Young, R. (1973) The historiographic and ideological contexts of the 19th century debate on man's place in nature, in Teich, M. and R. Young (eds), *Changing Perspectives on the History of Science.* Heinemann, London, pp. 344–438.

Zimmerman, E. W. (1964) *Introduction to World Resources.* (Ed., H. L. Hunker) Harper and Row, New York.

Zipf, G. K. (1949) *Human Behavior and the Principle of Least Effort.* Harvard University Press, Cambridge, Massachusetts.

Zweig, P. (1976) *Three Journeys: An Automythology.* Basic Books, New York.

Index

Aario, L., 268
Aay, H., 191, 275
Aberystwyth, 263
Åbo, see Turku
Ackerman, E. A., 187, 189, 191, 192, 194, 275, 288
Ackoff, R. L., 213
Adler, A., 228
Adorno, Th., 10, 190, 234, 275
Africa, 39, 71, 124, 127, 136, 144, 173, 269
Ahlenius, K., 262
Ahlmann, H. W:son, 153, 157, 158, 161, 265, 266
Ajo, R., 265, 268, 275
Alps, 31, 124, 169, 170
Alsace, 127, 169
America, 20–34, 40, 50, 52, 80–9, 97, 146, 186–95, 204, 231, 234, 244, 251, 262, 263, 264, 265, 266, 267, 269, 271, 274
American Geographical Society, 199
American geography/geographers, 62–5, 66–79, 80, 89, 98, 149, 186–95, 196–208, 233, 252, 266, 270, 271, 273
American West, 58, 64, 65, 85
Association of American Geographers, 83, 88, 189, 202, 262, 267, 274
   Annals of, 192, 197, 202, 263, 275
   Midwestern Field Conferences, 206, 265
   West Lake Conference, 201, 202
   Association of Pacific Coast Geographers, 88
   Economic Georgraphy, 266
   Geographical Review, 263
   Papers and Proceedings, 193
   Regional Science Association, 193
Ampferer, O., 171, 183n4
Amsterdam, university of, 18n11, 214, 215, 216, 261, 265, 266, 268, 270, 272
Ancel, J., 266
Ann Arbor, see University of Michigan
Antarctica, 262
Appelbaum, W., 191, 275
Arctic, 106, 113
Arendt, H., 6, 17n6, 275
Argentina, 267
Arnberger, E., 177, 178
Arnold, M., 47
Asby, 245
Asia, 20, 91, 262
   Central, 230, 262, 263, 264
   East, 30

South, 39
South-East, 39, 264, 266
Asia Minor, 25, 173, 180, 181, 230
Åström, S. E., 60n42
Auer, V., 106, 265, 267
Austria, 17n3, 25, 105, 167–85, 246, 262, 263, 268, 269, 277
   Atlas of, 177, 178, 181
   Academy of Sciences, 176
Ayer, A. J., 192, 275
Aylesbury, 44, 45, 47, 52

Baker, O. E., 71, 190, 275
Balandier, G., 129, 130
Baltic, 50, 80, 162
Banse, E., 264
Barrère, P., 133
Barrows, H. H., 63, 194, 275
Bartels, D., 17n3, 18n11, 195, 275
Barthes, R., 17n6
Bastide, R., 140n10
Bastié, J., 118, 144, 275
Bataillon, C., 3, 16n2, 114, 117, 119–140
Bates, M., 29
Baulig, H., 118, 128, 264, 266, 275
Baum, G., 7, 275
Bay Region, 28, 80
Beaujeu-Garnier, J., 114, 118, 132, 140, 141–152, 195, 271, 275, 278
Behrmann, W., 231, 232
Beijing, 20, 25
Benjamin, W., 5, 6, 275
Berdoulay, V., 18n17, 114, 115, 116, 275
Berelson, B., 191, 275
Bergen, 30, 155, 271
Bergmann, G., 192, 198, 199, 276
Bergsten, K. E., 246, 251, 252
Berkeley, University of California at, 22, 26, 28, 62, 65, 68, 69, 71, 72, 73, 74, 76, 78, 80, 82, 83, 84, 87, 88, 189, 190, 194, 199, 200, 201, 204, 227, 266
Berlin, 227
   University of, 86, 171, 172, 173, 174, 181, 226, 227, 229, 230, 231, 232, 263
Bernal, J., 7, 276
Bernard, A., 121
Berr, H., 116, 276
Berry, B. J. L., 193, 197, 204, 205, 269, 276
Bertalanffy, L. von, 194, 276
Bertrand, A. J. C., 271
Bingham, M. T., 7, 102n4, 276

Birot, P., 114, 117, 118, 119–40, 146, 267, 269, 276
Blache, J., 116, 131, 276
Black, M., 18n17, 276
Blanchard, R. 117, 122, 125, 126, 131, 134, 169, 262, 263, 276
Blaut, J., 191, 276
Bloch, M., 128, 132, 228, 266, 276
Blume, S. S., 7, 276
Blythe, R., 8, 276
Bobek, H. 167–85, 191, 230, 264, 268, 276
Boerman, W. E., 264
Bonn, 225
Boorstin, D. J., 49, 190, 276
Borchert, J., 70, 199, 200
Bordeaux, 114, 125, 126, 127, 130, 134, 271
Borsodi, R., 190, 275, 276
Boulding, K., 4, 189, 194, 276, 277
Bourdieu, 4, 7, 277
Boutroux, E., 116, 277
Boutruche, R., 129, 136
Bowden, M. J., 189, 277
Bowers, L., 190, 277
Bowman, I., 22, 92, 102n3, 277
Branch, M., 55, 61n48
Braudel, F. 117, 128, 140n9, 140n10, 277
Brazil, 129, 130, 134, 146, 151
Breese, G. W., 189, 277
Broc, N., 114, 277
Brodbeck, M. N., 192, 277
Broek, J. O. M., 199, 205
Brouwer, N., 214
Brown, E., 51, 60n24
Brown, R. H., 189, 199, 206, 277
Bruhat, J., 122, 129
Brunet, R., 271
Bruhnes, J., 116, 118, 194, 228, 246, 262, 277
Brun-Tschudi, A., 18n12, 35–43, 273
Brush, J., 269
Büdel, J., 268
Buffon, 31, 32, 33
Bultmann, R., 17n6
Bunge, W., 192, 195, 252, 277
Burke, K., 14, 277
Burtt, E. A., 23
Bury, J. B., 22
Busch-Zantner, R. 268
Buttimer, A., 17n7, 18n8, n10, n11, 114, 117, 205, 254, 277
Buursink, J., 272
Bylund, E., 271

Caen, 271
Cailleux, A., 146
Calef, W., 197
California, 21, 24
Calkins, R. D., 80, 81
Cambridge, University of, 50, 94, 98
Canada, 48, 50, 51, 52, 55, 106, 108, 110, 232, 267
Carcopino, J., 128, 131
Carpathians, 105

Carlstein, T. et al., 59n14
Cassirer, E., 18n17, 277
Caucasus, 25, 173
Cavaillès, H., 125
Cavell, S., 6, 277
Chabot, G., 118, 144, 147, 231, 276
Chadbourne, M., 59n7
Chamberlin, T. C. 81
Chargaff, E., 7, 278
Chase, S., 4, 278
Chicago, 17n3, 64, 66, 67, 68, 78
    University of, 26, 81, 82, 83, 189, 190, 192, 193, 194, 196, 197, 200, 201, 204, 205, 206, 262, 263, 271
    Northwestern, 193, 200, 201
Chicago Plan Commission, 192, 278
Chiltern Hills, 45, 46, 52, 59n5
China, 25, 30, 35, 36, 37, 38, 39, 40, 41, 91, 92, 94, 98, 171, 203
Chisholm, M., 93, 98, 252
Cholley, A., 117, 118, 122, 125, 126, 127, 131, 132, 134, 139, 140, 141, 142, 266, 267, 278
Chorley, R. J., 195, 252, 271, 278
Christaller, W., 7, 170, 172, 180, 246, 252, 264, 265, 266, 267, 268, 271, 273
Clark, T., 115
Clark University, 10, 11, 17n5, 18n13, 62, 63, 66, 74, 190, 265
Claval, P., 7, 17n3, 114, 116, 271, 278
Clayton, K. M., 59n5
Clozier, R., 136, 139
Colby, C. C., 192, 263, 278
Cole, G. 95
Cole, J., 252
Coleridge, S. T., 23
Collingwood, R. G., 7
Corn Belt, 65, 200
Croce, B., 17n6
Crowe, P. R., 266
Cvijik, J., 262
Cyprus, 25
Czechoslovakia, 105

Dacey, M. F., 278
Daguin, R., 130
Dahl, S., 63, 267, 278
Dalenius, T., 7, 278
Dallmayr, F. R., 7, 278
D'Almeida, C., 125
Darby, H. C., 51, 52, 60n28, 60n32, 63, 266, 278
Davies, G., 198
Davies, W. K. D., 191, 278
Davis, W. M., 62, 67, 81, 83, 85, 86, 89n4, 117, 134, 199, 201, 229, 262, 264, 267, 278
Deasy, G. F., 187, 278
de Bruyne, G. A., 272
De Dainville, A., 116, 278
Deffontainnes, P., 55, 60n39, 116, 117, 121, 267, 278
de Geer, S. 63, 83, 153, 157, 158, 161, 247,

262, 264, 278
de Jong, G., 63, 268, 278
de la Rué, A., 116, 278
Delobez, A., 147, 151
Delouvrier, P., 144
Demangeon, A., 63, 95, 117, 120, 121, 122,
    123, 125, 126, 127, 128, 132, 133, 137,
    231, 246, 262, 264, 266, 278
Demant Hatt, E., 1, 2, 11, 16
de Martonne, E., 118, 119, 120, 121, 122,
    123, 125, 126, 127, 128, 131, 132, 133,
    134, 136, 140, 142, 262, 266, 283
den Hollander, A. N. J., 265
Denis, P., 264
Denmark, 85, 244, 250
de Planhol, X., 273
Derrida, J., 17n6
Derruau, M., 129
Descartes, R., 7, 17n6
de Saussure, F., 17n6
de Smidt, M., 272
Detienne, M., 6, 278
de Vries Reilingh, H. D., 231, 266, 268, 270
de Vooys, A. C., 231, 265, 266, 268
Dickinson, R. E., 45, 59n2, 63, 266, 279
Dielemann, F. M., 272
Dietrich, B., 175
Dijkhuis, H. Th., 265
Dijkink, G. J., 272
Dilthey, W., 6, 8, 17n6, 279
Dion, R., 129, 140n9, 231, 266, 273, 279
Donau, 105
Dorpat, University of, 246
Dresch, J., 114, 117, 119–40, 269, 276
Dublin, 93, 94, 95, 96
Duncan, J. S., 63, 279
Duncan, O. D., 194, 279, 281
Duhamel, G., 47, 59n6
Durkheim, E., 116, 118, 129, 279
Dutch East Indies, 214

East, G., 269
Eddington, Sir A., 247, 248
Edinburgh, University of, 18n9, 90, 93, 97,
    250
Edison, Th. A., 241
Ehrendorfer, E., 184n27
Ellenberg, H., 174
Ely, R. T. 69, 71
Elzinga, A., 7, 279
Enequist, G., 63, 251, 269, 279
Engels, F., 138
England, 17n3, 25, 52, 53, 54, 84, 96, 97, 99,
    100, 148, 154, 155, 156, 250, 251
Enjalbert, H., 114, 119–40
Eriksson, G., 8, 279
Erixon, S., 263
Estonia, 103, 263, 267
Europe, 30, 41, 49, 50, 51, 52, 54, 64, 67,
    88, 91, 93, 94, 96, 98, 113, 156, 160,
    181, 188, 209, 225, 234, 246
European geography/geographers, 62, 63,
    187, 188, 190, 191, 193, 195, 198, 204,

252, 267, 270
Evanston, Ill., 193, 271

Far East, 30, 38, 94, 98, 188
Faucher, D., 124, 125
Fawcett, C. B., 96
Febvre, L., 116, 118, 128, 129, 140n9, 228,
    264, 279
Fellman, J. D., 187, 280
Fennemann, N. M., 263
Fesl, M., 178, 180, 185
Festinger, L., 190, 279
Finch, V. C., 64, 71, 199, 267, 279
Finland, 17n3, 25, 49, 50, 51, 52, 53, 54, 55,
    56, 57, 58, 59, 60, 61, 95, 103, 104,
    105, 106, 107, 109, 110, 114, 244, 252,
    261, 262, 263, 264, 265, 267, 268, 270,
    272
    Atlas of, 102n7, 261, 262, 264, 267, 268
    Fennia, 261
    Geographical Association (GA), 261, 262,
        263, 264, 265, 267, 270
    Geographical Society, 109, 111
    Society for the Geography of Finland
        (SGF), 261, 263, 264
    Winter War, 50
Finsterwalder, R., 170
Finsterwalder, S., 170
Florida, University of, 66
Fleck, L., 14, 18n16, 279
Fleure, H. J., 50, 64, 91, 93, 102n1, 264, 279
Forde, D., 264
Fosberg, R., 29
Foucault, M., 4, 6, 7, 15, 186, 279
France, 17n3, 25, 80, 84, 97, 114, 115, 116,
    119–40, 141–52, 228, 229, 232, 246,
    261, 262, 263, 264, 266, 267, 269, 271,
    273
    Annales de Géographie, 117, 128, 129, 133,
        139, 261
    Association of French Geographers, 132
    French geography/geographers, 114–18,
        119–40, 149, 214, 225, 228, 232,
        233, 262, 264, 265
Frankfurt
    University of, 225, 232, 234
    School, 234, 270
Frazier, F., 190, 279
Freeman, T. W., 62, 90–102, 279
Freiburg, University of, 174, 175, 176, 182
Freud, S., 17n6, 167, 228
Frödin, J., 265
Fromm, E., 10, 279

Gadamer, H.-G., 4, 6, 7, 16, 17n6, 279
Gale, S., 7, 279
Gallais, J., 271
Gallois, L., 120, 140, 262
Gamblin, A., 151
Garraty, J. A., 6, 279
Garrison, W. L., 193, 200, 201, 203, 205,
    269, 276, 279
Gautier, E. F., 264

Geddes, A., 93, 250
Geddes, P., 93, 98, 250
Geneva, 92, 227, 228, 229
George, P., 114, 118, 119–40, 144, 231, 267, 280
Germany, 17n3, 25, 65, 84, 93, 98, 134, 155, 231, 232, 234, 246, 262, 263, 268, 269, 272
German geography/geographers, 231, 234
Gilbert, G. K., 85, 102
Ginsburg, N., 197
Glacken, C. J., 20–34
Godlund, S., 246, 252, 253, 269, 271
Gombrich, E. H., 17n6
Göteborg (Gothenburg), University of, 157, 252, 262
Gottmann, J., 118, 186, 223, 252, 280
Gottschalk, M. K. E., 268, 270, 272
Gould, P., 252
Gouldner, A. W., 190, 280
Gourou, P., 266, 267
Granö, J. G., 46, 95, 246, 262, 263, 264, 265, 267, 268
Granö, O., 17n3, 18n11, 59n4, 254, 268, 280
Great Britain, 99, 261, 262, 263, 264, 266, 267, 269, 271, 273
British Association of Geographers, 97, 269
British geography/geographers, 57, 91, 97, 149, 233, 252
The Geographical Teacher/*Geography*, 50, 262
Institute of British Geographers, 266, 271
Royal Geographical Society, 262
Greece, 25, 268
Gregory, D., 192, 273
Grenoble, 122, 169
Groenman, S. J., 266
Groningen, University of, 216, 217, 219, 265, 266, 272
Grund, A., 262
Guilcher, A., 148
Gunn, J. V., 7, 280
Gurvitch, A., 129
Guyot, A. I., 85, 86

Habermas, J., 4, 6, 7, 17n6, 280
Hägerstrand, B., 245, 249, 250, 251
Hägerstrand, T., 9, 18n10, n11, n13, n15, 191, 193, 205, 223, 234, 235, 238–56, 262, 267, 269, 271, 272, 273, 277, 280
Haggett, P., 195, 252, 271, 278, 280
Haig, R. M., 158, 266
Hall, P., 252
Hamberg, A. H., 262
Hannerberg, D., 249, 251, 267, 269, 271
Hänninen, K., 263
Hard, G., 195, 280
Hardy, T., 53
Harris, C. D., 187, 192, 197, 233, 252, 280
Harrison, M. J., 60n27
Harrison, R. T., 18n17, 282
Hart, J. F., 64, 188, 280

Hartke, W., 18n12, 223, 225–37, 268
Hartmann, G. W., 193, 198, 280
Hartshorne, R., 63, 64, 71, 99, 102n13, 189, 191, 192, 199, 200, 205, 266, 267, 280
Harvard, 37, 39, 192, 262
Harvey, D., 195, 271, 273, 280
Hassinger, H., 175, 176, 177, 263
Haushofer, K., 262
Hedin, S., 261, 264
Hegel, G. W. F., 7, 138
Heidegger, M., 17n6, 281
Heidelberg, 170
Heinemeyer, A., 268, 270, 272
Heisenberg, W., 7, 281
Heklaakoski, A. R. 264
Hélias, P. J., 8, 281
Helmfrid, S., 17n3, 269, 281
Helsinki, 50, 53, 103
Helsinki School of Economics, 108, 109, 264, 267, 268
University of, 107, 108, 261, 263, 264, 265, 267, 268, 270
Helvig, M. 271
Henriksson, A. K., 187, 281
Herbertson, A. J., 95, 102n6, 262
Heslinga, 270
Hesselberg, J. 42–3
Hettner, A., 7, 63, 126, 170, 182, 244, 261, 264, 281
Hewes, L., 62, 65, 66–79
Higbee, E., 189, 281
Higham, J., 4, 190, 281
Hildén, K. 263, 264, 265
Hinderink, J., 272
Hjulström, F., 267
Hodgen, M., 23
Hoekveld, G. A., 270, 272
Hofmannsthal, H. von, 49
Hofstadter, R., 195, 281
Hofstee, E. W., 265, 266, 268, 270
Holland, *see* Netherlands
Hook, J. C., 193, 280, 283
Hoppe, G., 273
Horkheimer, M., 234
Horowitz, I. L., 7, 281
Horton School, 69
Hoskins, W. G., 52
Houston, E. J., 85
Howarth, O. J. R., 95, 97, 102n6
Hudson, D., 200
Hudson Bay, 106, 112n4, n11
Hult, R., 261
Humboldt, A. von, 62, 83, 149, 230, 281
Hunan, 35
Hungary, 67, 105, 107, 246
Huntington, E., 23, 41, 91
Hustich, I., 103–13, 265, 267, 268
Iceland, 48, 50
Ignatius, K. E. F., 261
Ile-de-France, 47, 144
India, 30, 92, 100, 173, 230
Information Géographique, 116, 187, 281

Innsbruck, 168, 169, 170, 171, 172, 180, 264
University of, 170, 174, 230
International Geographical Union, 90, 148
Commission on Agricultural Typology, 40
Commission on Applied Geography, 191
Commission on the History of Geographical Thought, 18n9, 90, 97
Commission on Medical Geography, 146
Commission on a World Population Map, 161, 166
Congress
in Amsterdam, 94, 172, 173, 180, 265
in Montreal, 97
in Moscow, 111
in New Delhi, 97
in Paris, 231
in Stockholm, 161, 162, 191, 270
in Tokyo, 144
at Uppsala, 166
in Washington, 146, 161
Iowa, University of, 62, 66, 73, 77, 189, 192, 193, 196, 198, 200, 201, 203
City, 271
Iran, 172, 176, 177, 181, 182
University of, 176
Ireland, 94, 95, 96, 98, 100, 101, 114, 188
Geographical Society of, 94, 96
Isachsen, F., 30, 38, 265, 266
Isard, W., 192, 193, 252, 281
Italy, 25, 173, 174, 232

Jaatinen, S., 55, 60n40, 61
Jackson, W. A., 191, 281
Jacoby, R., 10, 281
Jakobson, R., 7, 14, 18n17, 281
James, P. E., 62, 63, 64, 66–79, 189, 191, 205, 266, 281
James, W., 63
Jansen, A. C. M., 272
Japan, 25, 26, 27, 28, 30, 37, 39, 92, 121, 203
Joensuu, University of, 272
Johns Hopkins (University), 26
Johnson, R. J., 7, 17n3, 57, 61n55, 281
Jones, C. F., 62, 189, 263, 281
Jones, M., 55, 61n47
Jones, S. B., 191, 281
Journaux, A., 116, 281
Juillard, E., 129, 136, 139, 148
Jung, C. J., 10, 228, 282

Kain, R., 60n27
Kalm, P., 52, 54
Kan, C. M., 261
Kant, E., 191, 246, 247, 249, 254, 264, 267, 282
Kant, I., 164
Karelia, 55, 107, 108, 109
Kates, R. W., 194, 282
Kawir, 177, 182
Kazin, A., 7, 282
Kentucky, 82, 251

Keuning, H. J., 231, 265, 266, 268
Kiel, 18n11
Kimball, G. E., 186, 283
King, L., 193, 282
Kinzle, H., 170, 268
Kirby, D. G., 61n54
Kirk, R., 190
Kirk, W., 189, 282
Kish, G., 189, 196–208
Kjellén, R., 59n11
Kjellén-Björkquist, R., 59n11
Kleinpenning, J. M. G., 272
Kluckhohn, C., 190, 282
Klute, F., 264
Kniffen, F., 250
Knos, D., 189, 196–208, 283
Koestler, A., 49
Kohl, J. G., 246
Köhler, W., 231
Kohn, C., 62, 66–79, 189, 192, 282, 283
Kollmorgen, W., 195, 282
Köppen, W., 266
Korea, 25, 30, 200, 207
Kouwe, P. J., 268, 270
Kouwenhoven, J. A., 190, 282
Kovero, 103
Kranck, H., 105
Krebs, N., 63, 171, 227, 230, 282
Kristoffersson, A., 264
Kroeber, A. L., 23, 65, 73
Kropotkin, P., 93
Kruyer, G. J., 268, 270
Kruyt, J. P., 265
Kuhn, T. S., 7, 14, 18n16, 282
Kuklinski, A., 252
Kumamoto, 26
Kyushu, 26

Labasse, J., 118, 269, 282
Labrador, 103, 105, 106, 265
Lacoste, Y., 273
Lakatos, I., 7, 282
Lambooy, J. G., 270, 272
Langer, S. K., 14, 18n17, 116, 282
Lapland, 1, 103, 105, 106
Lapp, 1, 6, 16
Lasch, C., 10, 282
Lautensach, H., 63, 282
Lavedan, P., 266
LeBon, G., 228
Leeds, 91
Lehmann-Haupt, F., 169
Leighly, J., 80–9, 266, 282
Leiviskä, I., 262, 263, 264
Le Lannou, M., 117, 282
Leningrad, 18n9, 97
Leppard, H., 197
Levinas, E., 223n1
Lévi-Strauss, C., 7, 18n17, 56, 61n50, 114, 140n10, 282
Lewin, K., 192, 231, 248, 282
Lichtenberger, E., 17n3, 176, 179, 185
Lille, 143, 148

Lilley, S., 7, 282
Limousin, 120, 146, 148
Lincoln, Nebraska, 16n2, 65n1, 66, 79n1, 208n1
Lindståhl, S., 109
Linnaeus, C., 52
Linton, D., 93, 266
Liverpool, 51, 52, 53, 54, 60n30, 91
    University of, 53, 54, 263
Livingstone, D. N., 18n17, 282
Loesch, A., 268
Løffler, E., 7, 262
Löffler, H., 176
London, 48, 94, 154
    School of Economics, 50, 93
    University College, 55
Lord, A., 6, 282
Los Angeles, 201
Lösch, A., 192, 271, 282
Lotka, A. J., 248, 249
Lovejoy, A., 32
Lowe, J. L., 23
Lowenthal, D., 283
Lowie, R. H., 23
Lukermann, F., 116, 189, 196–208, 283
Lukkes, P., 272
Lund, 9
    University of, 18n13, 193, 242, 243, 246, 249, 251, 252, 253, 261, 267, 271
    Lund Studies in Geography, 267
Lundberg, G., 247, 248
Lundquist, M., 269
Lundsberg School, 154, 156, 163
Lut, 177, 182
Lynch, K., 189, 283

MacCarthy, T. A., 7, 278
McCarty, H. H., 193, 198, 205, 283
McDonald, J. R., 115, 283
Mackinder, H. J., 41, 50, 97, 98, 99, 262
MacIntyre, A., 6, 283
McMurry, F., 67
Madison, Wis., 78, 199
Malin, J. C., 64, 189, 283
Manchester, 96, 97
Manchuria, 37
Mannerfelt, C. M:son, 267
Mannheim, K., 14, 283
Markham, F., 50, 60n22
Marsh, G. P., 25, 29, 87, 88, 283
Martin, G., 62
Martin, W., 228
Marx, K., 17n6
Marxism, 138, 140, 270, 272
Marxist geographers, 136, 137, 138, 267, 269, 271
Masileh, 177
Massif Central, 126, 140, 146
Mather, E. Cotton, 62, 66–79, 206
Matzner, E., 254
Maury, M. F., 85
Mauss, M., 129
May, J., 146

Mayer, H. M., 191, 192, 197, 283
Mead, M., 190, 283
Mead, W. R., 44–61, 114, 252
Mediterranean countries, 25, 30, 31, 121, 124, 269
Medawar, P., 49
Mendelsohn, E., 7, 283
Merleau-Ponty, M., 14, 283
Merton, R. K., 191, 283
Metz, F., 170, 174
Mexico, 127
Meynier, A., 114, 119–40
Michelet, J., 116, 283
Michigan State University, 62, 67, 68, 70, 71, 72, 81, 82, 189, 196, 199, 201, 203, 204, 206, 265
    Land Economic Survey, 64, 67, 80, 82
    Central Michigan University, 80
Middle East, 144, 172, 173, 266
Midwest, 80, 83, 193, 197, 200, 203, 204
Mik, G. 272
Mikesell, M., 17n3, 62, 63, 65, 66–79, 185, 191, 194, 197, 204, 283, 287
Mill, H. R., 7, 283
Mills, C. W., 190, 283
Milton, J., 29
Minneapolis, 78
Minnesota, 56
    University of, 62, 66, 73, 74, 189, 190, 194, 196, 199, 201, 206
Monbeig, P., 114, 119–40
Mongolia, 265
Montpellier, 146
Moreno, J. L., 191, 248, 283
Morocco, 121, 130, 136, 268
Morrill, R. L., 189, 193, 196–208, 252
Morse, P. M., 186, 283
Morvan, 141, 142, 145
Mount Pleasant (Michigan), 80, 81, 82
Muir, J., 32
Mumford, L., 29, 88, 250, 283
München, 175, 225, 234
Murphey, R., 283
Musset, R., 133
Mycklebost, H., 17n3
Myrdal, G., 190, 283

Nakayama, S., 7
Nanjing, 25
Near East, 173, 180, 181
Nebraska, 62, 66
Needham, J., 7, 284
Nelson, H., 64, 244, 249, 263, 265, 284
Netherlands, 17n3, 211, 212, 216, 217, 231, 246, 252, 261, 263, 264, 265, 266, 268, 270, 272
    Dutch Geographical Society, 212, 231
    Dutch Geography, 246
    Geographisch Tijdschrift, 266, 272
    ISONEVO/SISWO, 215, 217, 268, 270
    Provincial Physical Planning Board, 214, 217
    Tijdschrift voor Economische Geographie (TEG), 263

New Deal, 190, 194
Newfoundland, 105
New Guinea, 231
Niemoeller, R., 226
Niermeyer, J. F., 261
Nijmegen, University of, 268, 272
Nisbet, R., 190, 284
Nonn, H., 133
Norborg, K., 191, 284
Nordenskjöld, A. E., 261
Nordfjord, 56
Norrland, 159
Norrlin, J. P., 261
Northern Ireland, 95
Norway, 17n3, 30, 31, 36, 37, 38, 39, 40, 47,
    52, 56, 58, 132, 244, 252, 261, 263,
    265, 267, 269, 271, 273
Nystuen, J., 201, 252

Oakland, 21, 24
O'Dell, A. C., 56
Odén, B., 4, 284
Oestreich, K., 261
Ogilvie, A., 93, 95
Okinawa, 26, 27, 30
Olsson, G., 7, 61n57, 271, 273, 279, 284
Ormsby, H., 50
Oslo, University of, 30, 39, 261, 265, 271,
    273
Otremba, E., 268
Oughton, M., 97
Oulu, University of, 269, 270
Oxford, 91, 92, 97

Pacific, 20, 22, 156
    Islands, 29
Palander, T., 246, 247
Palmén, J. A., 261, 262
Palmer, R. E., 4, 17n6, 284
Pannett, D. J., 60n27
Panzer, W., 171
Papy, L., 114, 119–40, 149
Paris, 32, 33, 39, 46, 67, 97, 114, 121, 127,
    129, 132, 135, 136, 137, 138, 142, 144,
    145, 146, 147, 229
    Atlas de, 144, 148, 149
    Ecole Normale Supérieure de la Rue
        d'Ulm, 137, 138, 140
    Ecole Normale Supérieure de Saint-Cloud,
        137, 138, 140
    basin, 124, 132, 146
    Sorbonne, Geography Institute, 114, 125,
        127, 128, 136, 141, 142, 143, 144
Parker, E., 197
Parker, R., 8, 284
Parkins, 263
Parkinson, C. N., 54
Parry, M., 6, 284
Pas-de-Calais, 148
Passarge, S., 126, 264, 266
Patagonia, 265
Pattison, W., 189, 191, 192, 193, 196–208,
    284

Peake, H. J. E., 91, 102n1
Pélissier, P., 271
Penck, A., 126, 169, 171, 227, 229, 262, 264
Penck, W., 84
Pennsylvania, University of, 69
Pepper, S. C., 18n17, 19n18n, 64, 117, 284
Perroux, F., 130
Peschel, O., 83
Pfeiffer, G., 201
Philbrick, A., 192, 197, 205, 284
Phliponneau, M., 135, 269
Pinchemel, Ph., 97, 133, 140, 273
Piveteau, J., 123
Platt, R. S., 191, 192, 196, 197, 205, 252,
    263, 284
Poincaré, H., 116, 284
Poland, 67, 105, 151, 211, 252
Pollock, N. C., 234
Popper, K. R., 7, 284
Portugal, 121
Potomac, 27, 188
Potter, D. M., 64, 284
Powell, J. W., 85
Prague, 174
    University of, 175
Pred, A., 193, 254, 276
Pribram, K. H., 14, 284
Princeton, 29
Prussia, 115
    East, 226, 231
Pyrenees, 122, 125, 127

Qualter, T., 59n13
Québec, 55, 106, 124

Rabinow, P., 6, 284
Rahner, K., 17n6
Rapp, A., 273
Ratzel, F., 116, 261, 262, 263, 285
Ravetz, J. R., 15, 285
Réclus, E., 93, 149
Redfield, R., 4, 285
Rennes, 114, 132, 136
Renouvin, P., 128
Restif de la Bretonne, 47
Rhine, 174
    Rhine-Danube frontier, 91
Rhône, 124
    lower, 124
Ricoeur, P., 14, 18n17, 116, 285
Riesman, D., 190, 285
Riga, 162
Ritter, C., 62, 83, 86, 261, 265, 285
Robbins, L., 49, 59n13
Robequain, Ch. 264
Rochefort, R., 117, 285
Roosevelt, F., 24
Rorty, R., 4, 7, 285
Rosberg, J., 261, 262, 263, 265
Rose, A. M., 4, 285
Rose, J. K., 187, 285
Rostlund, E., 31
Rotterdam, University of, 263, 264, 272

Roumania, 103, 105, 107, 170
Roupnel, G., 266
Roxby, P. M., 91, 93, 94, 98, 102n5
Ruellan, A., 121
Rühl, A., 229, 230
Russell, J., 68
Rutkis, J., 162
Ryukyu Islands, 27, 30

Saarinen, T. F., 285
Sack, R., 7, 192, 285
Sacramento, 21, 22, 24
  River, 24
Saint-Exupéry, A. D., 8
Salisbury, R. D., 64, 68, 81, 193, 206, 262
San Francisco, 24, 25
Sansom, Sir G., 26
Sauer, C. O., 4, 23, 27, 28, 29, 31, 63, 64,
  65, 67, 68, 70, 73, 78, 80, 82, 83, 85,
  87, 88, 189, 191, 201, 204, 263, 265,
  266, 267, 285
Sautter, G., 139, 271
Sauvy, A., 129
Savo, 53, 54
Scandinavia, 31, 35, 39, 49, 51, 53, 56, 57,
  60n41, 100
  Scandinavian connection, 56, 80
  Scandinavian geography, 111
Schaeffer, K., 192, 198, 201, 285
Schiff, M., 194, 285
Schlanger, J. E., 116, 285
Schlüter, O., 63, 169, 182, 262, 285
Schmieder, O., 7, 285
Schmithüsen, J., 174, 175
Schöller, P., 180, 181
Schott, C., 230
Schouw, J. F., 7, 285
Schultz, H. D., 63, 285
Schütz, A., 8
Schwartz, G., 63, 285
Schweitzer, A., 7, 285
Schweitzer, G., 185n29
Schwerin, H. H. von, 261
Scotland, 91, 99, 251
Scott, R., 262
Seamon, D., 18n8, n12, 277,
Seattle, University of Washington at, 114,
  189, 190, 192, 193, 196, 200, 201, 247,
  251, 271
Sederholm, J. J., 263
Ségala, 121, 126, 132
Sellars, J., 7, 285
Semple, E. Ch., 83, 263
Seoul, 26
Sestini, A., 191, 285
Shaler, N. S., 86, 87, 88
Sibelius, J., 49
Siebenbürgen, 107, 171
Siegfried, A., 7, 137, 263, 286
Sigtuna meeting, 10, 209, 222
Simiand, F., 116, 286
Simon, H. A., 194, 281
Sion, J., 114, 117, 286

Sjögren, A. J., 55, 61n48
Sjögren, O., 262
Småland, 238, 240
Smeds, H., 53, 55, 60n33, 61, 107, 108, 265,
  267
Smith, C. F., 198
Smith, N., 7, 273, 286
Snyder, W. H., 89n4
Sölch, J., 169, 170, 175, 181
Sorokin, P., 246, 247, 265
Sorre, M., 114, 116, 118, 132, 133, 141, 142,
  143, 146, 150, 191, 231, 246, 267, 286
South America, 27, 71, 127, 136, 230, 262,
  264, 265, 267
South Pole, 262
Soviet Union, 53, 98, 137, 161, 172, 203
Spain, 93, 268
Spate, O. H. K., 61n56
Spears, R. F., 191, 275
Spencer, J. E., 189, 190, 286
Spreitzer, H., 175
Sprout, H., 191, 286
Sprout, M., 191, 286
Stamp, L. D., 49, 96, 98, 99, 102n12, 266,
Stanford, 26
Steinacker, H., 169
Steinbach, J., 184n27
Steinbeck, J., 24
Steiner, G., 47, 48, 57, 59n8, 61n52, 191, 275
Steigenga, W., 231, 268, 272
Steinmetz, R., 231, 246, 261, 263, 264, 266,
  270
Stewart, G. R., 84
Stewart, H. D., 63, 285
Stockholm, 253, 265, 266
  University of, 39, 53, 55, 56, 83, 153, 156,
    157, 158, 161, 162, 163, 164, 249,
    251, 252, 264
  School of Economics, 162, 262
Stoddart, D. R., 18n9, 63, 64, 286
Stolz, O., 169
Stone, K. H., 187, 286
Stouffer, S. A., 192, 286
Strasbourg, 128, 134, 136, 148
Sullivan, W. M., 6, 284
Sundborg, Å., 270
Svensson, S., 249
Sweden, 9, 10, 17n3, 52, 56, 83, 84, 158,
  159, 161, 162, 205, 211, 241, 244, 247,
  249, 250, 251, 252, 253, 261, 262, 263,
  264, 267, 269, 270, 273, 274
  Geografiska Annaler, 63, 263, 280
  Svensk Geografisk Årsbok, 264
  Swedish geography/geographers, 149, 193,
    267
  Swedish Society for Anthropology and
    Geography (SSAG), 261
  Ymer, 162, 261, 274
Switzerland, 246
Szanton, D. L., 4, 286

Taaffe, E. J., 189, 193, 286
Taine, H., 22

Takes, Ch. A. P., 266
Tampere, University of, 270
Tamsma, R., 268
Tanner, V., 105, 106, 265, 267
Tansley, A. G., 62, 63, 286
Tanzania, 40
Tarr, R. S., 86, 89n4
Tartu, University of, 263, 264
Tawney, R. H., 49
Taylor, E., 96
Taylor, G., 187, 286
Taylor, G. R., 7, 50, 60n21, 286
Taylor, I., 84
Teggart, F. J., 22, 23, 31
Teich, M., 15, 286
Tennessee, 251
    Valley Authority Project, 64, 67, 266
ter Heide, H., 270
ter Veen, N. H., 264, 265
Theodorson, G. A., 64, 286
Thibault, A., 132
Thomas, E. N., 193, 286
Thomas, W. L., 29, 65, 189, 286
Thorarinson, S., 267
Thornthwaite, C. W., 68
Thünen, J. von, 246, 271
Tierra del Fuego, 265
Tillich, P., 238
Tobler, W., 193, 286
Tolvanen, V., 264
Topelius, Z., 54
Törnquist, G., 271
Toulouse, 114
Toynbee, A., 6, 287
Trewartha, G., 70, 189, 199, 287
Tricart, J., 140, 146, 267
Trinidad, 71
Troll, C., 186, 230, 231, 232, 268, 287
Tuominen, O., 267
Turi, J., 1, 2, 3, 6, 8, 16
Turkey, 172, 181
Turku (Åbo), University of, 18n11, 264, 265, 270
    School of Economics, 269, 270
Twin Cities, 200, 203
Tyrol, 169, 170, 171, 181

Ullman, E. L., 188, 189, 192, 193, 200, 201, 203, 252, 280, 287
Umeå, University of, 271
UNESCO, 7, 287
United States/USA, see America
Uppsala, University of, 56, 156, 157, 161, 163, 251, 262, 265, 270
Utrecht, 199
    State University of, 213, 215, 216, 261, 264, 265, 266, 268, 270, 272

Vallaux, C., 117, 118, 263, 287
van Amersfoort, J. M. M., 272
van Bath, B. H. S., 266
van den Berg, G. J., 18n12, n15, 209–24, 231, 268, 272

van der Knaap, W., 272
van der Valk, J. G., 265
van der Wusten, H., 272
van Doorn, J. A. A., 268
van Engelsdorp Gastelaars, R., 272
van Fraassen, Bas, C., 17n6
van Ginkel, G., 272
van Heek, F., 265, 268, 270
van Hise, C. R., 194, 287
van Hulten, M., 270
van Paassen, Ch., 17n3, 18n11, 213, 214, 215, 217, 231, 252, 254, 268, 270, 272
van Vuuren, L., 214, 217, 231, 264, 265, 268
van Westrhenen, J., 270
Varenius, B., 83
Varjo, U., 268
Vasa, 53, 55
Vendée, 125
Vernant, J. P., 6, 287
Vidal de la Blache, P. M., 62, 91, 93, 101, 102n2, n4, 114, 115, 116, 125, 129, 140, 214, 246, 261, 262, 263, 264, 287
Vienna, 167, 168, 175, 176, 263
    Master Plan of, 179, 181
    University of, 175, 183
    Circle, 192, 198
    School of Economics, 175
Viereck, R., 190, 287
Vilar, P., 122, 129
Virginia, 251
Visher, S. S., 263
Vogt, W., 27, 29

Wageningen, University of, 266
Wagner, Ph. L., 65, 185, 191, 194, 197, 204, 287
Wahlbeck, L., 109
Waibel, L., 63, 68, 71, 201, 264, 287
Wales, 90, 91, 99
Walter, E. V., 18n12, 287
Walter, J., 126
Warntz, W., 62, 85, 193, 269, 287
Washington, 71
Wastenson, L., 273
Watson, L. C., 4, 287
Watson, W., 51, 60n23
Wayne State, 206
Weaver, J., 199, 203, 205
Weber, A., 192, 271, 287
Weber, M., 169, 192, 287, 288
Weintraub, K. S., 4, 288
Werenskiold, W. W., 263
Wever, E., 272
Weulersse, J., 122, 266
Wheatley, P., 31
Whitaker, J. R., 189, 288
Whitaker, T. R., 194, 288
White, G. F., 194, 195, 197, 205, 288
White, H., 7, 12, 14, 18n17, 288
Whittlesey, D. W., 63, 64, 189, 191, 192, 269, 288
Whorf, B. L., 4, 288
Whyte, W. F., 190, 288

William-Olsson, W., 18n12, 53, 153–66, 201, 265, 267, 271
Williamson, A. V., 91
Wilson, L. S., 188, 288
Wilson, M. L., 275
Winsemius, J., 266
Wirth, E., 195, 288
Wisconsin, 56
    University of, 62, 68, 71, 199, 201, 202, 265
Wise, M., 252
Wissink, L., 268, 270
Wittfogel, K. H., 234
Wittgenstein, L., 4, 7, 15, 17n6, 288

Wood, W., 198
Wooldridge, S. W., 266, 269
Wopfner, H., 169
Wright, J. K., 62, 189, 190, 191, 288,

Young, R., 7, 14, 15, 286, 288
Yugoslavia, 174

Zagros, 176
Zeist, W., 176
Zelinsky, W., 70
Zimmerman, E. W., 64, 288
Zimmern, Sir A., 92
Zipf, G. K., 192, 288

298